T0136842

Studies in Systems, Decision and Control

Volume 255

Series Editor

Janusz Kacprzyk, Systems Research Institute, Polish Academy of Sciences, Warsaw, Poland

The series "Studies in Systems, Decision and Control" (SSDC) covers both new developments and advances, as well as the state of the art, in the various areas of broadly perceived systems, decision making and control–quickly, up to date and with a high quality. The intent is to cover the theory, applications, and perspectives on the state of the art and future developments relevant to systems, decision making, control, complex processes and related areas, as embedded in the fields of engineering, computer science, physics, economics, social and life sciences, as well as the paradigms and methodologies behind them. The series contains monographs, textbooks, lecture notes and edited volumes in systems, decision making and control spanning the areas of Cyber-Physical Systems, Autonomous Systems, Sensor Networks, Control Systems, Energy Systems, Automotive Systems, Biological Systems, Vehicular Networking and Connected Vehicles, Aerospace Systems, Automation, Manufacturing, Smart Grids, Nonlinear Systems, Power Systems, Robotics, Social Systems, Economic Systems and other. Of particular value to both the contributors and the readership are the short publication timeframe and the world-wide distribution and exposure which enable both a wide and rapid dissemination of research output.

** Indexing: The books of this series are submitted to ISI, SCOPUS, DBLP, Ulrichs, MathSciNet, Current Mathematical Publications, Mathematical Reviews, Zentralblatt Math: MetaPress and Springerlink.

More information about this series at http://www.springer.com/series/13304

Emil Pricop · Jaouhar Fattahi · Nitul Dutta ·
Mariam Ibrahim
Editors

Recent Developments on Industrial Control Systems Resilience

Editors
Emil Pricop
Control Engineering, Computers
and Electronics Department
Petroleum-Gas University of Ploiesti
Ploiesti, Romania

Jaouhar Fattahi
Department of Computer Science
and Software Engineering
Laval University
Quebec City, QC, Canada

Nitul Dutta
Computer Engineering Department
Marwadi University
Rajkot, Gujarat, India

Mariam Ibrahim
Department of Mechatronics Engineering,
School of Applied Technical Sciences
German Jordanian University
Amman, Jordan

ISSN 2198-4182   ISSN 2198-4190   (electronic)
Studies in Systems, Decision and Control
ISBN 978-3-030-31330-2   ISBN 978-3-030-31328-9   (eBook)
https://doi.org/10.1007/978-3-030-31328-9

© Springer Nature Switzerland AG 2020, corrected publication 2021
This work is subject to copyright. All rights are reserved by the Publisher, whether the whole or part
of the material is concerned, specifically the rights of translation, reprinting, reuse of illustrations,
recitation, broadcasting, reproduction on microfilms or in any other physical way, and transmission
or information storage and retrieval, electronic adaptation, computer software, or by similar or dissimilar
methodology now known or hereafter developed.
The use of general descriptive names, registered names, trademarks, service marks, etc. in this
publication does not imply, even in the absence of a specific statement, that such names are exempt from
the relevant protective laws and regulations and therefore free for general use.
The publisher, the authors and the editors are safe to assume that the advice and information in this
book are believed to be true and accurate at the date of publication. Neither the publisher nor the
authors or the editors give a warranty, expressed or implied, with respect to the material contained
herein or for any errors or omissions that may have been made. The publisher remains neutral with regard
to jurisdictional claims in published maps and institutional affiliations.

This Springer imprint is published by the registered company Springer Nature Switzerland AG
The registered company address is: Gewerbestrasse 11, 6330 Cham, Switzerland

Dedicated to our beloved families for supporting us all along.
Emil Pricop, Jaouhar Fattahi, Nitul Dutta and Mariam Ibrahim

Preface

Industrial control systems (ICS) have a critical place in the functioning and development of today's world. They are key components of every technical infrastructure around us, ranging from air conditioning in our homes and cars to the big factories, the energy production and distribution, the water distribution systems, and even the nuclear plants. The correct operation of the industrial control systems is essential for the functioning of our society, so they have to be designed to be resilient. This means that the ICS should be able to recover from various process faults and failures and to withstand emerging cyberattacks. These objectives can be achieved only by assuring both safety and security, being it physical or cybernetic. Also, a special interest is presented by predictive and preventive maintenance activities.

The main goal of the book is to collect valuable contributions of renowned researchers in the field of control engineering, Internet of Things, and cybersecurity. Some chapters are based on presentations and discussions that took place at the previous editions of the International Workshop on Systems Safety and Security (IWSSS, https://www.iwsss.org). The workshop, initiated in 2013, become a traditional annual scientific event in Romania. IWSSS is now a recognized venue for the exchange of experience and ideas in the field of systems safety, security, and resilience with the scope of stimulating joint work at a regional and international level.

The book comprises research based on theory, subsequent simulation and experimental results, numerous case studies, and practical implementations. Given the detailed discussion in the said context, the book offers profound insights on increasing the resilience of industrial control systems. Both fundamental and advanced topics are discussed, having the theoretical approaches sustained by practical examples.

The structure and chapters of the book are broadly grouped into core topics that address challenges related to safe operations of control systems, risk analysis and assessment, usage of attack graphs to evaluate and increase the resiliency of control systems, preventive maintenance, and malware detection and analysis. The resilience and cybersecurity of sensor networks and the Internet of Things devices, which are now an integral part of the various industrial control systems, are discussed in different chapters of the book.

Another notable contribution of this book is the inclusion of necessary and timely response to malicious attacks or hazardous situations. This topic will certainly help readers to decide the best approaches to handle such unwanted situations.

We believe, the contents of the book is essential readings for system engineers, researchers, and specialists. The topics discussed in the book are challenging and recent and we anticipate the book to represent a useful reference for all the professionals in the field of ICS resilience, safety, and security. Finally, the editors expect that this book will be a supportive auxiliary to undergraduate and graduate students, to academia and researchers trying to address security and safety issues related to the modern implementations of the industrial control systems.

Ploiesti, Romania   Dr. Emil Pricop
Quebec City, Canada   Dr. Jaouhar Fattahi
Rajkot, India   Dr. Nitul Dutta
Amman, Jordan   Dr. Mariam Ibrahim
August 2019

Contents

Editors and Contributors

About the Editors

Emil Pricop is currently with the Control Engineering, Computers and Electronics Department of the Petroleum-Gas University of Ploiesti, Romania. He holds the position of Senior Lecturer since 2018 and he is teaching Computer Networking, Software Engineering, and Human–Computer Interaction courses. He received his Ph.D. in Systems Engineering from Petroleum-Gas University of Ploiesti by defending in May 2017 the thesis with the title "Research regarding the security of control systems". His research interest is cybersecurity, focusing especially on industrial control systems security. Dr. Emil Pricop is co-editor of the book *Recent Advances in Systems Safety & Security (Springer, 2016)* and author or co-author of two national (Romanian) patents, five book chapters published in books edited by Springer, and over 30 papers in journals or international conferences. From 2013, Dr. Pricop is the initiator and chairman of International Workshop on Systems Safety and Security—IWSSS, a prestigious scientific event organized annually.

Jaouhar Fattahi is currently working with Defence Research and Development Canada (DRDC) at the Valcartier Research Centre as a defence scientist. He is also an adjunct professor with Laval University, Quebec City, Canada. He obtained his Ph.D. on the security of cryptographic protocols from Laval University in October 2015. He completed his postdoctoral fellowship at the Canadian Armed Forces Research Centre in the field of cybersecurity. He has also been a computer engineer since 1995. Dr. Jaouhar Fattahi is the author of *The Theory of Witness-Functions* for verifying security of cryptographic protocols. He now specializes in reverse engineering and machine and deep learning applied to security and cybersecurity. He is an IEEE member.

Nitul Dutta is a professor in the Computer Engineering Department, Faculty of Engineering (FoE), Marwadi University, Rajkot, Gujarat, since 2014. He has a total experience of 20 years. He received B.E. degree in Computer Science and

Engineering from Jorhat Engineering College, Assam (1995), and M. Tech. degree in Information Technology from Tezpur University, Assam (2002). He completed Ph.D. (Engineering) degree in the field of Mobile IPv6 at Jadavpur University (2013) and published 15 Journal and 30 conference papers. He has completed two AICTE sponsored research projects of worth Rs. 25 Lakhs (approx.) (Rs. Twenty-Five Lakhs only). His current research interests are wireless communication, mobility management in IPv6-based network, cognitive radio networks, and cybersecurity.

Mariam Ibrahim received her Bachelor's degree in Electrical and Computer Engineering from the Hashemite University, Jordan, in 2008, and M.S. in Mechatronics Engineering from Al-Balqa Applied University, Jordan, in 2011, and the Ph.D. in Electrical Engineering from Iowa State University, USA, in 2016. She was a lab supervisor with EE department at the Hashemite University (2008–2011). She joined the German Jordanian University (2011) as an RA, where she got a scholarship to pursue her Ph.D. studies; she is currently an assistant professor at GJU. Her research interests include discrete-event systems, stochastic systems, power systems, communication networks, healthcare systems, together with their control and resiliency analysis, and system model-based verification/attack graph generation using AADL. She is a member of Iowa Section IEEE Control Systems Society Technical Chapter. She serves as a scientific reviewer in the international scientific committee of the International Workshop on Systems Safety and Security—IWSSS since 2017, journal of *IET Cyber-Physical Systems: Theory & Application, 2018*, and *IEEE Network Magazine, 2018*.

Contributors

Qays Al-Hindawi Department of Mechatronics Eng, Faculty of Applied Technical Sciences, German Jordanian University, Amman, Jordan

Ahmad Alsheikh Department of Mechatronics Eng, Faculty of Applied Technical Sciences, German Jordanian University, Amman, Jordan

Horia Andrei SM-IEEE, Bucharest, Romania;
Doctoral School of Engineering Sciences, University Valahia Targoviste, Targoviste, Romania

Paul Cristian Andrei Department of Electrical Engineering, University Politehnica Bucharest, Bucharest, Romania

Alina-Simona Băieşu Automatic Control, Computers and Electronics Department, Petroleum-Gas University of Ploiesti, Ploiesti, Romania

Tomáš Bajtoš Faculty of Science, Institute of Computer Science, Pavol Jozef Šafárik University in Košice, Košice, Slovakia

Radoslav Benko Faculty of Law Institute of International Law and European Law, Pavol Jozef Šafárik University in Košice, Košice, Slovakia

Krishna Delvadia Chhotubhai Gopalbhai Patel Institute of Technology, Bardoli, Gujarat, India

Viktor M. Denisov "Flagman Geo" Ltd., Saint-Petersburg, Russia

Viorel Dumitru National Institute of Materials Physics, Magurele, Romania

Nirali Dutiya Department of Computer Engineering, Faculty of PG Studies, MEF Group of Institutions (MEFGI), Rajkot, India

Nitul Dutta Computer Engineering Department, MEF Group of Institutions, Rajkot, Gujarat, India

Marian Gaiceanu Department of Control Systems and Electrical Engineering, Dunarea de Jos University of Galati, Galati, Romania

Theodora Gaiceanu Gheorghe Asachi Technical University of Iasi, Iasi, Romania

Mariam Ibrahim Department of Mechatronics Eng, Faculty of Applied Technical Sciences, German Jordanian University, Amman, Jordan

Octavian Ionescu National Institute for Research and Development in Microtechnologies, IMT Bucharest, Bucharest, Romania

Nilesh Jadav Department of Computer Engineering, Faculty of PG Studies, MEF Group of Institutions (MEFGI), Rajkot, India

Dhara Joshi Department of Computer Engineering, Faculty of PG Studies, MEF Group of Institutions (MEFGI), Rajkot, India

Xinxin Lou Bielefeld University, Bielefeld, Germany

Ioan Marinescu Doctoral School of Engineering Sciences, University Valahia Targoviste, Targoviste, Romania

Terézia Mézešová Faculty of Science, Institute of Computer Science, Pavol Jozef Šafárik University in Košice, Košice, Slovakia

Stefan Pircalabu Cyberswarm Inc., San Mateo, CA, USA

Emil Pricop Petroleum-Gas University of Ploiesti, Ploiesti, Romania

Gabriel Rădulescu Control Engineering, Computers and Electronics Department, Petroleum-Gas University of Ploiești, Ploiești, Romania

Laura Rózenfeldová Faculty of Law, Department of Commercial Law and Business Law, Pavol Jozef Šafárik University in Košice, Košice, Slovakia

Pavol Sokol Faculty of Science, Institute of Computer Science, Pavol Jozef Šafárik University in Košice, Košice, Slovakia

Vasile Solcanu Dunarea de Jos University of Galati, Galati, Romania

Marilena Stanculescu Department of Electrical Engineering, University Politehnica Bucharest, Bucharest, Romania

Kajal Tanchak Computer Engineering Department, MEF Group of Institutions, Rajkot, Gujarat, India

Asmaa Tellabi University Siegen, Siegen, Germany

Andrey V. Timofeev LLP "EqualiZoom", Astana, Kazakhstan

Safety Instrumented Systems Analysis

Alina-Simona Băieşu

Abstract Operating most industrial processes, especially those in the oil and gas industry, involves an inherent risk due to the presence of dangerous/flammable substances. Therefore, using Safety Instrumented Systems (SIS) is mandatory. These systems are especially designed to protect personnel, equipment and environment by reducing the likelihood of an unwanted event to appear by reducing the severity of its impact. This chapter presents a comprehensive *Introduction* in the field of Safety Instrumented Systems, then the most important feature of a SIS is presented, *Safety Integrity Level of a Safety Instrumented System* and some *Practical Aspects Regarding Safety Instrumented Systems* are outlined. The chapter ends with some considerations regarding *IT Enabled Safety Systems*.

Keywords Safety instrumented systems · Risk analysis · IT enabled safety systems

1 Introduction

This paragraph outlines general aspects regarding the Safety Instrumented Systems (SIS) and their goal by presenting examples of such systems used in the industrial practice.

It also presents the evolution of the current in use standards that regulates the design, implementation and operation of SIS, focusing on IEC 61508 and IEC 61511 standards. The delimitation between the Control Systems (CS) and Safety Instrumented Systems (SIS) it is also highlighted by marking the major differences between the two types of automated systems.

A.-S. Băieşu (✉)
Automatic Control, Computers and Electronics Department, Petroleum-Gas University of Ploiesti, Ploiesti, Romania
e-mail: agutu@upg-ploiesti.ro

© Springer Nature Switzerland AG 2020
E. Pricop et al. (eds.), *Recent Developments on Industrial Control Systems Resilience*,
Studies in Systems, Decision and Control 255,
https://doi.org/10.1007/978-3-030-31328-9_1

1.1 Safety Instrumented Systems General Aspects

A Safety Instrumented System (SIS) aims to bring the process to a safe state when the normal operating conditions are violated, for reasons of safety and protection [1]. Therefore, the role of a SIS is to monitor potential hazardous conditions and to mitigate the consequences in case of an unwanted dangerous event appearance.

A SIS is a set of sensors, logic solvers and actuators [2].

A SIS does not improve production or efficiency but helps to reduce economic losses by reducing risks [1].

The structure of a SIS is presented in Fig. 1.

The sensors are used to measure process parameters (temperature, pressure, flow, etc.) and to use their measures to determine whether an equipment or process is in a safe or unsafe state. The sensors can be of various types, from simple pneumatic or electrical switches to intelligent sensors with diagnosis. These sensors are dedicated to SIS and use communication channels and power supplies, different from those of control system sensors.

The logic solvers have the role of establishing what decision should be made based on the information received from sensors. Typically, the logic solver is a Programmable Logic Controller (PLC) that receives as inputs the signals from sensors, runs a particular program according to these values in order to prevent possible dangerous situations and sends output control variables to actuators.

The actuators carry out the actions from the logic solver. Usually the actuators are two-way valves (opened/closed), pneumatically operated.

All SIS components must be designed so that they can safely isolate the process in the event of a hazardous situation [3].

In the following, a SIS example is presented to highlight its role, structure and operation [4].

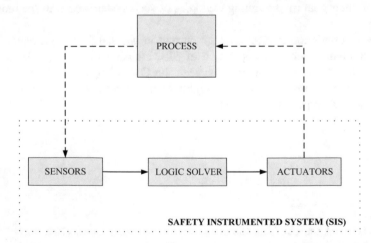

Fig. 1 A safety instrumented system (SIS) structure

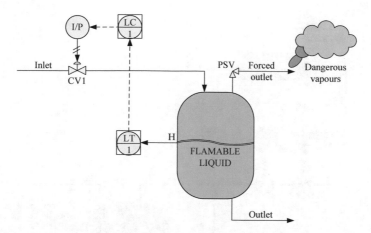

Fig. 2 A flammable liquid level control system: LC1—level controller, LT1—level transducer, I/P—electro/pneumatic converter, CV1—control valve, PSV—overPressure Safety Valve, H—liquid level

Figure 2 presents a vessel in which a flammable liquid is stored. A typical control loop must maintain the level (H) at 50%. An unwanted dangerous event can occur if the control loop fails for some reason and the vessel becomes full. The vessel has an overPressure Safety Valve (PSV) which evacuates the liquid out of vessel, but it will form a dangerous cloud of vapours [5].

Situations that can cause the control loop malfunction are:

– the Control Valve (CV1) cannot be operated;
– the Level Transducer (LT1) is faulty;
– the Level Controller (LC1) is manually operated and the Control Valve (CV1) is opened.

To prevent a possible damage, in the event of Control System (CS) malfunction, a Safety Instrumented System (SIS), as in Fig. 3, can be used. The CS equipment are marked with 1 and the equipment from SIS structure, with 2.

The SIS elements are symbolized using the ANSI/ISA 5.1 Specific Standard [6].

The SIS from Fig. 3 operation can be described as follows: the level transducer LZT2 has the role of detecting the vessel maximum liquid level (Hmax) and when this happens, the logic solver, for example a PLC, stops the vessel feeding through the UZV2 Safety Valve. Basically, the logic solver acts on the electromagnetic valve UZY2 that stops the air supply of UZV2, the air being ventilated outwards. The UZV2 will remain closed until all faults are removed. When the liquid level reaches a normal value, the operator can reset the UZV2 valve state, using HS2 (Human reSet) and the normal operation can continue.

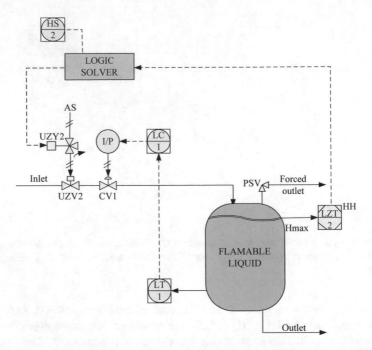

Fig. 3 Liquid level control system and safety instrumented system: LC1—level controller, LT1—level transducer, LZT2—high level transducer, UZV2 safety valve, I/P—electro/pneumatic converter, CV1—control valve, PSV—overPressure Safety Valve, UZY2—electromagnetic valve, AS—air supply, HS2—Human reSet, HH—high level alarm (High High), Hmax—high level

1.2 Safety Instrumented System Standards

The Safety Instrumented Systems (SIS) are receiving attention in the industrial sector due to the increasing environmental pressure in order to reduce the gas emissions and to get the most environmentally friendly products, but also because of the many accidents that occurred in different installations, through time.

Until the '80s, the industrial process safety issue was left to the decision of the various companies, who based on their own experience developed set of rules regarding SIS design and use. Subsequently, these sets of rules have been integrated into international standards and government regulations that require companies to comply with certain procedures.

IEC 61508 Standard

The IEC 61508 standard was developed by IEC (International Electrotechnical Commission) and covers a wide range of fields of activity and a multitude of SIS-associated equipment [7].

The standard applies for all steps which a SIS passes through from specification, design, operation, use, decommissioning and covers all the constituent parts of the SIS: sensors, logic solvers and actuators.

IEC 61508 standard has 7 parts [4]:

- Part 1, (December, 1998) presents some general specifications;
- Part 2, (May, 2000) presents the requirements for programmable electrical/electronic systems;
- Part 3, (December, 1998) presents requirements for software components;
- Part 4, (December, 1998) presents definitions, abbreviations and terminology to ensure a certain consistency;
- Part 5, (December, 1998) presents examples for determining the Safety Integrity Level (SIL);
- Part 6, (April, 2000) provides the appliance guide of Parts 2 and 3;
- Part 7, (March, 2000) presents a brief presentation of the techniques and methods relevant for Parts 2 and 3.

The most important part of IEC 61508 is the life cycle model of a SIS, shown in Fig. 4.

IEC 61511 Standard

IEC 61511 includes additional guidance for determining the Safety Integrity Level (SIL) to be imposed by the design team at the beginning of the SIS design phase and is structured in 3 parts [8]:

- Part 1, provides definitions, hardware and software requirements;
- Part 2, provides the appliance guide of Part 1;
- Part 3, provides guidance for determining the SIL.

The IEC 61511 standard is dedicated to the end user whose task is to design and operate the SIS in an industrial plant. The requirements are those imposed by IEC 61508, but modified to fit to practical situations from an industrial plant. 61511 standard does not cover the design and implementation of equipment used in safety applications, such PLCs, which remains standardized by IEC 61508.

1.3 Safety Instrumented Systems Versus Control Systems

A fundamental question is whether the Safety Instrumented Systems (SIS) and Control Systems (CS) systems should be combined or a clear delimitation between these two should be established.

According ANSI/ISA 84.01 [9], *Separation between control systems and safety instrumented systems reduces the probability that at some point in time either control and safety functions to be inactive, or some modifications to the control systems will lead to changes in the functionality of the safety systems. Therefore, it is generally necessary to separate the control systems from the safety instrumented systems.*

Several basic differences between the two types of systems support the idea that control and safety should be separated. Control operations are active, dynamic, and performance-oriented. Safety operations are passive.

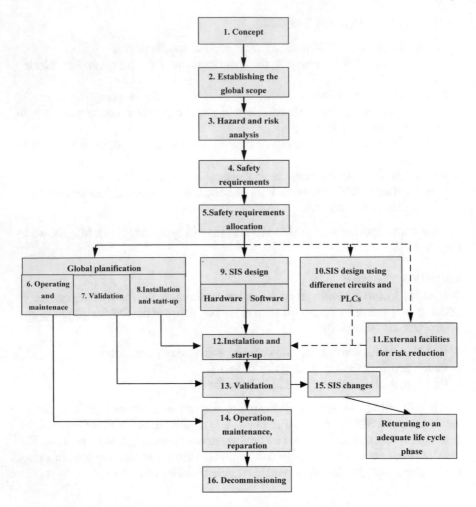

Fig. 4 Safety life cycle model according to IEC 61508 [7]

CSs act actively to maintain or change process conditions, helping to achieve the best performance of the process. They are often used to force the process to its limits, in order to obtain the required performance. They were not built to provide safety.

Since control system operation is continuous, it does not incorporate error diagnostics routines. A CS either work or not. There are no hidden defects. In these systems, operating errors are visible.

CSs are usually flexible and allow easy operation in the sense that, for example, operators can modify certain parameters or exclude certain part of the control system. This aspect has to be avoided in case of the safety systems.

Safety Instrumented Systems (SIS) are exactly the opposite of Control Systems (CS). They work only at certain times, their action must be restricted, and they must be reliable and act instantly when necessary. An example is the overPressure Safety

Valve (PSV) from Figs. 2 and 3, which must be closed as long as the pressure is below a certain value and opened when the pressure reaches a certain high limit. If the pressure never touches this limit, the valve should never open. Likewise, it may happen that this valve if stays closed for a long time, when a problem appears and must be opened, it cannot be opened. Therefore, these systems can hide defects that are not directly observed.

Because the SIS components stay for a long time in standby mode, it may not work if they must be put into operation. As such, these systems should be tested or should include techniques that provide the possibility of self-testing.

Safety instrumented systems are designed to involve human intervention as low as possible. The operator interacts with the control system. If it fails, the next step is to pass it on manually and the operator must act directly on the process. If this intervention also fails to make the necessary corrections, the last line of defence, the SIS, should work automatically and independently. The only human interference allowed is the start-up or maintenance of certain parts of the system.

2 Safety Integrity Level of a Safety Instrumented System

This paragraph describes the main methods for determining the Safety Integrity Level (SIL), which is the most important feature of a SIS.

The IEC 61508 standard defines four levels of SIL, marked with SIL1, SIL2, SIL3 and SIL4. SIL 4 denotes the highest level of safety integrity and SIL 1, the smallest. The Safety Integrity Level of a SIS can be expressed in terms of Probability of Failure on Demand (PFD) for systems/functions that operate with a low demand rate or in terms of Probability of dangerous failure Per Hour (PFH) for systems/functions that operate with a high demand rate or continuously.

There are several methods and tools for determining SIL, developed by different companies and organizations in order to provide support for assessing the process's risk and turning it into something palpable that has a certain meaning that is the required SIL level.

These methods can be grouped into four main categories: quantitative and semi-quantitative methods (e.g. LOPA—Layer of Protection Analysis), qualitative (e.g. risk matrix, risk graph) and semi-qualitative methods (e.g. calibrated risk graph).

A system risk is a function of the frequency of an unwanted dangerous event and the severity of the consequences of that event/hazard. Risk can influence personnel, production, environment etc. [10].

$$\text{RISK} = \text{FREQUENCY} \times \text{HAZARD CONSEQUENCE}$$

According to [5], hazard is an inherent feature of a system/process that has the potential to cause damage to individuals, processes or the environment.

In the case of chemical processes, hazard is the combination of a hazardous material, an operating environment with problems and some unplanned events that can cause accidents.

Depending on how the Risk Reduction Factor (RRF) is expressed, quantitatively or qualitatively, what is the scope and purpose of the risk analysis a particular type of method is chosen. Although the method used is a qualitative one, a number always quantifies the determined SIL (SIL1, SIL2, SIL3 or SIL 4).

The SIL level is a measure of the performance of the safety system and not a direct measure of process risk. The higher the risk of a process is, the higher the safety level will be and the number that follows the SIL increases as value.

Usually, the qualitative methods are used in the design phase of the SIS and the quantitative methods are used more in the SIL verification and validation phases.

2.1 Quantitative Methods

For systems with a low demand ratio, after determining the risk level, the next step is to determine, using (1), the Risk Reduction Factor (RRF) required to meet the tolerable risk level. This is achieved by dividing the number of times per year when a SIF function fail to function to the number of demands per year. The result is the acceptable number of times when a SIF may not operate at a demand per year, named the Probability of Failure on Demand (PFD) [8].

The Risk Reduction Factor (RRF), in frequency, is given by [11]:

$$RRF = \frac{F_{np}}{F_t},$$ (1)

where F_{np} is the frequency of the risk without protection and F_t is the tolerable risk frequency.

The Probability of Failure on Demand (PFD) is:

$$PFD = \frac{1}{RRF} = \frac{F_t}{F_{np}}.$$ (2)

Further, SIL is determined using Table 1.

The Probability of Failure on Demand, PFD of a Safety Instrumented System (SIS) is obtained by summing the PFD for transducers (ZT), logic solvers (PLCs) and actuators (UZ) [3]:

$$PFD_{SSP} = PFD_{ZT} + PFD_{PLC} + PFD_{UZ}.$$ (3)

Also, PFD for transducers (PFD$_{ZT}$) is obtained by summing PFD for sensors and adapters:

Table 1 Safety integrity level and required values for the instrumented system performance in case of low demand rate system [7]

Safety integrity level, SIL	Probability of failure on demand, PFD	Availability, 1-PFD (%)	Risk reduction factor, RRF = 1/PFD
4	10^{-4} to 10^{-5}	99.99–99.999	10^4 to 10^5
3	10^{-3} to 10^{-4}	99.9–99.99	10^3 to 10^4
2	10^{-2} to 10^{-3}	99–99.9	10^2 to 10^3
1	10^{-1} to 10^{-2}	90–99	10^1 to 10^2

Table 2 Safety integrity levels (SIL) and required values for the safety system performance in case of a high demand rate/continuous operation systems [7]

Safety integrity level, SIL	Probability of failures per hour, PFH	Mean time to failure, MTTF
4	10^{-9} to 10^{-8}	10^4 to 10^5
3	10^{-8} to 10^{-7}	10^3 to 10^4
2	10^{-7} to 10^{-6}	10^2 to 10^3
1	10^{-6} to 10^{-5}	10^1 to 10^2

$$PFD_{ZT} = PFD_{adaptor} + PFD_{sensor}, \tag{4}$$

and the PFD for actuators (PFD$_{UZ}$) is obtained by summing the PFD for electromagnetic valve UZY and the safety valve UZV:

$$PFD_{UZ} = PFD_{UZT} + PFD_{UZV}. \tag{5}$$

For systems with a high demand ratio or continuous operation, the Safety Integrity Level (SIL) is determined using the Probability of a hazardous Failure per Hour (PFH) or the Mean Time To Failure indicator (MTTF), according to Table 2.

The two measures (PFD and PFH) of the SIS performance are related, the dependence between them being expressed through equation [12]:

$$PFD = \frac{T}{2} \cdot PFH = \frac{T}{2 \cdot MTTF}. \tag{6}$$

PFD is the Probability of Failure on Demand, T represents the test or replacement interval, PFH Probability of a dangerous Failure per Hour and MTTF is the Mean Time To Failure.

10 A.-S. Băieşu

2.2 Risk Matrix

The risk matrix is one of the most popular method for determining the Safety Integrity Level (SIL), due to its simplicity. The risk matrix takes into account the frequency and severity of an unwanted event, based on a classification of the risk parameters.

The frequency of an event to occur can be quantified in terms such Small (S), Medium (M), High (H) or any other suggestive terms, Table 3.

Consequences and its severity can be quantified based on various risk factors such personnel, environment, production, equipment, capital, etc.

Table 4 provides an example of the severity of risk values, associated with personnel, environment and production.

Table 3 Risk frequency values [13]

Level	Frequency	Qualitative interpretation
3	High (H)	An unwanted event may occur more than once in the predicted life time of the plant
2	Medium (M)	An unwanted event can occur once in the predicted life time of the plant
1	Small (S)	An unwanted event may appear with a low probability over the predicted life time of the plant

Table 4 Example of severity values/risk consequences [13]

Level	Severity/consequences	Personnel	Environment	Production/equipment
III	Catastrophic (C)	More fatalities	Escapes of dangerous substances outside the plant perimeter	Losses greater than $1,500,000
II	Serious (S)	Single death or injuries requiring recovery time	Releases of non-hazardous substances outside the perimeter of the plant or leakage of dangerous substances into the perimeter of the plant	Losses between 100,000 $ and 1,500,000 $
I	MINor (MIN)	Injured that requires medical treatment or first aid	Releases of hazardous substances to a restricted area of the plant perimeter or without leakage	Losses up to $100,000

Fig. 5 Global risk (risk matrix)

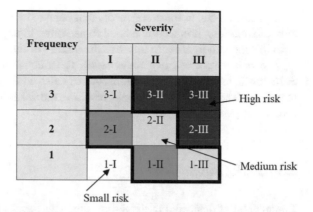

According to Table 4, the consequences may be MINOR (MIN), SERIOUS (S) or Catastrophic (C), according to severity level. Categories can be selected either qualitatively or quantitatively by attaching some economic figures, deaths, etc.

By joining the values of the two properties, the frequency and severity of the risk, the risk matrix is obtained (Fig. 5).

If the identified risk is high, then changes are recommended. If the identified risk is medium, it is necessary to add additional safety levels. If the identified risk is low, no changes or additions of protection levels are required.

Given that, if each risk matrix cell has an associated SIL level, then the process of determining the SIL level is simple [13].

Figure 6 shows a typical modified risk matrix chart for determining the SIL level.

Severity-minor consequence (I), low frequency (1) leads to unnecessary SIL. This means that the risk is considered tolerable.

Severity-minor consequence (I), average frequency leads (2) to a low SIL level, while catastrophic consequence, high frequency leads to a high SIL level. In case of a SIL 3 or SIL 4 required level, additional studies should be conducted as a single SIF may not provide sufficient risk reduction [14].

Fig. 6 The modified risk matrix for determining the SIL level

Frequency	Severity		
	I	II	III
3	SIL 2	SIL 3	SIL 4
2	SIL 1	SIL 2	SIL 3
1	No SIL	SIL 1	SIL 2

Sometimes, due to the assessments that need to be made, the result may be unrealistic. Therefore, it is recommended to use other tools and methods in conjunction with this method in order to improve the quality of the determination [8].

The risk matrix has two dimensions, the frequency and severity of an event. Sometimes, in practical applications, there is also added a third dimension that takes into account additional safety levels, resulting the risk matrix of safety and protection levels.

2.3   Risk Graph

The method of determining the Safety Integrity Level (SIL) using the risk graph is based on the methods written in the German publication DIN 19250 [15].

Table 5 lists the risk parameters classification suggested in IEC 61511.

The risk graph method is a qualitative method that takes into account the consequence and frequency of a dangerous event, but also the likelihood that the personnel could avoid the danger [14].

For the Consequence parameter (C), in relation to personnel's risk, four categories of consequences are suggested, ranging from minor injury to multiple deaths. C1 is the least severe category. In general, the consequences are measured by the degree of injuries of individuals but also by environmental or financial measures [8].

Occupancy Frequency (F) shows the fraction of time in which the hazardous area is occupied by personnel. F2 shows a higher risk than F1 because the area is occupied more frequently. Usually, in accordance with IEC 61511, F1 can be selected if the hazardous area is occupied less than about 10% of the time.

Table 5   Risk parameters classification according to IEC 61511

Risk parameter	Notation	Classification
Consequence (C)	C_1	Minor injuries
	C_2	Serious injury of one or more persons
	C_3	Death of one person
	C_4	Catastrophic effect. Many deaths
Occupation frequency of the affected area (F)	F_1	Rare to frequent (<0.1)
	F_2	Frequent to continuous (>0.1)
The probability of avoiding the consequences (P)	P_1	Possible in certain conditions (>90%)
	P_2	Almost impossible (<10%)
The probability of occurrence of an unwanted event (A)	A_1	Very unlikely to appear (F < 0.01/year)
	A_2	It is unlikely that an unwanted event will appear (F > 0.01/year)
	A_3	Relatively large probability that an unwanted event to appear (F > 0.1/year)

The possibility of personnel to avoiding the danger is incorporated into parameter P. This parameter shows which methods must be identified by personnel in order to escape the danger. Additionally, the rate of development of the dangerous event is taken into account. Two categories are suggested, P1 and P2, P2 indicating the highest risk. In order to select P1, a list of statements must be validated. Such lists of statements are suggested in IEC 61511.

The final parameter is the probability of occurrence (A), which is the frequency of the occurrence of an unwanted event without SIF (Safety Instrumented Function).

Figure 7 is a typical chart of personnel risk. Similar graphs can be obtained for the risk associated with equipment, production losses or environmental impact.

The path from left to right is determined by the selected risk parameter values. The selected result, the occupancy frequency and the probability of avoidance leads to a certain output line, O. The output line leads to three values of the parameter A. Choosing the parameter A is the last step in determining the SIL level. Choosing a higher A parameter results in a higher SIL level [14].

If the probable consequence is assessed to be serious injury to one or more persons, C2 is selected. If the area could be exposed to personnel rarely to more frequent, F1 is chosen. It is possible that under certain conditions the unwanted events can be avoided, which means that parameter P1 should be chosen. The combination of these risk parameters leads to the O2 output line. Considering a high probability that an unwanted event to occur, A3 is selected. According to Fig. 7, SIL 1 is required.

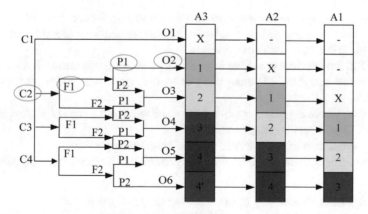

Fig. 7 Risk graph: C—consequences (C1—minor injuries, C2—serious injuries, C3—one death, C4—more deaths); F—frequency (F1—rare to frequent, F2—frequent to continuous); P—probability of avoidance (P1—sometimes possible, P2—almost impossible); A—appearance probability (A1—very small, A2—small, A3—relatively high); X—SIS is not necessary, 1, 2, 3, 4—safety integrity level, SIL, 4'—one SIS is not enough

3   Practical Aspects Regarding Safety Instrumented Systems

This paragraph outlines some general aspects regarding SIS implementation by listing the main requirements of the components, the available implementation technologies and ways to connect them.

The performance of a SIS depends on each element from its structure (sensor, logic solver or actuator), so a special attention must be given starting from the stage of choosing the type of each sensor and actuator to be used, to the implementation of the logic solver phase or to the way that these parts are interconnected.

The field devices that are parts of the Safety Instrumented Systems (SIS) must be selected and installed to meet the performance requirements expressed through the Safety Integrity Level (SIL), for each SIF (Safety Instrumented Function).

Any SIS associated field device must be separate and independent of those of the control system. With few exceptions, sharing equipment with the control system can cause procedural and maintenance problems.

In order to achieve the necessary independence, the following components should be separated from the control system [16]:

– the field sensors, safety and electromagnetic valves, pulse lines;
– wires, panels and junction boxes;
– voltage sources.

If the SIS signals are to be sent to the control system for comparison, the signals between the two systems should be isolated (using optical insulators) to prevent that one single failure affecting both systems.

The SIS field devices should be fail safe. In most cases, this means that the devices are normally energized. De-energizing or losing power will initiate the unit/plant shutdown.

Systems where the shutdown function is activated in the case of alarms are used to reduce false errors due to power failures.

Only tested technologies should be used in safety applications.

When installing sensors, the following general requirements must be considered [17]:

– contacts must be normally closed and normally energized;
– sensors must be directly connected to the logic system;
– smart transducers are more commonly used for SIS due to improved diagnostics facilities and rigorous reliability. When using such transducers, procedures should be established to ensure that they cannot be left in forced outputs;
– the contacts of the electric circuit should be hermetically sealed for greater reliability;
– the SIS associated field sensors should be differentiated in a certain way by the sensors of the control system by a single tag, numbering or colour;
– when using redundant sensors, a discrepancy alarm should be provided to indicate the failure of a single sensor.

In case of flow measurement, although diaphragm transducers are primarily used in most safety applications, the vortex and magnetic flow transducers can offer some advantages such ease of installation and improved performance.

In the case of temperature sensors, the main failure mode of thermocouples is burning; therefore, detection and alarm systems should be used.

In the case of pressure sensors, the main precautions to be taken are when selecting the range and that the condensate accumulation will not cause calibration problems.

Level transducers with air blowers and nitrogen purge have proven to be reliable, requiring low maintenance.

When installing actuators, they should be set to maintain their status after a shut-down function, until manual reset. They should be allowed to return to their normal state only if the variables that generated the stop have returned to their normal operating values.

The following aspects should be considered when selecting the actuators:

- closing/opening speed;
- leaks;
- fire resistance;
- suitability/compatibility of the material;
- diagnostic requirements;
- the safety of the valve;
- the need for a position indicator or limiter;
- on-line maintenance capacity.

The bypass valves must be considered for each safety valve that fails in the closed state. Limiters can be used to initiate an alarm if a bypass valve is opened.

Generally, safety valves should be dedicated to safety applications and separate from the control valves.

Electromagnetic valves have a low reliability and therefore can be one of the most critical (weaker) components in the whole system. A common cause of the failure is burning the coil causing a false stop. It is important to use a tap like to withstand high temperatures, including heat generated by its own coil, heat generated by furnaces, exposure to sunlight, etc. Double ball valves or redundant valves can also be used.

24 V powered electromagnetic valves seem to be more reliable due to low energy consumption and low heat output.

When installing a field equipment, consideration should be given to issues related to:

- environment (temperature, vibration, shock, corrosion, humidity etc.);
- on-line testing, if necessary;
- maintenance requirements;
- accessibility;
- local indication;
- protection against frost, if required.

Common wiring defects include grounding, noise and induced voltages.

General recommendations to minimize connection issues:

- each field device must have its own set of connections to the logic system;
- it is not recommended to connect multiple discrete inputs to a single input channel of a logic system, in order to reduce the number of wires and the expenses related to the input modules. A disadvantage of such an arrangement is that it will be much more difficult to troubleshoot and diagnose problems.
- it is also not advisable to connect a single logic solver output to multiple valves;
- each input and output field variable must be limited to the electric current; This can be done either as an integral part of the input/output modules of the logic element or by using external fuses;
- SIS wiring and junction boxes must be delimited from all other instruments and/or wiring of the control system;
- all equipment, wires etc. must be clearly labelled.

4   IT Enabled Safety Systems

Information Technology (IT) is a technology that has the fastest rate of development and application in all areas of different industrial fields and requires adequate protection in order to provide high security. The goal of the safety analysis for an IT system is to identify the main threats, vulnerabilities and safety features in order to protect the information stored electronically with implications regarding data integrity, availability and confidentiality [18].

As the complexity of different type of projects increased, the advances in information technology and data acquisition equipment and tools, offered the possibility to significantly increase the ability to store, retrieve, transmit and manipulate data and information during the entire steps of a project, Fig. 8. A precise real-time control of the equipment, materials, construction methods and work environment is needed in order to ensure the safety risk prevention and emergency response [19].

IT can reduce the errors rate in three ways: by preventing the main errors, by offering a rapid response after an unwanted dangerous event has occurred and by tracking and providing feedback about that event. The main methods for preventing errors and dangerous events appearance are based on tools that can improve communication, make information more accessible, assist with calculations, perform real time checks, provide monitoring and decision support.

In order to minimize losses, it is necessary to use risk management and risk assessment concepts regarding the area of the information technology [20].

The IT enabled systems risk management consists of analysis, planning, implementation, control and monitoring of implemented measurements. Using risk management the risk level is identified and measures to reduce risk are taken in order to maintain the risk on an acceptable level.

The IT enabled systems security modules are grouped into generic aspects (organization, personnel and data backup policy), infrastructure (buildings, server room

Fig. 8 Transmitting the information between the different layers of an industrial system

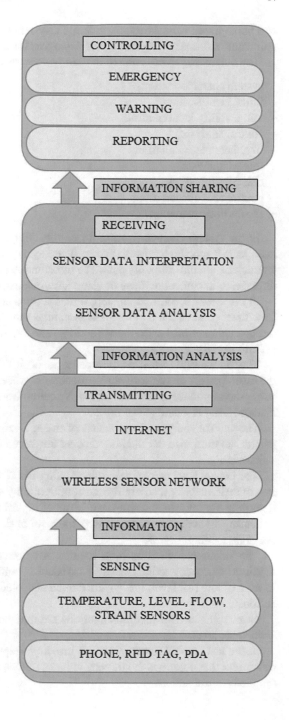

and home-based workstation modules), IT systems (servers, clients), networks, and applications (e-mail, web server and databases for modeling modules) [18].

The risk management and methodology consists of [18]:

- system characterization,
- threat identification,
- vulnerability identification,
- control analysis,
- likelihood determination,
- impact analysis,
- risk determination,
- control recommendations,
- results documentation.

In the system characterization phase the information relevant to the IT system is collected: specific hardware and software system interfaces, data and information characteristics and criticality data. The information is collected using questionnaires, interviews, document reviews or using automated scanning tools.

In the thread identification step, threat actions and threat sources are identified. The threat sources can be classified as natural threats (floods, earthquakes), human threats (unintentional or deliberate actions) and environmental threats (power failure, pollution) [18].

The vulnerability identification phase identifies the IT system vulnerabilities of the assets (hardware and software), procedures, processes and information transfer by using security requirements check list. Vulnerabilities can be identified by verifying if the security standards are fulfilled.

The control analysis step consists of the analyzing of the controls implemented in order to minimize the likelihood of an unwanted event, which affects the system vulnerability.

The impact level can be determined using the IT system and data sensitivity, i.e. loss of their integrity, availability and confidentiality and can include some estimation regarding the frequency occurrence, costs of repairing and assumed damage factor.

During the risk determination step, the level of risk is assessed. The risk level can be: high, medium or low.

The values classified as high require fast corrective measures. In the case of medium values, corrective measures are required within a reasonable period of time. In case of low risk level, the system can be accepted with or without any corrective action.

Next, some control recommendations are made in order to reduce the risk to an acceptable level.

In the last step all results from all previous steps are documented in an official risk report that describes the threats, vulnerabilities, measured risk level and recommended control actions.

References

1. USPAS: Controlling Risks Safety Instrumented Systems. http://uspas.fnal.gov/materials/12UTA/15_safety_instrumented_systems.pdf (2012). Accessed 25 June 2019
2. Honeywell Industrial Measurement and Control: Safety Instrumented Systems (SIS), Safety Integrity Levels (SIL), IEC61508, and Honeywell Field Instruments. Phoenix, AZ (2002)
3. Emerson Process Management—Fisher Controls International LLC: DVC6000 SIS Training Course 1—Basic Fundamentals of Safety Instrumented Systems. Marshall-town, Iowa. http://www.documentation.emersonprocess.com/groups/public_valvesprodlit/documents/training_info/sis_training_course_1.pdf (2005), Accessed 15 May 2019
4. Macdonald, D.: Practical Industrial Safety—Risk Assessment and Shutdown Systems. Newnes, Boston (2004)
5. Technologies, I.D.C.: Overview of Safety Instrumented Systems. IDC TECHNOLOGIES, West Perth, Australia (2012)
6. Instrument Society of America: ANSI/ISA-5.1-2009 Instrumentation Symbols and Identification. Research Triangle Park, North Carolina (2009)
7. International Electrotechnical Commission: IEC 61508—Functional Safety of Electrical/Electronic/Programmable Electronic Safety-Related Systems, 2nd edn. Geneva, Switzerland (2010)
8. International Electrotechnical Commission: IEC 61511—Functional Safety—Safety Instrumented Systems for the Process Industry Sector. Switzerland, Geneva (2018)
9. Instrument Society of America: ANSI/ISA 84.01—Application of Safety Instrumented Systems for the Process Industries. Research Triangle Park, North Carolina (1997)
10. Center of Chemical Process Safety: Guidelines for Chemical Process Quantitative Risk Analysis. American Institute of Chemical Engineers (AIChE), New York (1989)
11. Torres-Echeverria A.C: On the use of LOPA and risk graphs for SIL determination. J. Loss Prev. Process Ind. **41**, 333–343 (2016)
12. Gulland, W.G.: Methods of determining safety integrity level (SIL) requirements—pros and cons. In: Redmill, F., Anderson, T. (eds.) Practical Elements of Safety, pp. 105–122. Springer, London (2004)
13. Lassen, C.A.: Layer of Protection Analysis (LOPA) for Determination of Safety Integrity Level (SIL). The Norwegian University of Science and Technology, Department of Production and Quality Engineering (2008)
14. Marszal, E., Scharpf, E.: Safety Integrity Level Selection—Systematic Methods Including Layer of Protection Analysis. The Instrumentation, Systems and Automation Society (ISA), Research Triangle Park, NC (2002)
15. Baybutt, P.: An improved risk graph approach for determination of safety integrity levels (SILs). Process Saf. Prog. **26**, 66–76 (2007)
16. ABB Safety Lead Competency Centre: A Methodology for the Achievement of Target SIL. Cambridgeshire, UK (2013)
17. Gruhn, P., Cheddie, H.: Safety Instrumented Systems—Design, Analysis and Justification, 2nd edn. The Instrumentation, Systems and Automation Society (ISA), Research Triangle Park, NC (2006)
18. Nikolić, B., Ružić-Dimitrijević, L.: Risk Assessment of information technology systems. Issues Inf. Sci. Inf. Technol. **6**, 595–615 (2009)
19. Mirosław, J.: Information technology applications in construction safety assurance. J. Civ. Eng. Manage. **20**(6), 778–794 (2014)
20. International Organization of Standards: ISO/IEC 13335-2—Information Technology—Guidelines for the Management of IT Security—Part 2: Managing and Planning IT Security. Switzerland, Geneva (1997)

Risks Assessment of Critical Industrial Control Systems

Gabriel Rădulescu

Abstract When we deal with the risks associated with the industrial control systems (ICS), we have to frame them into a more general problem: the risk management (RM) techniques. In fact, organizations always manage risk in order to fulfill their (business) tasks and objectives, and we speak here about economic and financial risk, personnel physical risk, equipment failure, ICS malfunctioning and so on. Normally, such organizations evaluate their business risks, determining how to deal with them in the frame of their priorities, taking into account internal and external constraints. In fact, RM is regarded as an interactive process, this being in permanent connection with usual (technical) processes. At the same time, when using ICS it is normal to have some sort of good engineering practices and safety compulsory rules. These safety assessments are formulated as regulatory requirements, this being a part of the official operating procedures. This is why we consider that RM (in general) and ICS associated risks (in particular) may be regarded as an added/complimentary dimension to any plant operation. Based on a comprehensible literature study, this chapter will concentrate on how ICS associated risks (usually formulated at the global system information level) are identified, expressed and (if possible) quantified. At the same time, like any other RM activity, we will indicate that dealing with these risks usually impacts the other system levels. It is our intention to show how extending the concepts here emphasized (for the control level) provides ICS-specific rules to be integrated in specific system/plant operating procedures. Finally, some conclusions will be presented.

1 Introduction

As global systems are engineered and operated, their productivity, safety and security are the major reasons why ICS operators always make decisions. Productivity is a measure of the system's efficiency in converting inputs into useful outputs [1]. At

G. Rădulescu (✉)
Control Engineering, Computers and Electronics Department, Petroleum-Gas University of Ploiești, Ploiești, Romania
e-mail: gabriel.radulescu@upg-ploiesti.ro

© Springer Nature Switzerland AG 2020
E. Pricop et al. (eds.), *Recent Developments on Industrial Control Systems Resilience*,
Studies in Systems, Decision and Control 255,
https://doi.org/10.1007/978-3-030-31328-9_2

the same time, safety may be seen as "*freedom from conditions that can cause death, injury, occupational illness, damage to or loss of equipment or property, or damage to the environment*" [2]. Likewise, system's security is the area of concern involving the systems safeguarding by keeping the integrated hardware devices and software applications under strict control [1]. The major generic risk in managing all these three requirements is that sometimes serious conflicts may arise between them. For instance, if safety requirements conflict with good security practice, or if productivity conflicts with safety regulations, a difficult question has to be answered: how will the organization decide?

The availability of ICS and their services is another major task for ICS operators, as ICS may be part of critical infrastructure (our case), where continuous and reliable operations are always needed. In this case, ICS comply with strict requirements in order to provide this high availability and low time recovery. Subsequently, the system operating procedures should contain strict references on such assumptions, constrained by system requirements, otherwise the organization may manage the risks in a way that could cause unintended consequences.

Another key feature of ICS is that, when a risk assessment is done, there may be extra-considerations that are not taken into account when doing a risk assessment for a traditional IT system. In this case, risk assessments have to incorporate both physical and digital effects due to the impact of a cyber-incident in an ICS.

To sum up, by using a literature study, we have to perform a relevant analysis of the identification, formulation and quantification of risks associated with ICS, taking into account the impacts on process safety assessments. We will consider here the physical impact of a cyber-incident on an ICS and the impact of non-digital control components incidents within ICS, as a whole approach. At the same time, we will try to see how the impact of an incident from the ICS could propagate to a connected system (ICS or a physical one) as such failures can easily cascade to other internal or external systems.

2 Risk Assessment—An Overview on How to Proceed

The actual technical systems imply various communication devices and technologies in order to communicate with various external actors to support their business objectives. ICS is one such system which plays a major role in the monitoring, controlling and sometimes even optimization of industries (oil and natural gas extraction and processing, automotive, aerospace, electric power, nuclear energy and so on [3–5]. The ICS are part of critical infrastructure elements, as their provided services failure usually leads to adverse effects on system security, economic development, public health and safety (or any combination among them) [6]. These systems have to be secured and highly protected, because a lot of people increasingly rely on the critical infrastructures in order to provide essential services on a day-to-day basis (taking into account the continuous advancement in technology) [3, 7].

In the past, the ICS were almost perfectly isolated from other systems, because they use particular pieces of hardware, dedicated communication networks, special channels and used standards. Of course, all these peculiar elements were completed by a total lack of connection to the Internet. However, at present, smart devices, open sensors, standard operating procedures and even open software have exposed the system to the Internet network. The major reason for that is the growing integration of 'cost-effective and emerging technologies' such as 'commercial-off-the-shelf' devices, exposing the assisted systems to the real and cloudy world [3, 4, 8, 9]. These open technologies and their dedicated devices bring together a great variety of vulnerabilities, some of them being very difficult to be detected and/or removed.

Stuxnet, for instance, is one of the strongest attacks that have been directed to world ICS. This computer worm, which appeared in June 2010, was estimated to take total control of programmable ICS implemented by Siemens AG [4]. A Stuxnet infection caused the systems under control to quietly malfunction, offering a perfect illusion of normal operation. As proven by specialists, the worm used to introduce false data directly to the system monitor facilities [10].

These state-sponsored attacks put at risk the government infrastructure and industry frameworks, especially for Nordic European countries, as publicly proven by a top cyber security company, FireEye Inc. Another example is that, in 2014 (and probably in 2016 too), we witnessed a systematically unlawful retrieval of intelligence (both political and military) by a Russia-based Advanced Persistent Threat group. Their actions have been placed also in Nordic countries, consisting in phishing emails and fake login pages [11, 12].

Day by day, the ICS—especially the critical ones—become more susceptible to cyber-attacks due to their inherent critical nature, combined with the subsystems and components vulnerabilities. These attacks are carried out not only by individual attackers, but also from strong-organized and well-funded groups [3, 8]. This is why the protection and security of the nationwide facility systems, resources and information are of real importance to the governments and their subsidiary organizations.

But several questions have to be put: 'how do we know a system is safe and secure?', 'how much do we pay for this security?', and 'does a single and system-wide measure of security exists?' [4, 13–17]. There are several researchers that have attempted to answer these questions by adopting a measure of security, speaking here about a common metric or framework of metrics.

Many approaches exist in this continuous effort to organize the knowledge in this field, but this book chapter concentrates only on the use of risk analysis techniques to present this security posture of the system. We have also to take into account that risk analysis techniques are extensively used to reduce the risks effects for ICS, as a measure of safety and security [3].

This chapter will firstly explore the concept of security measure by metrication (to a small extent). Furthermore, it will highlight the major concepts in Risk Analysis, which is an explicit subset of RM methods. Risk Analysis is a wide concept, as it may be differently performed depending on a specific subject. In our case, generic ICS are used to present a systems overview and how they differ from the conventional Information and Communication Technology systems. Their specific threats and vulnerabilities are also presented. The next step is to identify different available Risk Analysis methods in order to classify them and develop a generic framework applicable to ICS. The chapter ends with appropriate concluding remarks.

2.1 Problem Statement Description

Recently, we have witnessed the introduction of countless vulnerabilities in the ICS, making them an easy target for attacks (both internal and external) due to currently used technologies (the wireless sensors, control smart devices together with old legacy devices, the Internet network and so on) [4]. We sustain that ICS may be victims of potential vulnerabilities and risks to the operations, as they use high-tech technologies for interconnection and inter-connectivity [6, 7]. In addition, the situation is aggravated by the critical nature of the ICS which increases these wider and more complicated attacks. From passive unfriendly users experimenting their skills to active attacks originating from other states and terrorists, all have the same target: ICS [7, 8, 12]. But there is one big difference: these attacks can cause physical damage to the ICS, their controlled systems and the humans, unlike in traditional Information Technology systems [4, 10].

2.2 The Research Interest

With the progress of time and technology, critical ICS became an interesting issue to investigate for many people and organizations. Our research interest was inspired by Joseph Mukama, who, in his thesis "Risk Analysis as a Security Metric for Industrial Control Systems", shows the following aspects [3].

> A review by the Department of Homeland Security (United States) reported a continued increase in the frequency and sophistication of cyber-threats against the country's CI with over 145,000 cyber security incidents in the latest year [7]. This trend can destabilize national services and cripple the national economy and therefore protecting the ICS is of utmost importance for both the economy and stability of a nation [18].

> According to the National Institute of Standards and Technology (NIST), security facilitates decision making, determining the efficiency and performance as well as the security posture of an organization [19]. The risk-based approach is an effective approach against cyber-attacks in the ICS [18]. The NIST Special Publication 800-82r1 further emphasizes the need for assessing and rating risk of possible vulnerabilities [5].

> Risk analysis, as an approach for security [evaluation], guides organizations in their approach to security and helps them manage the security in a systematic, repeatable and formal way. Furthermore, risk analysis empowers the management with vital information including: justification for cost-effective countermeasures and mitigation approaches, highlighting areas that need to be made more secure and those that can be less secure, and increases security awareness by assessing and reporting the strengths and weaknesses of the security posture an organization.

ICS have the potential of being much better if the security is enhanced and handled using a more proactive approach. The risk analysis helps in identifying the probable consequences/risks associated with the vulnerabilities and prevent information leakage and security attacks.

2.3 A Deeper Examination—The Research Procedure Explained

Due to the existing methodologies of providing and applying security measures in various fields of activity, research in this field proves to be very vast and quite difficult to summarize. An extensive papers' list related to critical ICS was at first collected, all containing references about risk analysis methodologies. We have mainly used Web of Science (Clarivate Analytics) and Google Scholar tools (with a few IEEE Xplore paperwork included) and most of them were reviewed. These scientific tools provided access to both the latest and the most relevant articles, journals and conference proceedings in the ICS security field.

The applicability of Web of Science was motivated because it gives the best results when searching for most commonly used indexing approaches which serve the technical research. Furthermore, the general Google engine was adopted in order to supplement the data already gathered. In many situations we have used key words like 'Industrial Control Systems', 'ICS', 'Risk Analysis', but 'Risk Assessment' and 'Risk Frameworks' were also used.

This systematic research methodology overview is represented in Fig. 1.

We processed the (primarily) chosen papers, mainly searching for the terms described above, looking for potential a match between them and the words in abstract, introduction, methodology or implementation and the conclusion sections.

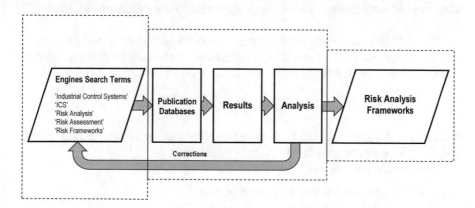

Fig. 1 The research technique. Author's own illustration

Only some papers have qualified after his stage further to which a thorough analysis was performed on them. This process was iteratively applied until we identified the risk analysis frameworks that serve this chapter.

The chosen frameworks were then used to fill in the next categories:

- their global target;
- their type (qualitative or quantitative);
- the type of RM addressed;
- their capacity to cover multiple issues;
- their relevance to different fields of activity;
- the amount of detail offered by the approach;
- their advantages and disadvantages.

Based on this structured information for a general ICS, the appropriate risk model was formulated into its convenient Risk Analysis Framework.

3   Risk Management—How to Do It in a Systematic Manner

RM, as part of the global company management, is broadly used in operational (technical) environments. This section offers some definitions concerning RM, the compatible standards and ground rules applicable for RM, the processes implicated in RM and risk assessment frameworks and associated techniques [20, 21].

3.1   Concepts and Broadly Used Definitions

Risk, as a general component, produces effects in all performed activities, even when people are not aware. A particular risk is frequently associated with the uncertainty that an event apparition may or may not take place and the expected effect, severity, outcome and consequence in case it actually occurs, at a very fundamental level [22, 23].

This definition is usually context-specific and broad variations arise in how an organization deals with the result [3]. For instance, we usually have to consider the financial risk in terms which concentrate it into a single value calculated by (1):

$$E_R = P_H \times C_S \tag{1}$$

where E_R is the expected risk, P_H is the probability that a particular hazard appears and C_S is the cost reflected in the system's functioning.

Additionally, the risk may be split into four components:

- event (an uncertain situation that could occur);
- asset (the direct or indirect target of an event);
- outcome (the event impact);
- probability (the necessary measurement in risk assessment) [24].

RM means managing risks of organizational operations (like mission, functions, appearance or own reputation), benefits, impact to individuals and/or other organizations resulting from the operation or use of an information system [3, 25]. Usually, RM includes:

- the conduct of a risk estimation;
- the application of a risk effects remission method;
- the operation of techniques for the continuous monitoring of the information system security;
- the clear documenting of the global RM program [25].

We have to keep in mind the distinction between risk analysis and assessment. Risk analysis refers to the information analysis in order to identify the risk to an information system [25]. In contrast, risk assessment means identifying risks to organizational operations (including mission, functions, image, reputation), organizational assets, individuals and/or organizations resulting from the information system employment [3, 4].

3.2 Administrative Features of RM

There are organizations developing standards, guidelines and recommended methods for carrying out the specific RM duty. The expected result should be the efficient application of these RM rules. This section concerns the main organizations and their standards/guidelines necessary when reading this book chapter.

3.2.1 The Use of Standards and Their Advantages

An international standard is a document keeping together some practical information and best practices that normally describes an acceptable way of doing something in order to solve a problem [26].

The standards are beneficial to all partners in a specific field of activity, in many ways, and these include:

- compatibility—because, by using standards, different organizations in the world make products that fit and work well with each other;
- cost reduction—because organizations are able to pay more attention on what they produce best (instead of trying to produce "everything");
- increased client satisfaction—because all standards users achieve a high degree of comfort when using standardized products and services;
- wide market—because the organizations using the same standards produce compatible items and so products manufactured in one area may be sold in a totally different area of the globe;

- better organizational collaboration—because all are part of the same production system;
- improved products' safety—because standardization also brings about the same solutions in the field of products' safety, keeping in mind this is also an international problem;
- better ideas and solutions propagation—because the best concepts from an organization's experts are shared all over the world to every other organization, regardless of their employees' competence level [3].

3.2.2 Organizations and Standards

The following organization list was suggested by Mukama, which in the same quoted thesis "Risk Analysis as a Security Metric for Industrial Control Systems" states the following aspects [3].

The Federal Information Security Management Act (FISMA) is an information security framework, part of the U.S. e-government act, put in place to secure information systems as well as manage the risk associated with information resources in federal government agencies. It sets requirements for the U.S. federal agencies to develop, implement and document an information security program to protect the organization assets as well as support the operations and processes.

The National Institute of Standards and Technology (NIST) is a U.S. organization that promotes the national economy and public welfare by providing guidance and leadership on the national measurement and standards infrastructure [4]. It provides the 800 series security-specific special publications to assist the governments, industries and academia follow standardized best practices. With the exception of SP 800-53, each 800 series SP gives guidance on a specific subject area. For example, the Special Publication 800-30 gives guides and best practices for conducting risk assessments of federal information systems conforming to the statutory requirements of the FISMA. The publications are updated on a continuous basis with the changing technology advancements.

The International Organization for Standardization (ISO) and the International Electro-Technical Commission (IEC) work in collaboration to form the specialization system for worldwide standardization. ISO, in particular, is an independent and non-governmental body that comprises of experts who develop innovative international standards to address the global challenges [26]. With a wide portfolio of standards and regulations in various industries, the following are the most commonly used in our field:

ISO/IEC 17799:2005 (Code of practice for information security management) is the international standard that established the guidelines and general principles used throughout the lifetime (initiation, implementation, maintenance, and improvement) of information security management in an organization.

The ISO/IEC 17799 was later revised and replaced by the ISO/IEC 27002:2005 with a change in the reference number but having the same title. ISO/IEC 27002:2005 contains the recommended best practices of control objectives and controls intended to meet the risk assessment requirements during information system management. The current version of the standard is the ISO/IEC 27002:2013 that further enables not only implement commonly accepted information security controls but also develop organization-specific information security management guidelines [26].

The BS7799 was an information security management standard, developed in the United Kingdom to secure the confidentiality, integrity and availability of information assets through security controls used within the organization. It is the foundation of some risk analysis methods such as the CRAMM [27]. BS7799 was used as a basis for certifying organization against its Information Security Management System (ISMS) but later revised and superseded by the ISO/IEC 27001:2005 and ISO/IEC 27001:2013.

The ISO/IEC 27001 (Information security management) family of standards focuses on asset management to help organizations keep their information assets secure [26].

ISO/IEC 27005 (Information security risk management) on the other hand provides guidelines for information security risk management applicable to all organizations and supports the general concepts, models and processes from both the ISO/IEC 27001 and the ISO/IEC 27002 [26].

3.3 Risk Management Processing Stage

All partners in an organization, from the junior staff to top management, should be a part of this very complex RM mechanism. Less experienced staff operates the systems and so they support the business functions. On the other hand, top management are involved in identifying the business plan, target and objectives. But most importantly, both of them have to provide the support throughout the RM process, from both executive and managerial point of views.

The NIST SP 800-30 standard presents a basic RM process consisting of four parts:

- framing risk;
- assessing risk;
- responding to risk;
- monitoring risk [4].

Figure 2 shows how the RM process proposed by NIST is aggregated. As suggested, an effective RM process means a continuous information exchange between its individual components. Each RM (sub)system may address all of these four parts [3].

3.3.1 Framing Risk

This first step in RM technique refers to how organizations define risk and how they create the risk background. This stage in RM, if successful, has to result in an exhaustive strategy for the organization. This strategy addresses the risk assessment methods, how to respond and how to permanently audit the risk.

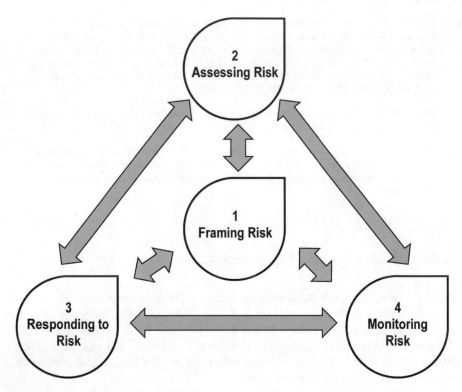

Fig. 2 Risk Management in the NIST view. Author's own illustration inspired by National Institute of Standards and Technology [4]

3.3.2 Assessing Risk

This is certainly a very critical step in the RM process. In the context of the organization risk frame already stated in the first step, this stage indicates how an organization assesses the risk. According to NIST SP800-30, this risk assessment is especially used to specify threats and vulnerabilities directly addressed to organizations. Their possibility to appear and their impact practically determine the risk. As the main subject in this book chapter, risk assessment and analysis rely on the data quality and expert skills. Relevant data can be collected through system inspection, system documentation and interviews with teams of analysts, having members of distinct work experiences [3].

3.3.3 Responding to Risk

After precisely determining and accurately assessing the risk, the organization has to respond according to the obtained conclusions. This RM stage ensures a consistent, repeatable, organization-wide response as determined and stated by the organizational risk frame. Paying attention to the organizational risk tolerance rules, the current step helps to establish and evaluate the different courses of action for risk reaction tasks. Moreover, this component has to simplify the risk responses implementation based on the selected courses of action [3].

3.3.4 Monitoring Risk

This step involves the risk monitoring (over quite a long) time in order to:

- determine the performance offered by the three steps described above;
- find significant changes (from the whole organization to some small technical systems) that can reduce the produced plan effectiveness or intensify the risk effects;
- ensure that the risk responses are based on the industry's regulations, the organization's business goals, standards and best operational practices [3].

4 Identifying Threats and Vulnerabilities in Critical ICS

4.1 System Threats

A threat may be represented by any event or circumstance having a potential negative impact on the organization's operations, its mission, provided functions, image, reputation, organizational resources, individuals and other organizations. In our view, this threat has to interfere with the system via unauthorized access, destruction, disclosure, modification of information, and/or denial of service [25].

Although it be caused by naturally occurring acts, we will only take into account the attacks from humans, and this is why in our case a threat actor is the human creating a specific threat, in an active or passive manner.

The critical ICS may be assaulted by different attackers and the threats may originate from any motivated human having any number of reasons. The threats may originate from the following sources:

- the people inside the attacked system, having access privileges to express their authority and to use the system;
- the people inside the system having no intrinsic authority, but gaining it from external sources;
- the people outside the system, from new and random individuals to big organizations and well-funded attackers;
- the people outside the system, but being part of another system connected in cascade with the attacked system [3].

The threat actors may attack the critical ICS for the following reasons:

- receiving financial gains when detaining the stolen data;
- disrupting and weakening the victims' services;
- posing the greatest risk to the ICS in order to affect the global production system;
- posing the greatest risk to governments in order to influence their decisions;
- obtaining state secrets, personal data and intellectual property in order to benefit from their owners [12].

The literature studied also shows several groups of critical ICS attackers, trying to additionally explain why such a system may be a point of interest to the given threat agents.

The first category are the insiders, seen as users with authorized system access privileges, who are triggered by job dissatisfaction, money gain or simple revenge [8]. The next one are the professional vendors which support the malware and virus/worm writers. They intend to permanently invest in development of malware, viruses and worms, and are also interested in botnets management. Such a group's interest is to gain control of devices and the system and use them contrary to intended purpose. Many of the people involved also try to sell or rent out the botnet [3, 10, 12].

One of the most dangerous group of attackers is organized crime, acting through gangs and cyber criminals. Usually, they are involved in debit and card fraud. Such a group steal personal identities in order to gain system access, intending to extort money from the affected owners [8]. Next, we are talking about state sponsored attacks achieved by using foreign intelligence techniques. The attackers are well-funded and legally protected, using also any information obtained to their own benefit. Usually, they retrieve information regarding national secrets, intellectual property, technologies used and security strategies [12].

Terrorists, activists or any motivated groups and individuals in general are part of the following group. In their case, people are joined together by common ideologies, and they are able to gather enough resources to cause attacks on various systems. Such a group publicizes their cause by sabotaging critical ICS assets, the public sector and government services. Likewise, we witness thefts of system plans and various layouts and strategies intended to bring harm to national security [3, 8, 12].

On a separate plan, we should mention the industrial espionage performed by mercenaries. They are explicitly hired to attack specific corporate assets. Such a group is usually interested in information regarding intellectual property, production and security strategies [12]. On the same level, the quoted literature positions the competitors, seen as actors producing competitive products and services for similar purpose. They fight to steal intellectual property for financial gain, in order to finally reduce competition in the marketplace.

The last category of attackers is formed by beginners, the so-called "the script kiddies". They are commonly people who are fairly new to programming and scripting, but their "experiments" may use such tools that could affect ICS assets [8].

To conclude, many attacks lead to severe implications on the socio-economic development, military security and even political influence of a country, taking into account the inherent nature of ICS [3, 12].

4.2   System Vulnerabilities

A vulnerability is regarded as an information system weakness. This also regards the system security strategies, internal system controls or internal/external implementation. Everything that could be exploited by a threat source is a vulnerability [25].

In order to understand the vulnerabilities of a system, we need a detailed review of it together with the applied security plan, as long as the system is secure, in order to detect its weakest possible connection. The ICS vulnerabilities are classified into three main categories, as it is shown by the quoted literature (having also been inspired by the NIST SP 800-82r1) [5]:

- protocol and plan vulnerabilities;
- platform vulnerabilities;
- network vulnerabilities.

4.2.1   Procedure and Plan Vulnerabilities

Humans are doing their task according to well-known good practices, being guided by procedures and plans. Whenever the security documentation is missing, is affected by inaccuracies or it is inappropriate, at least one vulnerability is issued in ICS [3, 5].

Starting a security plan needs at first the top management's commitment and support, because the required documentation has to be generated in the context of the system users' training (both operational and non-operational).

At the same time, these security procedures and plans have to be proactive to prevent and reduce any possible system vulnerabilities. For instance, such a plan may be having a good procedure for system logon and critical file access, for all ICS client systems.

4.2.2   Vulnerabilities Induced by Platform Abnormalities

Vulnerabilities may appear when weakness, lack of configuration and low maintenance characterize the platforms used in ICS. They usually consist in hardware infrastructure, the resident operating system and application software [5].

The NIST SP 800-82r1 standard announces four categories for these platform vulnerabilities, based on the following items [3, 5]:

- configuration key;
- hardware troubles;
- software problems;
- malware protection.

4.2.3 Vulnerabilities Induced by Network Abnormalities

As components of an ICS represent a network, vulnerabilities may also appear when weakness, lack of configuration and low administration/maintenance characterize the network used in ICS and its connection to other entities.

The NIST SP 800-82r1 states four categories for these network vulnerabilities [5]:

- hardware vulnerabilities;
- configuration vulnerabilities;
- perimeter vulnerabilities;
- monitoring and logging vulnerabilities;
- communication vulnerabilities;
- wireless connection vulnerabilities.

5 Characterizing the Risk Analysis Methods

The academia, research institutions and also the industry proposed many risk assessment methods and approaches, each of them being driven by different factors: the industry productivity and safety, the security based on regulations, the need that the services at hand be available and so on. In this part of our work we analyze the main (existing) risk analysis methods, making it possible anytime to issue a convenient framework for (critical) ICS.

5.1 The Current Risk Analysis Methods

Some of the current risk analysis methods of importance to our study have been published in the quoted literature, disseminated through workshops, standards rules, control materials and printed papers. As shown below in this section, these approaches were briefly described in [3], the main ideas being gathered from the technical literature [4, 27–36].

5.1.1 CCTA Risk Analysis and Management Method (CRAMM)

CRAMM is an approach based on the qualitative risk assessment methodology developed by the Central Computer and Telecommunications Agency (CCTA, UK) to carry out information systems security reviews within the government departments [3]. CRAMM is used for all types of organizations in order to justify the security investment in information systems and networks at a managerial level. At the same time, CRAMM demonstrates the compliance with the British standard for information security management (BS7799, superseded by the ISO/IEC 27001 series) during a certification process [3, 27]. The tool can also be used for benchmarking risk in organizations. The CRAMM approach review has four stages:

- identifying and valuing assets;
- identifying threats and vulnerabilities;
- calculating associated risks;
- identifying and prioritizing countermeasures.

The approach uses a risk matrix, so for each identified group it calculates the risks against the level of threats and vulnerabilities [3, 27].

5.1.2   Hazard and Operability Studies (HAZOP)

The HAZOP method is a highly systematic, formal and structured hazard identification method that helps in identifying possible hazards in complex systems, shows the current defenses and makes appropriate recommendations to avoid accidents [3]. This method was initially developed for the chemical plants and uses a multidisciplinary team of experts. HAZOP mainly uses the 'guide words' in order to identify the hazards and operational problems. For each of them, some causes and consequences are identified, making also additional recommendations [3, 29].

5.1.3   Failure Modes and Effects Analysis (FMEA)/ Failure Modes, Effects, and Criticality Analysis (FMECA)

These are a systematic and highly structured approach used to investigate how a system or subsystems can lead to performance problems and even system failure. It helps in identifying potential failure modes of systems processes and products, assess the risk associated with each failure mode, rank the risk and carry out corrective and preventive actions [3]. It evaluates the risk priority numbers by using three factors:

- risk occurrence;
- risk severity;
- detection possibilities [30].

FMEA can be modified to FMECA when more information is needed. The quantitative frequencies, their estimation and possible rankings of the consequences (when using a formal procedure) may be used in this case. Any well-defined system (for instance electrical or mechanical plants) are ready for both FMEA and FMECA to be used, from component to system-wide level [3].

5.1.4 Risk Assessment for Safety Critical Systems (CORAS)

CORAS is a research and development project set up to provide method and tools for an efficient and unambiguous risk assessment for security critical systems. The CORAS approach was modeled using the Reference Model for Open Distributed Processing (RM-ODP) as a reference model and the Unified Modelling Language (UML) to show the interactions and dependencies between the users and the environment [3]. It has four defining elements:

- the system documentation framework;
- the risk management process;
- the system development process;
- the platform for tool-integration [3, 31].

5.1.5 Operationally Critical Threat, Asset and Vulnerability Evaluation (OCTAVE)

The CERT Coordination Center (CERT/CC) developed the OCTAVE approach in order to manage information security risks by considering both organizational and technological issues. It examines how people use the organizational computing infrastructure on daily basis and focuses on mainly two aspects: operational risk and security practices [3]. OCTAVE exists in different versions, used independently or in several combinations:

- OCTAVE method for large organizations;
- OCTAVE-S approach for small organizations.

The OCTAVE approach is divided into three phases:

- phase 1—for building the asset-based threat profile;
- phase 2—for identifying the infrastructure vulnerabilities;
- phase 3—for developing the security strategy and action plans [3, 32].

5.1.6 Harmonized Risk Analysis Method (MEHARI)

The MEHARI approach, created in 1996 by the Club de la Sécurité de l'Information Français (CLUSIF), is designated to enable company executives to manage their information security, IT resources and to reduce the security related risks [3]. It is added to the implementation of the ISO/IEC 27005 standard. The approach complies with the ISO 13335 Risk Management standard and is suitable for the ISO 27001 Information Security Management System (ISMS) process. The MEHARI approach manages information security through four phases:

- analysis and classification of the potential vulnerabilities and risks;
- evaluation of security services for vulnerabilities;
- concrete risks analysis;
- definition of security plans [3, 33].

5.1.7 Common Safety Method for Risk Evaluation and Assessment (CSMRA)

Initiated by the European Railway Agency (ERA) and the European Commission, CSMRA aims to encourage diversity and competitiveness in the railway market without compromising the safety levels, and/or improving them when reasonably practicable. It was put in place to harmonize the risk assessment and evaluation processes and provide documentation during the application of the processes [3]. When any technical, operational or organizational change is proposed, CSMRA is used for changing the management process in the railway system. In this case, significant focus is placed on the RM impact on the system safety. The methodology specifies where it will be applied on the system. At the same time, it specifies the key personnel to lead and undertake this process [3, 34]. Figure 3 depicts a flowchart of the CSMRA methodology.

5.1.8 Hierarchical, Model-Based Risk Management of Critical Infrastructures (HMRM-CI)

Baiardi et al. developed a quantitative model-based risk management framework that considers a sequence of hierarchical models which describe the dependencies among infrastructure components [3].

This approach implies two independent graph representations:

- an infrastructure hypergraph. It shows the interdependent system components in terms of their internal states and operations carried out on them.
- an evolution graph. This one represents an attack that can be composed into more complex threats. In order to design the evolution attack graphs for risk analysis and automated reduction measures, this method uses mathematical computations and software tools [3, 35].

5.1.9 NIST Special Publication 800-30 (NIST SP 800-30)

The NIST SP 800-30 proposes a guide for conducting risk assessments for federal IT systems and organizations with processes similar to those given by the ISO/IEC. It provides best practices of carrying out a risk assessment [3]. The approach given is a general guideline for information systems. It can always be applied and adapted to general needs in a specific industry [4].

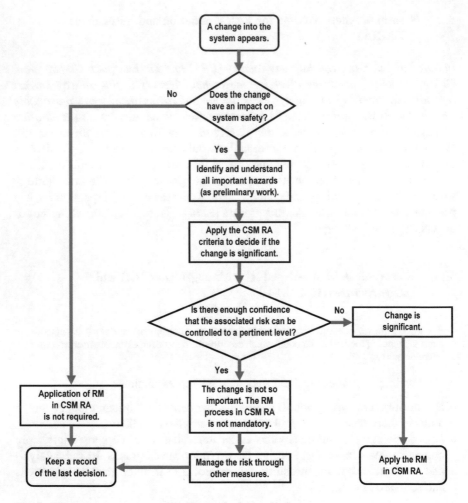

Fig. 3 Illustration adapted from CSMRA methodology [34]

5.1.10 Automated Risk Management System (ARMS)

Henderson et al. proposed ARMS, an automated risk modelling framework to address the deficiencies left by the traditional risk modelling frameworks such as CRAMM and OCTAVE. The framework identifies and addresses four main aspects of timeliness, granularity, accuracy and comprehension that are not efficiently addressed by the traditional threat and risk assessment methods [3, 36].

ARMS has to be applied in large environments for the following reasons:

- it uses automated dependency modelling to supervise the system linkages;
- it shows the risk posture using a series of continuous-based metrics;
- it can discover the cascading failures, possibly caused by a given fault [3].

5.2 Interpreting the Risk Analysis Methods

To better overview the risk analysis approaches presented above, it is a good idea to clearly formulate their definitions, to express each analysis goals and then classify them in a very constructive manner including:

- the approach goals;
- the qualitative or quantitative nature;
- their scope;
- the stages involved in the RM process.

5.2.1 Definition and Methods Goals

The following short comments start by these risk analysis approaches definitions, keeping in mind their intended purpose and main objectives. The definition may include them into one of the following category:

- comprehensive risk analysis approaches;
- frameworks;
- guidelines;
- risk assessment platforms;
- model-based frameworks.

On the other hand, the main goal of all methods is identified risk recognition, study and possible reduction. This produces a wide range of methods, which may lay the foundation for a suitable framework (or frameworks) for ICS.

Summarizing, CRAMM is a comprehensive tool for identifying security requirements. It has to justify the emphasis put on necessary and corrective measures taken especially when an IT operation is performed [3, 27]. At the same time, OCTAVE is only a method for identifying, prioritizing and managing the risks involved when dealing with the information security [32]. CRAMM and OCTAVE are surpassed by ARMS, which is a model-based approach to counter the organizations' impediments to identify and react to fundamental vulnerabilities and their inherited impacts [3, 36].

MEHARI is an approach for large organizations. It tries to identify risks, quantifying their level, reducing strong risks to an acceptable point. The risk tracking is also possible [33]. Besides that, CSMRA applies when any technical, operational or organizational change is proposed. This model was designed for railway systems but may be extended to other fields of activity [34]. Regardless of any change in the system, there is another method, FMEA framework, which tries to determine all possible failures in a designing and/or manufacturing process [3, 30].

HAZOP may be a good approach for identifying potential hazards in the system. It tries to find the potential operability problems. Additionally, it attempts to identify the causes for operational disturbances to the system [29]. It may be completed (in a way) by the HMRM-CI method, because this model-based approach is used to provide a formal definition of a risk reducing plan. It is also very interesting that HMRM-CI tries to achieve an evaluation of the infrastructural robustness [35]. On the other hand, the CORAS platform may be useful for a precise, unambiguous, and efficient risk assessment of security critical systems [3, 31], enhancing the HAZOP and HMRM-CI results.

Finally, the NIST SP 800-30 is a general guideline to manage the risk assessments in the information systems and organizations, however it may be applied only in conjunction with the SP 800-39 standard [4].

5.2.2   One Question Remains: A Qualitative or a Quantitative Method?

There are two main risk analysis methods, the quantitative and the qualitative one.

The most common type of risk analyses are the qualitative approaches. They are broadly used and operated in industry, regardless of the activity field. They follow an applied procedure which addresses the steps recommended by NIST. This is a variation of Threat and Risk Assessment, consisting in the following stages [3]:

- determining value's assets;
- identifying threats associated with the determined assets;
- evaluating existing controls' ways;
- determining the risks to assets;
- making recommendations to address the risks.

Noticeably, risks are approached in a given priority order. Their toughness, probability to appear and existing repercussions are judged as key parameters, so the qualitative methods classify the risks (in terms of non-numeric levels) such as low, medium and high.

On the other hand, the quantitative methods use two *metrics* for the risk model:

- occurrence probability of an event;
- the loss suffered when an event takes place.

These two metrics are separately evaluated. Next, they are multiplied in order to calculate the expected loss for a given time period. Risk is evaluated for several events appearance and then it has to be classified (in order to make decisions and reduce the risk).

The big problem when trying to apply such a quantitative method is the accurate and reliable measurement of the potential loss appearance probability and the produced effect. This dispute was explicitly mentioned when reviewing the three quantitative methods—FMECA, HMRM-CI and ARMS [3].

In conclusion, it is not at all facile to make a clear decision in choosing either a qualitative or a quantitative approach, but we believe that a quantitative one may be a better candidate when dealing with critical ICS.

5.2.3 The Risk Analysis Methods Scope

These methods were also organized in groups by using the following details:

- their scope;
- the granted degree of detail;
- the key issues addressed (organizational, technical or directly operational);
- their flexibility (applying the method to a different field of activity or be taken as it is).

The results of this analysis are shown in Table 1, which can also be found in the quoted literature [3].

The three issues discussed by the approaches can be expressed as the following:

- organizational issues (management decisions);
- technical issues (technological decisions);
- operational issues (process flow decisions).

We have also considered the level of detail offered by the methods, including the related risk analysis activities.

The CORAS, HMRM-CI and ARMS approaches are converged on the technical aspect of the risk analysis. Here the organizational or operational issues are considered on the secondary level. At the same time, these methods are not very flexible, as they are designed in a specific way. Moreover, HMRM-CI and ARMS benefit from a mathematical approach and this is why they are suitable to a narrow system scope.

Table 1 Scope of the risk analysis methods

Approach	Issues addressed	Adjustability	Offered detail
CRAMM	Organizational Technical	Flexible	Low
FMEA	Organizational Technical	Flexible	Low
CSMRA	Organizational Technical Operational	Not flexible	Low
MEHARI	Organizational Operational	Flexible	Low
HAZOP	Organizational Operational	Flexible	Low
NIST SP800-30	Organizational Operational	Flexible	Medium
OCTAVE	Technical Operational	Flexible	Low
CORAS	Technical	Not flexible	Low
HMRM-CI	Technical	Not flexible	Medium
ARMS	Technical	Not flexible	High

Apparently, this fact seems to lower their applicability for ICS, but the research in this field is still developing.

CSMRA takes into account all the three aspects (organizational, technical, and operational), but it is a hard method to deal with, since it is not flexible at all. It is mainly suitable for a significant system change, and this change has to influence people's safety and security, and also the ICS components normal functionality.

MEHARI, HAZOP and NIST SP800-30 take care of the operational and organizational aspects. They prove to be adaptive and extensible, applicable and modifiable from one ICS to another. Among them, NIST SP800-30 is currently adopted for its full detailed features and results, better sustained by other special publications in the field of risk analysis.

Finally, a combination of HAZOP and FMEA may be a good partnership for organizational, technical and operational issues, as a customizable approach in the hazard and risk identification for usual ICS.

5.2.4 The Risk Management Phases

Lastly, these methods were analyzed taking into account how they address the RM mandatory steps:

– risk identification;
– risk analysis;
– risk evaluation;
– risk treatment [3].

As expected, all the methods address the risk identification and risk analysis steps. At the same time, only HMRM-CI misses the risk evaluation step, while the only methods that are not taking into account the precise risk treatment stage are CRAMM and CSMRA.

Table 2 marks these methods' risk model. The ARMS method performs well when talking about its documentation, but it does not offer satisfactory details regarding the application context.

HMRM-C, CORAS and FMEA do not explain well the risk evaluation and, because they are quantitative methods, they apply different sets of metrics in order to appreciate the system security.

On their side, CORAS, FMEA and NIST SP800-30 adequately treat all steps in the RM process. FMEA seems to be very strong when identifying the risks (determining at the same time the system's various failure modes). The method helps to constitute the attack routes, because it has a special set of risk factors when evaluating the risk model—different to the traditional ones used by CORAS and the NIST SP800-30.

Table 2 The method's risk model

Method	Risk model
CRAMM	Threat level System vulnerability level
CORAS	Risk probability Risk consequences Active threats
OCTAVE	Risk assets System threats System vulnerabilities System security requirements
MEHARI	Risk exposure Risk impact
CSMRA	System changes System hazards Risk impact on safety
FMEA	Risk severity Risks frequency of occurrence Problem detectability
HAZOP	System hazards
HMRM-CI	Security dependency between components Strategies of attack
NIST SP800-30	System threats System vulnerabilities Risk probability Risk impact
ARMS	System interdependencies System intradependencies System cascading impact of events Problem granularity Results accuracy

5.2.5 Advantages and Disadvantages of the Presented Methods

All the risk analysis approaches have their relative good points and bad ones. The following remarks try to reveal these considerations, with the purpose of offering a guideline on how to finally choose the appropriate method.

CRAMM proves to be very fast, especially in conducting security reviews. It is of great importance that this approach offers consistent results for similar risk profiles, but it needs qualified operating personnel and experienced users At the same time, CRAMM may be time-consuming when performing full reviews. Furthermore, another disadvantage is that it does not provide an entire system risk measurement.

CORAS simplifies the connection between the actors involved by risk analysis, improving the accuracy at the same time. Its weakness is that this method does not offer any relation between risks, which can be critical when such a correlation is needed [3].

OCTAVE is well known as a low cost approach, encouraging the institutions to self-assess their own practices. Its main disadvantage is that it addresses key security issues rather than tactical ones. It also shows no relation between risks, even if they are inter-dependent.

MEHARI has the only advantage of being a free, open-source approach. However, it is intended to be used at high organizational level, as this is a strategic risk analysis method. MEHARI does not seem to be interesting for ICS, since it does not address (in a satisfactory manner) the technical risks [3].

FMEA and FMECA methods are flexible and appropriate for safety and reliability analysis, since they benefit from a knowledge base (with failure modes connected with the corrective actions). Although they use elementary tools to record the analysis results, these approaches are not quite accurate. Moreover, the risk factors are difficult to be estimated. Offering no relationship among failure modes, FMEA and FMECA are totally inefficient for big systems with many components (which may be a characteristic of ICS) [3].

The HAZOP method proves to be a systematic, easy manageable and well documented approach. It also covers the equipment's safety and human operators' security. This approach always considers system parts as individual components, therefore it depends greatly on the research team members' competence and experience.

HMRM-CI is a completely automated approach, producing and optimal risk reduction plan. Like HAZOP, it is designed to be applied by qualified and experienced users. Although HMRM-CI seems to be appropriate for ICS, the highly formal approach makes it sometimes difficult to implement in technical environments.

ARMS is also an automated method, perfectly fitted for large environments (including the technical plants). It addresses timeliness and offers a more accurate results granularity. Like HMRM-CI and HAZOP, it always needs qualified and well-trained users [3].

Finally, the NIST SP 800-30, although applicable to all industries, proves to be more vulnerability-centric. When such a feature is hard to identify for the studied ICS, the NIST method is almost impossible to apply.

6 Conclusions

The industrial systems' productivity, safety and security are the most important aspects when talking about the efficient way to produce goods. This is why the systems' associated ICS are considered as critical components, for which the operators always make decisions.

The availability of ICS and their provided services is a major task for ICS technical personnel, this is why they should comply with strict requirements in order to provide this high availability (and low time recovery in some cases). Another feature of ICS is that there may be views (both technical and economical) that are hard to integrate in the global organizations' policies.

According to the reasons stated above the ICS normal operating procedures contain strict references on such facts and their constraints. At the same time, the risk analysis and management is a distinct operating policy, especially when the organization intends to deal with them in a way that never causes unintended consequences.

By using a literature study, we performed an analysis of the identification, formulation and (sometimes) quantification of risks associated with ICS. We used the risk analysis treated as a measure of safety and security in ICS, this is why we explored and analyzed several existing risk analysis methods. Our analysis tries to maintain focus on the impacts they may have on process safety assessments. The results at this chapter's end may be considered promising, taking into account the ICS diversity.

We believe that, in addition to this analysis method, special risk-depending parameters need to be identified, in order to produce a more precise measure of risk, depending on the ICS feature that is attacked.

References

1. Stouffer, K.A., Falco, J.A.: Guide to Supervisory Control and Data Acquisition (SCADA) and Industrial Control Systems security. Recommendations of the NIST. National Institute of Standards and Technology. Gaithersburg, MD, USA (2006)
2. Department of Defense (DoD): MIL-STD-882E Standard Practice—System Safety. https://www.dau.mil/cop/armyesoh/DAU%20Sponsored%20Documents/MIL-STD-882E. pdf (2012). Accessed 7 Apr 2019
3. Mukama, J.: Risk Analysis as a Security Metric for Industrial Control Systems. Master's Thesis. Department of Computer Science and Engineering, Chalmers University of Technology, Gothenburg, Sweden (2016)
4. National Institute of Standards and Technology: Guide for Conducting Risk Assessments. NIST Special Publication 800-30 Revision 1. https://nvlpubs.nist.gov/nistpubs/legacy/sp/nistspecialpublication800-30r1.pdf (2012). Accessed 7 Apr 2019
5. Stouffer, K.A., Falco, J.A., Scarfone, K.A.: SP 800-82. Guide to Industrial Control Systems (ICS) Security: Supervisory Control and Data Acquisition (SCADA) Systems, Distributed Control Systems (DCS), and Other Control System Configurations such as Programmable Logic Controllers (PLC). Revision 1. Technical Report. National Institute of Standards & Technology, Gaithersburg, MD, USA (2013)
6. National Institute of Standards and Technology: Framework for Improving Critical Infrastructure Cyber-security (2014)
7. National Cybersecurity and Communications Integration Center, Industrial Control Systems Cyber Emergency Response Team: NCCIC/ICS-CERT Year in Review. https://ics-cert.us-cert. gov/sites/default/files/Annual_Reports/Year_in_Review_FY2015_Final_S508C.pdf (2015). Accessed 7 Apr 2019
8. Christiansson, H., Luiijf, E.: Creating a European SCADA security testbed. In: Goetz, E., Shenoi, S. (eds) Critical Infrastructure Protection. ICCIP 2007. IFIP International Federation for Information Processing, vol. 253, pp. 237–247. Springer. Boston, MA, USA (2008)
9. Knapp, E.D., Langill, J.T., Samani, R., Cruz, M.I.: Industrial Network Security: Securing Critical Infrastructure Networks for Smart Grid, SCADA, and other Industrial Control Systems, 2nd edn. Syngress, Waltham, Massachusetts, USA (2015)

10. Langner, R.: Stuxnet: dissecting a cyberwarfare weapon. IEEE Secur. Priv. **9**(3), 49–51 (2011)
11. Baezner, M., Robin, P.: Hotspot Analysis: Stuxnet. Risk and Resilience Team Center for Security Studies (CSS). ETH Zürich, Switzerland (2017)
12. FireEye Threat Intelligence: Report: Cyber Threats to the Nordic Region. https://www.fireeye.com/content/dam/fireeye-www/global/en/current-threats/pdfs/rpt-nordic-threat-landscape.pdf (2015). Accessed 7 Apr 2019
13. Applied Computer Security Associates: Information System Security Attribute Quantification or Ordering. Workshop on Information Security System Rating and Ranking (WISSRR), Williamsburg, Virginia, USA (2001)
14. Hallberg, J., Hunstad, A.: Towards Quantifying Computer Security: System Structure and System Security Models. Workshop on Information Security System Rating and Ranking (WISSRR), Williamsburg, Virginia, USA (2001)
15. Kahn, J.: Certification of Intelligence Community Systems and Measurement of Residual Risks. Workshop on Information Security System Rating and Ranking (WISSRR), Williamsburg, Virginia, USA (2001)
16. Hoo, K.J.S.: How Much Is Enough? A Risk-Management Approach to Computer Security. https://fsi-live.s3.us-west-1.amazonaws.com/s3fs-public/soohoo.pdf (2000). Accessed 4 Mar 2019
17. Brotby, W.K., Hinson, G., Kabay, M.E.: Pragmatic Security Metrics: Applying Metametrics to Information Security, 1st edn. CRC Press, Boca Raton, FL, USA (2013)
18. Zhang, Q., Zhou, C., Xiong, N., Qin, Y., Li, X., Huang, S.: Multimodel-based incident prediction and risk assessment in dynamic cybersecurity protection for industrial control systems. IEEE Trans. Syst. Man, Cybern. Syst. **46**(10), 1429–1444 (2015)
19. Swanson, M., Bartol, N., Sabato, J., Hash, J., Graffo, L.: SP 800-55. Security Metrics Guide for Information Technology Systems. Technical Report, Gaithersburg, MD, USA (2003)
20. Chew, E., Swanson, M., Stine, K., Bartol, N., Brown, A., Robinson, W.: Performance Measurement Guide for Information Security. NIST Special Publication 800-55 Revision 1. Maryland, USA (2008)
21. Premaratne, U., Samarabandu, J., Sidhu, T., Beresh, B., Tan, J.-C.: Application of security metrics in auditing computer network security: A Case Study. In: 4th International Conference on Information and Automation for Sustainability (ICIAFS 2008), pp. 200–205. Sri Lanka (2008)
22. Foroughi, F.: Information security risk assessment by using Bayesian learning technique. Lect. Notes Eng. Comput. Sci. **2170**(1), 91–95 (2008)
23. Hayden, L.: IT Security Metrics: A Practical Framework for Measuring Security and Protecting Data, 1st edn. McGraw-Hill, USA (2010)
24. Talabis, M., Martin, J.: Information Security Risk Assessment Toolkit: Practical Assessments Through Data Collection and Data Analysis. Newnes Edition (2012)
25. Committee on National Security Systems, CNSS Instruction No. 4009: National Information Assurance (IA) Glossary. https://www.ecs.csus.edu/wcm/cias/pdfs/4009.pdf (2010). Accessed 7 Apr 2019
26. International Organization for Standardization: ISO/IEC JTC 1/SC 27–IT Security techniques. https://www.iso.org/committee/45306.html (1989). Accessed 7 Apr 2019
27. Yazar, Z.: A Qualitative Risk Analysis and Management Tool—CRAMM. SANS Institute Information Security Reading Room, Philadelphia, PA, USA (2002)
28. Broder, J.F., Tucker, E.: Risk Analysis and the Security Survey, 4th edn. Elsevier, Butterworth-Heinemann (2012)
29. Australian Standard IEC: Hazard and Operability Studies (HAZOP Studies)—Application Guide. AS IEC 61882:2017, IEC 61882, Ed. 2.0. https://infostore.saiglobal.com/preview/293323916298.pdf?sku=99523_SAIG_AS_AS_209229 (2016). Accessed 7 Apr 2019
30. Filip, F.-C.: Theoretical research on the failure mode and effects analysis (FMEA) method and structure. recent advances in manufacturing engineering. In: Proceeding of the 4th International Conference on Manufacturing Engineering, Quality and Production Systems, pp. 176–181 (2011)

31. Aagedal, J.O., den Braber, F., Dimitrakos, T., Gran, B.A., Raptis, D., Stolen, K.: Model-based risk assessment to improve enterprise security. In: Proceedings of the IEEE's Sixth International Enterprise Distributed Object Computing, 20–20 Sept., Lausanne, Switzerland, pp. 51–62 (2002)
32. Alberts, C.J., Dorofee, A.J., Stevens, J.F., Woody, C.: Introduction to the OCTAVE approach. Software Engineering Institute, Carnegie Mellon University, Pittsburgh, PA, USA (2003)
33. Club de la Sécurité de l'Information Français—CLUSIF: MEHARI 2010: Fundamental Concepts and Functional Specifications. https://clusif.fr/publications/mehari-2010-fundamental-concepts-and-functional-specifications/ (2010). Accessed 7 Apr 2019
34. Office of Rail and Road: Common Safety Method for risk Evaluation and Assessment. https://orr.gov.uk/rail/health-and-safety/health-and-safety-laws/european-railway-safety-legislation/csm-for-risk-evaluation-and-assessment (2019). Accessed 7 Apr 2019
35. Baiardi, F., Telmon, C., Sgandurra, D.: Hierarchical, model-based risk management of critical infrastructures. Reliab. Eng. Syst. Saf. **94**(9), 1403–1415 (2009)
36. Henderson, G., Sawilla, R., Matwin, S., Bacic, E., Tremblay, L., Sayyad-Shirabad, J., de Souza, E. N.: Automated risk management system. Decision making support for continuous improvement of IT mission assurance. Defence R&D Canada—Ottawa, Technical Report DRDC Ottawa TR 2012-060 (2012)

Machine Learning Based Predictive Maintenance of Infrastructure Facilities in the Cryolithozone

Andrey V. Timofeev and Viktor M. Denisov

Abstract This chapter provides some practical aspects and peculiarities of the use of Machine Learning based Predictive Maintenance for the infrastructure facilities in the cryolithozone. Some mathematical models of Machine Learning based Predictive Maintenance are described, which have shown their practical effectiveness. The solutions of several important problems of Predictive Maintenance for pipelines located in cryolithozone are considered, including: problem of leak detection from pipelines taking into account the possible damage to the pipeline foundation due melting of permafrost; problem of automatic classifying of defects that led to leaks; problem of prompt corrosion spot detection in the pipelines as well as problem of identifying the current state of the corrosion process in the pipeline. The problem of optimizing the procedure for incident tickets processing in the Predictive Maintenance system for oil pipelines was also considered.

1 Introduction

Predictive maintenance (PdM) [1–4] is a technology to recognize the condition of an equipment to identify maintenance requirements to maximize its performance. The PdM system's output is probabilistic prediction of the future equipment state. This prediction is directed towards preventing equipment from future breakdowns, failures, or outages with usage of continuously monitoring diverse data related to performance and efficiency of a given asset. The use of PdM allows to predict when and which equipment needs maintenance. In industry [5], predictive maintenance serves to organize optimum utilization of industrial assets with conditional monitoring of data from different types of sensors. PdM also plays an important role to detect existing problems before scheduled inspection [4]. Apart from these, it plays a crucial role in Timely Repairing of Machines and Hazardous Outage Preventing.

A. V. Timofeev (✉)
LLP "EqualiZoom", Astana, Kazakhstan
e-mail: timofeev.andrey@gmail.com

V. M. Denisov
"Flagman Geo" Ltd., Saint-Petersburg, Russia

© Springer Nature Switzerland AG 2020
E. Pricop et al. (eds.), *Recent Developments on Industrial Control Systems Resilience*,
Studies in Systems, Decision and Control 255,
https://doi.org/10.1007/978-3-030-31328-9_3

In short words, the main goal of PdM is to predict at a particular time moment, using the various type data, which were collected up to this time moment, whether the equipment will fail in the close future [6]. According to PricewaterhouseCoopers [7], there are four levels of PdM:

- **Level 1**. Visual inspections: periodic physical inspections; conclusions are based solely on inspector's expertise.
- **Level 2**. Instrument inspections: periodic inspections; conclusions are based on a combination of inspector's expertise and instrument read-outs.
- **Level 3**. Real-time condition monitoring: continuous real-time monitoring of assets, with alerts given based on pre-established rules or critical levels.
- **Level 4**. PdM 4.0: continuous real-time monitoring of assets, with alerts sent based on statistical and AI-based predictive techniques including regression analysis and various automatic classification methods (ANN, SVM, XGBoost etc.).

In the following text, instead the name "PdM 4.0", we will use the name "**Machine Learning-based Predictive Maintenance (ML PdM)**" [8, 9], since it is greater extent corresponds to the essence of the present study. Namely ML PdM, which is a part of Industry 4.0 concept, will be in focus of this research.

To date, a number of industries are objectively ready for a large-scale implementation of ML PdM: thousands of sensors monitor equipment (phase: condition monitoring); data from these sensors are collected in special data banks (phase: data collection). This collected data can be leveraged for better maintenance practices, but it is not being fully leveraged or in many cases, it is ignored. The only way to exploit this high volume of data is using machine learning (ML) and other mathematical methods, which have been developed to solve various predictive problems. With special learning approaches, ML-algorithms are trained to detect abnormal and correlated patterns of abnormal sensor data. Based on this analysis, the ML-algorithms identify machine degradation or fault before they occur. Advanced ML-algorithms do not require rules or simplistic threshold setting, because it is looking at behavioral patterns. Vast amounts of data can be analyzed in real time without the need of human involvement, the more that people are simply not able to process such a huge amount of data. In simply words, ML PdM can be formulated in one of the four ways: (a) regression approach (predicts how much time is left before the next failure); (b) classification approach (predicts whether there is a possibility of failure in next n-steps); (c) flagging anomalous behavior; (d) survival models for the prediction of failure probability over time. In this study, which is very limited in scope, we will use only approaches (b) and (c). The application of ML PdM to objects located in the cryolithozone has certain specific features. First of all, these features are associated with the need to take into account the influence of instability of the foundations on the state of the controlled objects (buildings or equipment). Also, during the analysis of the state of objects of control it is necessary to take into account weather conditions that are extremely harsh in the cryolithozone. In addition, a critical feature of the practical operation of various metal structures (including pipelines) in the cryolithozone is the increased risk of the development of corrosion processes, which lead to accidents and loss of efficiency.

The paper discusses several very important practical problems related to the maintenance of the pipelines systems. It will also show how the ML-methods are used in the maintenance of pipelines systems in the cryolithozone. In particular, the following will be considered: new methods for reliable detection of leaks in pipelines; methods for the operational classification of the type of defect that led to a leak; new method for timely detection and evaluation of the stage of corrosion processes in pipeline, based on the joint use of high-precision methods for analyzing the pipeline vibroacoustic field (photon counting technology) and ML-methods for processing measurement data. There will also be considered a practically effective solution based on use of ML-methods and designed to optimize the incident tickets processing in the oil pipelines control systems.

2 Research Objectives

The aim of the study is to create a group of math-methods based on statistical and AI-approaches, which intended to solve following problems of the PdM for infrastructure facilities located in the cryolithozone:

- Reliable detection of leaks in pipelines located in cryolithozone using fiber optic monitoring systems. The solution to this problem is absolutely necessary for the early planning of the resources that are required to solve it.
- Operational classification of the type of defect that led to a leak. Solving this problem is necessary for optimal planning of the resources required to eliminate the leak.
- Timely detection and evaluation of the stage of corrosion processes in pipeline. This problem is relevant for the conditions of the cryolithozone, where corrosion processes are developing very intensively. Solving this problem will make it possible to predict and prevent the occurrence of leaks, which sooner or later will occur as a result of the development of local corrosion spots due to metal loss. It is necessary to monitor the rate of corrosion metal loss, which will allow implementing preventive measures in time.
- Optimizing the sequence of handling incident tickets in PdM systems for oil pipelines. This problem is extremely important for the Predictive maintenance of oil pipelines in cryolithozone conditions, since the promptness in servicing the most important incident tickets ensures the minimization of the risks of accidents, the elimination of which is extremely expensive in the permafrost zone.

3 General Concepts and Notations

- $o \in O$ is a maintenance object: pipeline, bridge, structure being in stress-strain state. Here O—denotes the entire set of service objects. Abbreviated: **MO**.

- $F(o)$. The volume of space actually occupied by the MO. For simplicity, we assume that $F(o)$ is a convex region.
- $p(o)$ is approximate value of the object perimeter $o \in O$.
- $t_1 < t_2 < \ldots < t_k < \ldots$. The increasing sequence of time points, in each of which a decision is made on the state of the MO.
- $\Delta t_k = (t_{k-1}, t_k)$: **$k$-th time interval**: the time interval between the moments t_{k-1} and t_k. During this interval, information is accumulated that is needed to make a decision about the state of the MO at the time t_k. Otherwise, Δt_k is called the **k-th monitoring interval**.
- **Monitoring** the MO current state is one of the basic components of PdM system. Monitoring is carried out using a network of different types of sensors located on the object.
- **Monitoring Task**. For different types of MO, monitoring tasks are different. The work addresses the following tasks:

  - Leak detection in pipelines. Task code: "LD"
  - Automatic classification of the damage type through which pipeline leakage occurs. Task code: "CL".
  - Detection of pipeline corrosion processes. Task code: "CP".
  - Intelligent diagnosis of corrosion degradation state of a pipeline. Task code: "IDC".

- $\Theta_{v_o}^{(B)}$ is a finite set of all possible values of status parameters characterizing MO ($\Theta_{v_o}^{(B)}$ is determined a priori). The identification of these parameters is the goal of the specific **monitoring task**. The composition of the set $\Theta_v^{(B)}$ depends on the object type (v) and on the monitoring task type (B—task code). In this chapter, one type of MO is considered: "pipeline". Object code: "PL".
- $\Omega(x|o, x^c)$ is a **monitoring point**. $\Omega(x|o, x^c)$ defines a convex subdomain of $F(o)$ domain, the current state of which can be estimated within the existing technical capabilities of the monitoring system. $\Omega(x|o, x^c) \subseteq F(o)$. Here, $x^c \in F(o)$ is the geometric center of the $\Omega(x|o, x^c)$, x is the set of minimal projections of the point x^c on the object's o surface (in general, x is a set). The coordinates of x^c are tied to the $F(o)$ in accordance with strict rules that are defined at the design stage of the monitoring system for the MO. The dimensions of the $\Omega(x|o, x^c)$ determine the spatial resolution of the monitoring system. By definition, for any real $o \in O$, the number of possible monitoring points is limited and equal to the number of information channels of the monitoring system. In general, $\bigcup_k \Omega(x_k|o, x_k^c) \neq F(o)$. Further, for convenience (if this does not cause ambiguity), instead of $\Omega(x|o, x^c)$, we will simply write x. Example: MO is pipeline. Monitoring system: C-OTDR [10]. In this case, monitoring point $\Omega(x|o, x^c)$ is a cylinder "stretched" on a real pipeline, whose height is equal to the spatial resolution of the C-OTDR-system, x^c is a certain point in the pipeline center that coincides with the geometric center of the cylinder, x is a set of minimal x^c projections on the cylinder surface. In this case, the number of monitoring points is equal to the integral part from dividing the fiber optic sensor length by the value of spatial resolution of the monitoring

system. The number of monitoring points in this case is equal to the number of the F-OTDR system channels. The set of monitoring points is denoted by the symbol **x**.

- X is the space domain that the foundation of the object o occupies and which is monitored by the monitoring system. For example, for objects of type PL, the domain X is a straight parallelepiped with linear dimensions $M_L \times M_W \times M_D$. On the upper face of this parallelepiped is located a part of the object's structures (or the whole object together with its foundation) so that the boundaries of the object's basement recede from the boundaries by the values of $\pm M$ meters. Here M_L, M_W are the linear dimensions of the edges of the parallelepiped, bounding its upper face, which approximately coincides with the surface of the Earth, M_D is the linear size of the "vertical" edge of the parallelepiped, which determines the depth of its immersion into the ground.

- $\Delta(X, o) = \{(\delta_i, t_i, x_i)_i \mid i = 1, \ldots N_\Delta\}$ are soil shifts that occurred within domain X and which were detected by the monitoring system. Here i—number of shift within the set $\Delta(X, o)$; δ_i is the absolute value of the soil shift; $t_i \in \Delta t_k$ is the time moment of the shift (if the shift was gradual, this parameter is assumed to be t_k), $x_i^{(P)} \in X$ projection of coordinates of the shift center on the surface of the MO; N_Δ is the number of shifts in the monitoring domain X. In $\Delta(X, o)$ not only the soil shifts recorded during the current monitoring period Δt_k are saved, but also those that occurred earlier, as they continue to affect the foundation of the MO. Over time, this effect decreases, which is taken into account by the coefficient $\gamma(x_i, t, t_i \mid \alpha)$, which is described in Sect. 6.1.

- $k(o)$ is seismic resistance coefficient MO $o \in O, k(o) \in [0, 1]$.

- $\phi(x, t \mid \eta_\delta)$ is a generalized function of soil shift effect on the value of the probability of damage to the MO structure at a point $x \in X$ on its surface, at time t. Here $\eta_\delta = (\Delta, \alpha, a, b, A_0)$ is some parameters set (for more details, see in Sect. 6.1). The set $\{\alpha, a, b, A_0\}$ is determined as a result of a machine learning training or empirically.

- $E_o = \left\{e_i^{(o)}\right\}$ is the set of MO structural member (element) $e_i^{(o)}$, $o \in O$. For extended objects (for example, for oil pipelines), the element $e_i^{(o)}$ is otherwise called a **section**; $F(e_i^{(o)})$ is the volume of space occupied by the $e_i^{(o)}$.

- $\lambda\left(e_i^{(o)}\right)$ is a parameter that, in the case of an extended object, indicates the frequency of accidents in section e_i.

- $S(t, x \mid \Delta t_k)$ is the so-called set of objective parameters (OP) of monitoring, defined for the subinterval of time $t \subseteq \Delta t_k, \cup t = \Delta t_k$, for a specific structural member $e_i^{(o)} \in E_o$, at the monitoring point $x \in \mathbf{x}, x \subseteq F(e_i^{(o)})$. This set consists of various parameters directly measured by the sensor system either once or repeatedly during the interval Δt_k. The specific set of parameters depends on the MO type. Examples of objective parameters: the angles of inclination of the structural elements, the natural frequencies of these elements, the parameters of the vibroacoustic field of the structural elements, and so on. Based on the analysis of the OP values dynamics, the main monitoring tasks are effectively solved, the essence of which is to estimate the MO status parameters.

- **The MO vibroacoustic field parameters** are included in the set of objective parameters $S_{E_o}(t, x \mid \Delta t_k)$. In the process of monitoring with usage of a point microphone network, as well as data from fiber-optic monitoring systems, a group of parameters characterizing the frequency responses of the MO vibroacoustic field is calculated. These parameters are calculated at each monitoring point $x \in \mathbf{x}$ (channel), for the time interval $t \subseteq \Delta t_k$. This group includes:

  - $B_t(x, \omega_h)$ is the Fourier coefficients vector for the frequency range of $[0, \omega_h]$ (the value ω_h is determined by the frequency bandwidth of the monitoring system and lies within ranges from a few hundred hertz to several kilohertz);
  - $P_t(x \mid \omega_h)$ is the vector of cepstral coefficients;
  - $EV_t(x)$ is the vector of heuristic characteristics of the stochastic dynamics of the oscillatory process in each monitoring system channel, during the interval $t \subseteq \Delta t_k$, which takes into account the macrodynamics of these oscillations and the irregularity of their amplitude in time.

The parameters of the vibroacoustic field form a subset $s(x, t) = (EV_t(x), B_t(x \mid \omega_h), P_t(x \mid \omega_h))$ of the main set of objective parameters $S(t, x \mid \Delta t_k)$, which is used to solve the problems "LD", "CL" and "CP". Denote by $\mathbf{s}(x \mid \Delta t_k) = \{s(x, t) \mid t \in \Delta t_k\}$, $x \in \mathbf{x}$, the sequence of such parameter subsets repeatedly measured during each monitoring cycle by the subsystem of fiber-optic monitoring based on the count of single photons.

4 Specialties of ML PdM Processes in Cryolithozone Conditions

The PdM-processes for infrastructure facilities in the cryolithozone are significantly affected by the soil instability factor at the base of the foundations of these objects. This is due to the influence of freezing–thawing processes of foundation soils. These processes are intensified by the process of global warming. To take into account these circumstances, information of soil shifts in the infrastructure foundations location area must necessarily be included in the general information contour of PdM system. You should also consider the factor of low temperatures, changes in atmospheric pressure, as well as sharp daily fluctuations in air temperature, which are so characteristic of the cryolithozone. For example, the operability of trunk pipelines in conditions of low climatic temperatures is mainly determined by their cold resistance. Major damage to trunk pipelines in the cryolithozone occurs under static loading and leads to brittle, quasi-brittle and ductile fracture. In addition, corrosion processes are significantly accelerated in the cryolithozone. All of these factors can be taken into account by using of appropriate types of sensors, as well as by using of joint data processing peculiarities. In Fig. 1 schematically shows the principle of data processing in a ML PdM system of pipeline system management in the cryolithozone. This diagram displays the following levels of data processing: Sensors-Data Level, Sensors Data Transfer Level, Data Collection and Storage Level, ML PdM Decision Support System Level. To ensure the effective functioning of the ML PdM system for pipelines in the cryolithozone, it is necessary to collect and take into account a wide variety of data that fully describe both the state of the object itself and its environment. For this it is necessary to solve the following monitoring tasks:

- **The technological parameters monitoring of pipeline elements**. To solve this monitoring task the next sensor types are used: gauges, thermometers, vibrometers and etc. In Fig. 1 these sensors have number "1".
- **Monitoring of soils and foundations**. To solve this problem, strings of inclinometers are used, which burrow into the ground near the foundations of objects. These

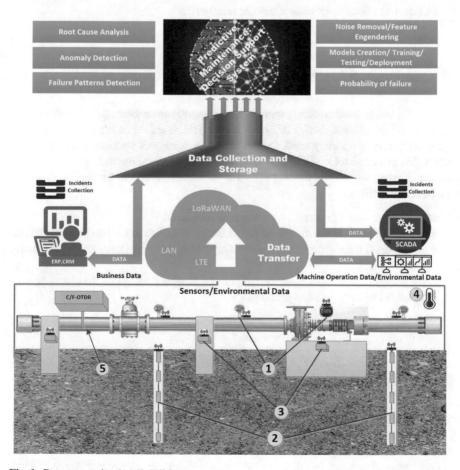

Fig. 1 Data processing in ML PdM systems

sensors make it possible to detect even minimal ground shifts in the foundation area. In Fig. 1 these sensors have number "2".

- **Monitoring of facilities and structures**. Networks of inclinometric sensors, as well as sensors for estimating the natural frequency of structural elements and the logarithmic decrement of these vibrations, are placed on the surface of objects, including its foundation and supports. The readings of these sensors must clearly correspond to the intervals of admissibility, determined at the stage of designing objects. In Fig. 1 these sensors have number "3".
- **Climate conditions are monitored** using temperature, humidity, pressure, and wind speed sensors. This data is used in the intellectual unit of ML PdM as a supplement to the feature set that characterize the targeted incidents occurring on the object. In the diagram, these sensors have number "4".
- **Monitoring of corrosion** processes can be done by various methods. This study will focus on vibroacoustic methods based on fiber-optic monitoring systems [10]. In Fig. 1 the fiber optic sensor has the number "5".

5   Models and Methods of ML-Based PdM Systems

Currently, many mathematical models and methods have been developed that are used in ML PdM processes. For example, models based on Markov processes and queuing theory are widely used. On the other hand, various variants of optimization strategies are used for optimal planning in PdM systems. Due to the limited scope of the chapter, we will focus only on some of these models. In particular, they will be briefly described:

- an approach based on analysis of targeted incidents;
- a method of interval estimation of the moment, at which a random process parameters changed (this approach has shown high efficiency in a number of practically important cases);
- some methods for multi-class classifying of stochastic objects that have proven themselves in solving PdM problems.

5.1   A Incident-Based Approach to the Intellectualization of PdM Processes

An **incident** is a collection of situations registered by the sensory system and bearing signs of a **pre-failure condition**. For example, an incident is a short-term increase above the normal temperature of a component of a system or a short-term increase above a normal pressure in a pipeline system or a short-term increase in the maximum allowable equipment voltage. Not every incident that carries part of the signs of a pre-failure condition actually corresponds to a failure. The fact is that real industrial systems are very complex and multi-connected. When trying to change the operating parameters of these systems, for example, in the process of repair or maintenance, some of the system's sensors may well give abnormal readings even if the system as a whole is in good condition. Predicting this kind of alerts is pretty hard. ML PdM systems provide special mechanisms for optimal response to these cases, which will be discussed in the following sections. The sequence of states of the incident life cycle is simple: (1) "new"; (2) "eliminated"; (3) "completed" and (4) "closed". In the process of managing an object, the corresponding process control systems constantly form a database of historical data of incidents. Each incident, in manual mode, is classified by the system operators upon completion of its life cycle. In Fig. 1 shows that incident data comes from both production control systems (SCADA) and business systems (ERP, CRM). As a result, the Incident Historical Data Base (IHDB) is formed, which underlies the intellectualization system of PdM processes. IHDB also retains all sorts of characteristics of incidents, which are otherwise called incident feature. The set of features, naturally corresponding to the incident of a particular type, will be called the patterns of this incident. Incident feature set include: sensor system data, time and place characteristics, climatic environment parameters etc. The IHDB also stores information about the feature dynamics which correspond to the time intervals preceding each specific incident. Thus, this database contains all the necessary information for training incident automatic classification system. This approach is the key to intelligently solving the following PdM problems: "Predicting Repairs" and "Predictive Maintenance to the Rescue". The PdM approach based on IHDB analysis using Machine Learning methods will be called the incident approach. In addition to this approach, there are other ways to PdM system organize. Especially common is the approach based on the use of semi-Markov degradation models, described below.

5.2 Semi-Markov Models of Parametric Degradation Description

Most of the degradation processes of MO target parameters are continuously monotonous. Despite this, the transition from one parametric status to another is quite adequately described by the semi-Markov model [11], the number of states of which is countable. A semi-Markov process is a random process that passes from one state to another in accordance with specified probability distributions. The residence time of the process in any state is a random variable. The distribution of this quantity depends both on this state and on the state to which the next process transition will take place. This model adequately approximates the processes of the MO structure components degradation. This adequacy is due to the fact that the procedure for identifying the current parametric status of MO is rather long. During the entire

identification process, from the monitoring system view point, the current parametric status of MO remains unchanged. The parametric status of MO, with a certain
probability, can be discretely changed only after the completion of the identification
procedure. And status MO can remain the same. An example of the adequacy of
the semi-Markov model in describing the degradation: the semi-Markov model well
approximates the processes of monotonous metal loss during the development of
corrosion processes in pipelines, as well as the processes of gradual changes in the
natural frequencies of structures, which is due to its slow destruction.

Let $X(t)$ be a semi-Markov process with a finite set of states $N = \{\mu_1(t_1), \mu_2(t_2), \ldots.\}$, which has stepped trajectories with jumps at times $0 < t_1 < t_2 < t_2 < \ldots$. In most practically interesting cases, the set N is (not strictly) mono-
tonic sequence of scalar quantities. For example, in the case of monitoring the devel-
opment of a local corrosion center in a pipeline, we have:

$$N_{1,n}(t_1, t_n) = (\mu_1(t_1), \mu_2(t_2), \ldots, \mu_n(t_n)), \mu_1(t_1) > \mu_2(t_2) > \cdots > \mu_n(t_n),$$

here $\mu_1(t_1)$ is a pipeline wall thickness at the beginning time of the corrosion spots
observation, $\mu_2(t_2)$ is the pipeline wall thickness which is correspond to the corrosion
process next stage, that led to such the loss of metal $\delta(t_1, t_2)$ in the wall of the
pipeline so that $\mu_2(t_2) = \mu_1(t_1) - \delta(t_1, t_2)$. Here $(\delta(t_1, t_2), \delta(t_3, t_4), \ldots \delta(t_{n-1}, t_n))$ is
a deterministic sequence of scalar quantities that determine the quantitative difference
between different pipe corrosion states. This sequence is determined a priori, at the
PdM system setup stage, for each type of pipe. The values of the semi-Markov $X(t_n)$
process at the moments of "jumps" form a Markov chain with transition probabilities:

$$p(\mu_{n-1}(t_{n-1}), \mu_n(t_n)) = P(X(t_n) = \mu_n(t_n)|X(t_{n-1}) = \mu_{n-1}(t_{n-1})).$$

In turn, distributions of jumps moments $\{t_n\}$ are described as follows:

$$P(t_n - t_{n-1} \leq x, X(t_n) = \mu_n(t_n)|X(t_{n-1}) = \mu_{n-1}(t_{n-1})) =$$
$$p(\mu_{n-1}(t_{n-1}), \mu_n(t_n)) \cdot F_{\mu_{n-1}(t_{n-1})\mu_n(t_n)}(x)$$

Here $F_{\mu_{n-1}(t_{n-1})\mu_n(t_n)}(x)$ is a distribution function. For most technical applications,
the functions of $F_{\mu_{n-1}(t_{n-1})\mu_n(t_n)}(x)$ and $p(\mu_{n-1}(t_{n-1}), \mu_n(t_n))$ are a priori unknown,
since they depend on many factors, some of which cannot be taken into account
either at the analysis stage or at the stage of system operation. In this regard, the
main task of the semi-Markov process state tracking system, is the timely evaluation
of consecutive points of $T^{(n)} = \{t_2, t_2, \ldots, t_n\}$, at which the parametric status of
this process changes. This problem can be solved by various methods. In practice,
the process of $X(t)$ is almost never observed directly. Only the $z(t) = X(t) + \varepsilon(t)$
process is directly observed, where $\varepsilon(t)$ is a random process with known statistical characteristics that describes measurement errors. The method for this problem
solution from the standpoint of interval estimation is briefly described in Sect. 5.3.
An alternative to this approach is the use of methods for **automatic classification**

of the MO current parametric status. For example, classical binary classifiers like SVM, RVM, logistic regression and others can be used to solve this problem. The classification decision is made according to the observations of the $X(t)$ -process during the time interval $\Delta t = [t_s, t_f] \subseteq \Delta t_k$. Let the MO be in the state μ_k at the time moment $t = t_s$. Then, during the interval Δt the observations of process $X(t)$ form a data set $\mathbf{X}(\Delta t) = \{X(t)|t \in \Delta t\}$. The hypothesis of which class corresponds to this set is tested: class μ_k or the class of the next state μ_{k+1}? At the stage of training the system, each of the classes $\mu_k \in N_{1,n}$, must be represented by the corresponding training set $\mathbf{X}^{(k)}(\Delta t)$. These data sets are collected either at the stage of pre-setting of the PdM system in a test site, or in actual operating conditions (if possible). Thus, for a sequence of parametric states $N_{1,n}(t_1, t_n)$, for each μ_k, the corresponding binary classifier $C_{k,k+1}: \mathbf{X}^{(k)}(\Delta t) \rightarrow \{\mu_k, \mu_{k+1}\}$ must be created and trained. To consistently solve classification problems $N_{1,n}(t_1, t_n)$ it is necessary to have a sequence of binary classifiers $(C_{1,2}, C_{2,3}, \ldots C_{n-1,n})$. Instead of a sequence of binary classifiers, we can use one multiclass classifier $C(N_{1,n})$ such that $\underset{k}{\forall} C(N_{1,n}): \mathbf{X}^{(k)}(\Delta t) \rightarrow N_{1,n}$.

This approach is very convenient, but in case of insufficiently representative training sets, the multi-class classifier can be significantly inferior to the sequence of binary classifiers in reliability. This approach, as well as the incident one, makes it possible to effectively solve the "Predicting Repairs" and "Predictive Maintenance to the Rescue" tasks. Some examples of such solutions will be given below.

5.3 Interval Estimation of the Random Process Change-Point

Let the semi-Markov process $X(t)$ change its parametric status at a priori unknown moments of time $T^{(n)} = \{t_2, t_2, \ldots, t_n\}$. Observations are described by the following model $z(t) = X(t) + \varepsilon(t)$, where $\varepsilon(t)$ is a noise process with known statistical characteristics p_ε. In [10], a method was proposed for sequential interval estimation of the $t \in T^{(n)}$, which makes the decision according to the following rule:

$$\begin{cases} \text{If } Y(X(t)|\alpha, \beta) \geq b(p_\varepsilon) \text{ Then } t \in \Delta \mathbf{t} \\ \text{If } Y(X(t)|\alpha, \beta) < b(p_\varepsilon) \text{ Then } t \notin \Delta \mathbf{t} \end{cases}$$

Here $\Delta \mathbf{t}$ is the time interval with prescribed length $\delta = |\Delta \mathbf{t}| > 0$, $Y(X(t)|\alpha, \beta)$ is some statistic, which depending on observations and predetermined values $0 < \alpha, \beta < 1$. Here α is a predetermined upper bound for the probability of making type I errors; β—is a predetermined upper bound for the probability of making type II errors; $b(p_\varepsilon)$ —decision threshold. At the same time, the decision reliability is guaranteed in the following form:

$$\mathbf{P}(Y(z(t)|\alpha, \beta) \geq b|t \notin \Delta \mathbf{t}) \leq \alpha;$$
$$\mathbf{P}(Y(z(t)|\alpha, \beta) < b|t \in \Delta \mathbf{t}) \leq \beta.$$

Under certain conditions, which, as a rule, are carried out in practice, the properties of this procedure are strictly proved (Theorem 4.2 [10]). In general case, observations are not scalar. In this situation, the described procedure is implemented for each component of the observation vector independently. The decision that $t \in \Delta t$ is made when, at least for one of the single components of the vector of observation, have place inequality $Y(\bullet|\alpha, \beta) > b$. For convenience, we denote the procedure of interval estimation of the moment t as follows: $\mathbf{IE}(\{z(t)\}|\mathbf{PR})$. Here $\mathbf{PR} = (z, \alpha, \beta)$ is the parameters set, that define the properties of this procedure. Also valid entry: $\mathbf{IE}(\{z(t)\}|\mathbf{PR}) = \Delta \mathbf{t}(z)$, where $\Delta \mathbf{t}(z)$ is the time interval of a given length z, for which the assertions of Theorem 4.2 will be satisfied [10]. Thus the interval $\Delta \mathbf{t}(z)$ is the confidence interval for the parameter t.

5.4 Some Methods for Classifying Objects

In context of the PdM, different approaches to classification problems solutions are in demand. Many classification algorithms are relevant for PdM due to the fact that in various practical cases we have significantly different conditions and restrictions. There are situations when we have only training sample of small volume, and vice versa: there are situations when training sample volume is large, but it is littered with false data or statistical outliers. In this connection, we must use different methods to solve the different classification problems in PdM. In this section we shortly describe some basic classification methods which provide the acceptable effectiveness in various PdM practical cases. In order to save space, we will not give a complete mathematical basis of the used classification methods, especially since the study of the mathematical details of various classifiers is not the goal of this work. We confine ourselves to the description of the main features of the application of some classification methods to the tasks of the PdM.

Multiclass SVM (Support Vectors Machine). This is a well-studied and proven in various practical applications, which has already become a classic method. The SVM mathematical basic is fully described in [12]. The positive features of SVM include the following properties: accuracy; SVM works well on smaller cleaner datasets; SVM works well with data of high (more than 100) dimensions; SVM has a minimal set of hyper-parameters (regularization parameters and others), the essence of which is clear and understandable; SVM is defined by a convex optimization problems (no local minima) for which we have many efficient methods. The main SVM disadvantage is that it's not very suited to larger datasets as in this case the SVM training time can be unacceptable high. Also SVM is less effective on noisier datasets with overlapping classes. We have good experience of usage multi-class SVM in next classification problems of the **PdM** context: class leak identification, corrosion process stage identification in frame of the RrM-procedures for pipelines, incident class identification.

XGBoost (Extreme Gradient Boosting). XGBoost is a very popular implementation of Gradient Boosting. Ever since its introduction in 2014, XGBoost has been lauded as one of the most efficient classification algorithms: various teams repeatedly became winners of machine learning competitions using XGBoost [13]. XGBoost is an ensemble learning method. This algorithm provides the best trade-off between bias-related errors and variance-related errors. In contrast to SVM, XGBoost has quite a bit of hyper-parameters, which must be identified very carefully in the learning process. If these seven hyper-parameters determined non-optimal the model will be overfitted. The main XGBoost advantages: extremely fast (due parallel computation); very effective even on large datasets; versatile (can be used for classification and regression); do not require feature engineering (missing values imputation, scaling and normalization). The main disadvantages: XGBoost only works with numeric features (XGBoost cannot handle categorical features by itself, it only accepts numerical values); if hyper-parameters are not tuned properly its leads to overfitting. Despite some shortcomings, XGBoost shows excellent results in analyzing big data, including in the incident type classification tasks (context PdM). The classical **Gradient Boosting** (GB) may be used too, but GB always will show worse performance in comparison with XGBoost.

ANN (Artificial Neural Net). We mention this widely discussed method simply because even those who have never practiced intellectual data processing have heard about it. The mention of this method in scientific and especially in popular science literature has become so frequent, although in many cases at the same time inappropriate, that it has generated a whole wave of unhealthy speculation around this technology. At some point it might even seem that the ANN is almost a panacea and is able to effectively solve all tasks in the field of intellectual data processing. In actual practice, this is not quite true. Indeed, in the field of image classification, as well as in some other applications, the ANN shows very good results. On the other hand, in those problems where it is necessary to work with training datasets of relatively small volume, ANN does not have an advantage over other classification methods, for example, over the XGBoost. But at the same time, ANN learns much more slowly than, for example, modern ensemble methods (XGBoost, CatBoost, LightGBM). Therefore, for each specific problem, in practice, the most convenient and adequate method of classification is chosen. And not always the model of ANN has advantages in this choice. In our practice of solving **PdM** problems, as a result of testing and comparing, ANN-model always lost to other methods of classification. Probably, this situation is due to the fact that we never had large data sets for training: in **PdM** applications, the size of the training sample rarely exceeds 300–500.

6 ML-Based PdM of Oil Pipelines in the Cryolithozone

This section describes solutions to some specific ML PdM tasks that are relevant to the maintenance of oil pipelines in the cryolithozone.

6.1   Leaks Detection Task in Cryolithozone Context

If PdM is successfully implemented on the pipeline, then one of its main objectives will be to minimize the leakage from the pipeline. In this regard, as part of the implementation of PdM procedures, a whole range of measures should be implemented to prevent the occurrence of a leak. However, the likelihood of a leak will remain non-zero even if PdM is used. For example, a leak may occur due to force majeure (sabotage, technological incident) or due to hidden factory defect of pipeline equipment. Therefore, pipelines should be provided with leak detection systems, which mainly play a controlling role. When deciding on leak detection, the likelihood of errors of the first and second kinds should be below the limits set by the relevant industry standards. Therefore, leak detectors will always be part of PdM systems for pipeline monitoring. The peculiarity of the solution of the leak detection problem from pipeline systems in the cryolithozone is the influence of the instability of the pipeline foundation. If in the area of the location of the pipeline foundation there was a significant soil shift due to thermokarst processes, this shift can significantly deform the pipeline design and conditions will be created for its destructive changes to occur. And this, in turn, may cause a leakage of the transported agent from the pipeline. In these circumstances, to correctly solve the leak detection problem, it is necessary to take into account information on the dynamics of soil shifts in vicinity of the pipeline construction foundation.

A system for detecting leaks will be referred to as a "leak detector" and will be denoted as follows: $D_L(\mathbf{s}(x|\Delta t_k))$. The source data for the leak detector $D_L(\mathbf{s}(x|\Delta t_k))$ is a sequence of subsets $\mathbf{s}(x|\Delta t_k)$ collected using a single-photon fiber optic monitoring system. Ordinary C/F-OTDR system can also be used, but the spatial resolution will be slightly worse (1–5 m vs. 2–3 cm).

Denote by τ_L—the leakage occurrence moment. The accuracy of the leak localization is determined by the size of the monitoring point $\Omega(x|o, x^c)$. For simplicity, we assume that the hole through which leakage of the transported agent is realized belongs to a single monitoring point, which actually determines the geometric dimensions of the monitoring system channel. Thus, the leak must be detected in one (or in several) channels $x \in \mathbf{x}$ (monitoring points) of the monitoring system. General requirements for systems of this type are set forth in the standard practice of the American Society for Testing and Materials [14], as well as in the recommendations of the European Committee for Standardization E1211-97 [15]. The proposed approach allows evaluating the status of the vibroacoustic field throughout the object. In fact, this is a continuous monitoring of whole physical body of the object. This provides spatial resolution in linear coordinates from 0.5 m to 5 m, depending on the system settings. This feature is a significant advantage due to the use of a fiber-optic monitoring system based on the single-photons counting technology. When deciding whether there is a leak, we work not with individual points in time, but with finite time intervals $u_z \subseteq \Delta t_k$, with length z. In this case, the decision is made according to the results of observations of the data measured during the interval $u_z \subseteq \Delta t_k$. According to Sect. 3, this data set is $\mathbf{s}(x|\Delta t_k)$. A priori, there are two hypotheses:

- **First hypothesis** (status: **no leak**): $L_0(u_z, x)$: **no leak** in the interval u_z, in monitoring point (channel) $x \in \mathbf{x}$: event $\omega_{0L} : \tau_L \notin u_z$.
- **Second hypothesis** (status: **leak**): $L_1(u_z, x)$: **leak** in the interval u_z, in monitoring point (channel) $x \in \mathbf{x}$: event $\omega_{1L} : \tau_L \in u_z$.

Here $\Theta_{PL}^{LD} = \{L_0(u_z, x), L_1(u_z, x)\}$ is a set of status parameter values. According to the results of the analysis of the set of objective parameters $\mathbf{s}(x | \Delta t_k)$, collected during the k-th monitoring interval Δt_k, at the time t_k of this interval completion, the leak detector, $D_L(\mathbf{s}(x | \Delta t_k))$, decides which of the hypotheses of the set Θ_{PL}^{LD} is most likely. In the general case $u_z \subseteq \Delta t_k$, but in practice we often have $u_z = \Delta t_k$. Anyway, $\max\{t' | t' \in u_z\} = t_k$. The leak detector $D_L(\mathbf{s}(x | \Delta t_k))$ generates a solution in two phases. The first phase is implemented on the principle of interval estimation, described in Sect. 5.3. This approach is very economical computationally. In the case when the signal-to-noise ratio is small (less than 2–3 dB), for this simplicity and efficiency we will have to accept a significant width of the confidence interval, which contains the moment τ_L. The length of this interval can be from several seconds to several minutes. Based on the readings of the sensors placed directly on the surface of the object, the probability of the event $\omega_{1L} : \tau_L \in u_z$ is calculated at the monitoring point $x \in \mathbf{x}$ (actually, the LD task).

At the same time, a priori information about the state of the object, statistics of past malfunctions and leaks, as well as data on soil shifts in vicinity of the object's construction are **not taken into account**. The output of the detector $D_L(\mathbf{s}(x | \Delta t_k))$ at this stage is a two-dimensional vector $\mathbf{P}_{LD}^{(I)}(t_k, x) = (P(\omega_{0L}), P(\omega_{1L}))$, whose components are estimates of the probabilities (reliabilities) of hypotheses from the set Θ_{PL}^{LD} under the condition of observations $\mathbf{s}(x | \Delta t_k)$ (without taking into account various a priori information about the state of the object). Naturally, that $P(\omega_{0L}) + P(\omega_{1L}) = 1$.

Taking into account the notation introduced in Sect. 5.3, the interval estimate of the moment τ_L will be obtained in the form of a time interval $u_z = \mathbf{IE}(\mathbf{s}(x | \Delta t_k) | \mathbf{PR}) \subseteq \Delta t_k$ with length z. Here $\mathbf{IE}()$ is an interval estimation procedure, $\mathbf{PR} = (z, \alpha, \beta)$ is a set of quality parameters of the procedure $\mathbf{IE}()$.

In the case when the signal-to-noise ratio for each point $x \in \mathbf{x}$ significantly different (by more than 3–5 dB), for each $x \in \mathbf{x}$ you can choose different values of the width of the confidence interval z_x. The processing principles remain the same, but $\forall x1 \neq x2 \in \mathbf{x} : z_{x1} \neq z_{x2}$. In view of Theorem 4.2 [10], it is not difficult to see that $\mathbf{P}(\tau_L \in u_z) > 1 - \alpha$. Accordingly, $\mathbf{P}(\tau_L \notin u_z) \leq \alpha$. Therefore, taking into account the fact that $u_z \subseteq \Delta t_k$, the following entry is valid: $\mathbf{P}_{LD}^{(I)}(t_k, x) = (\alpha, 1 - \alpha)$.

Thus, the **first phase of detection is completed** and the **second phase begins**, the purpose of which is to correctly account for a priori information within the framework of the **Bayesian paradigm**. In the absence of any priori information about the object state, the hypothesis $L_0(u_z, x)$ and $L_1(u_z, x)$ are equally probable.

Therefore, a priori probabilities of hypotheses from the set Θ_{PL}^{LD} are equal, that is: $P_A(L_0(u_z, x)) = P_A(L_1(u_z, x)) = 0.5$. A priori information, which can significantly affect leak detection performance, is of various types. Including:

- data on soil shifts that occurred near the base of the MO foundation;
- $\lambda_i = \lambda\left(e_i^{(o)}\right)$ is the frequency of accidents in the pipeline section $e_i^{(o)}$ to which point $x \in \mathbf{x}$ belongs, provided that for the whole object o the average intensity of accidents $\tilde{\lambda}_o$ is determined;
- When point $x \in \mathbf{x}$ (PL type object) belongs to the so-called risk zone. Examples of risk zones: gas pipeline sections after compressor stations (5 km, risk factor: non-stationary dynamic loads); sections of gas pipelines in the connection points; sections of underwater crossings; pipelines sections with high anthropogenic activity.

There may also be significant other a priori information that characterizes the $e_i^{(o)}$ section, for example, the frequency of passage through this section of the cleaning projectile (PIG). The passage of PIG, in some cases, can provoke the occurrence of a hole through which leakage can occur, for example, in the presence of a factory marriage or in the presence of a developed corrosion hearth. Also important is the life of the pipeline and life of its structural elements, as well as brand of the structural elements manufacturer.

As part of this work, data on soil shifts, which are so frequent in the cryolithozone, as well as data on accident statistics at pipeline sections, are especially important for us. It will be show how this information is taken into account when a leak is detected.

So, a priori information about the presence of soil shifts that occurred near point $x \in \mathbf{x}$ (PL type object) is taken into account using the value of influence function $\phi(x, t|\eta_\delta)$. The purpose of this function is to smoothly approximate the effect of soil shift on the likelihood of destructive processes in the object's body. The main requirements for this function are as follows: the closer to the point $x \in \mathbf{x}$ was a the soil shift (determined by the value of $\|x - x_i\|$, x_i—is the i-th shift coordinate), the larger the amount of this shift $\delta_i > 0$, and the less time occurred since its inception (determined by the value of $t - t_i > 0$, here t_i is the moment of the i-th shift), the higher should be the a priori probability of destruction of the object's structure at the point $x \in X$ at the moment of time t. And vice versa: the larger the values of $\|x - x_i\|$, $(t - t_i)$ and the smaller the magnitude of the shift δ_i, the lower the prior probability of destruction of the object's structure to the point $x \in \mathbf{x}$ at the moment of time t. The general view of the function $\phi(x, t|\eta_\delta)$ is shown in Fig. 2. Considering the fact that the soil shift sensors are located in the immediate vicinity of the foundation, the distance from the shift center to the foundation can be neglected. The function $\phi(x, t|\eta_\delta)$ must be differentiable in all its arguments, and $\underset{x,t}{Max}(\phi(x, t|\bullet)) = 1$, $\phi(x, t|\eta_\delta) > 0$.

There are many functions that satisfy these requirements, including, for example, the following function:

$$\phi(x, t|\eta_\delta) = N^{-1} \sum_{i=1}^{N} W(\delta_i|A_0, o)\mu(x - x_i|a, b)\gamma(x_i, t, t_i|\alpha),$$

$$\eta_\delta = (\Delta, \alpha, a, b, A_0), \mu(x - x_i^{(P)}|a, b) = \left(1 + \left(\left\|x - x_i^{(P)}\right\| \cdot a^{-1}\right)^{2b}\right)$$

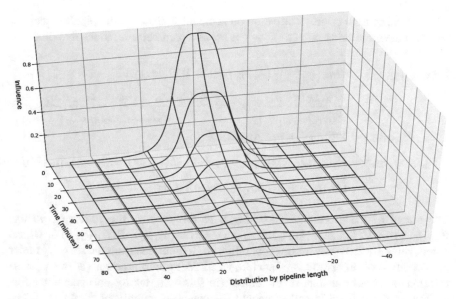

Fig. 2 The influence function $\phi(x, t|\eta_\delta)$

$$\gamma(x_i, t, t_i|\alpha) = \begin{cases} \exp(-\alpha|t_i - t|), & t \geq t_i \\ 0, & t < t_i \end{cases},$$

$$W(\delta_i|A_0, o) = 1 + \ln(1 + A_0 \cdot \delta_i^2)k^{-1}(o)$$

Here, the parameters α, a, b, A_0 are determined at the stage of the system setting based on the data of the field tests, expert estimates and the results of computational experiments.

Consider the matrix function:

$$\Gamma(x, t_k|\eta_\delta) = diag\left(\frac{P_A(L_0(u_z, x))}{L(x, u_z)}, \frac{P_A(L_1(u_z, x)) \cdot \phi(x, u_{max}|\eta_\delta)}{L(x, u_z)}\right),$$

where $L(x, u_z) = P_A(L_0(u_z, x)) + P_A(L_1(u_z, x))\phi(x, u_{max}|\eta_\delta)$, $u_{max} = \max\{x \in u_z\}$.

In the case when no additional information was involved, the prior probabilities of the hypotheses from the set Θ_{PL}^{LD} are equal, the form of this function is simplified:

$$\Gamma(x, t_k)|\eta_\delta) = diag\left((1 + \phi(x, u_{max}|\eta_\delta))^{-1}, \phi(x, u_{max}|\eta_\delta) \cdot (1 + \phi(x, u_{max}|\eta_\delta))^{-1}\right).$$

Accounting for information on the soil shifts contained in the set $\mathbf{\Delta}(X, o)$, as follows: $\mathbf{P}_{LD}(t_k, x) = P_{LD}^{(I)}(t_k, x)\Gamma(x, t_k|\eta_\delta)$. Components of vector $\mathbf{P}_{LD}(t_k, x)$ carry information about the effect of the soil shifts that occurred at the instants of time preceding the time t_k in vicinity of $x \in \mathbf{x}$ on the magnitude of the leakage probability.

Accounting for a priori information on accident statistics at the pipeline section, which at the monitoring point $x \subseteq e_i^{(o)}$ is set as a parameter λ_i, looks like this:

$$\Gamma_i(x, t_k)|\eta_\delta, \eta_\lambda) = diag\left(\varphi_\delta^{(1)} \cdot W\left(\tilde{\lambda}_o|\eta_\lambda\right), \varphi_\delta^{(2)} \cdot W(\lambda_i|\eta_\lambda)\right),$$

$$\varphi_\delta^{(1)} = \frac{P_A(L_0(u_z, x))}{P(\tilde{\lambda}_o, \lambda_i, \eta_\delta, \eta_\lambda)}, \varphi_\delta^{(2)} = \frac{P_A(L_1(u_z, x)) \cdot \phi(x, u_{\max}|\eta_\delta)}{P(\tilde{\lambda}_o, \lambda_i, \eta_\delta, \eta_\lambda)},$$

$$W(\lambda_i|\eta_\lambda) = 1 + A_2 \cdot \ln(1 + A_1 \cdot \lambda_i),$$

$$P(\tilde{\lambda}_o, \lambda_i, \eta_{\delta\lambda}) = P_A(L_0(u_z, x))W\left(\tilde{\lambda}_o|\eta_\lambda\right) + P_A(L_1(u_z, x))\phi(x, u_{\max}|\eta_\delta)W(\lambda_i|\eta_\lambda)$$

$$\eta_{\delta\lambda} = (\Delta, \alpha, a, b, \lambda_i, A_0, A_1, A_2, o), \eta_\lambda = (A_1, A_2).$$

The output of the detector $D_L(\mathbf{s}(x|\Delta t_k))$, in this case, is the following vector: $\mathbf{P}_{LD}(t_k, x) = \mathrm{P}_{LD}^{(I)}(t_k, x)\Gamma_i(x, t_k)|\eta_\delta, \eta_\lambda) = (P(L_0(u_z, x)), P(L_1(u_z, x)))$, whose components are estimates of the probabilities of hypotheses from the set Θ_{PL}^{LD} under the condition of observations of $\mathbf{s}(x|\Delta t_k)$ and taking into account various a priori information about the state of the object (in this case, taking into account information on the soil shifts and accidents frequency in vicinity of $x \subseteq e_i^{(o)}$). For simplicity, we assume that among the components $< \mathbf{P}_{LD}(t_k, x) >_k$ of vector $\mathbf{P}_{LD}(t_k, x)$ there is a single maximal component. The final solution is as follows: $\theta = Arg \underset{k}{Max} < \mathbf{P}_{LD}(t_k, x) >_k$—is the index of the maximum component. Accordingly, the hypothesis $L_\theta(u_z, x) \in \Theta_{PL}^{LD}$ will be chosen as true. A priori information of other types can be taken into account in a completely similar way.

6.2   Automatic Classification of the Damage Type Through Which Pipeline Leakage Occurs

The type of leakage and its intensity fundamentally determine the response of the PdM system to this incident. To optimize costs and properly plan mitigation actions for this incident, it is very important to determine the type of leak. For these purposes, a system is used which we will call the leakage classifier and denoted as follows: $C_L(\mathbf{s}(x|\Delta t_k))$. As a source of data, this system uses a set of objective parameters $\mathbf{s}(x|\Delta t_k)$. The set of status parameters for this task (CL code) is as follows: Θ_{PL}^{CL} = {"H", "S_Cr", "M_Cr", "G_R"}. The set of status parameters for this task (CL code) is as follows: W. Here "H" –hole, "S_Cr" is a small crack, "M_Cr" is a middle crack, "G_R"- guillotine pipe rupture.

Total, we have four main classes of leakage. The validity of just such a composition of the set Θ_{PL}^{CL} is confirmed in a number of works, for example, in [16, 17]. Consider this issue in more detail. Let L be the characteristic linear size of the defect, m; D be the nominal diameter of the pipeline; S_0 be the cross-sectional area of the pipe, m^2; S_e be the equivalent area of the defective hole, m^2; f_L be the conditional probability of occurrence of a defective hole with a characteristic size L. Table 1 [16,

Table 1 Defective holes parameters

Defective hole parameters	Defective hole type			
	"H"	"S_Cr"	"M_Cr"	"G_R"
L/D	$\ll 0.3$	0.3	0.75	1.5
S_e/S_0	$\leq 10^{-4}$	0.0072	0.0448	0.179
f_L	0.7	0.165	0.105	0.030

17] shows the results of field studies that prove the existence of a stable statistical relationship between a set of values $(L/D, S_e/S_0)$ and the probability of occurrence of a certain type (class) defect. This is an extremely important result, which significantly simplifies the system training procedure, and also potentially improves the efficiency of the classifier, by providing the possibility of taking into account the value of a priori probabilities $\{f_L\}$ for each class of leakage. The very mechanism of the account of this type of a priori information may be various, for example, it may correspond to the scheme set out in Sect. 6.1.

Let the event "leak" be characterized by the following two features, forming a pair $(L/D, S_e/S_0)$. The information presented in the table, in fact, postulates that if in a given feature space an infinite stream of "leakage" events will be realized (corresponding to physically realizable situations on real pipelines), then three clusters with centers $d_{S\_Cr} = (0.3, 0.0072)$, $d_{M\_Cr} = (0.75, 0.0448)$ and $d_{G\_R} = (1.5, 0.179)$ in this space will be formed. These clusters will correspond to three types of defect, respectively: "S_Cr", "M_Cr", "G_R": All other events will correspond to relatively low-power and more frequency type defects "H".

This information provides opportunities for effective solution of the classification problem. In this case, the solution will be to simulate on the test site only four types of defects (all types from Θ_{PL}^{CL}), leading to a leak. For each of the above-mentioned clusters, a set of realizations of the defect $\{d_i^j | i \in \Theta_{PL}^{CL}; j = 1, \ldots N_i\}$, is formed, with the powers $N_i = 1 \ldots 50$ for each class (the value of N_i depends on the type of class). For classes "S_Cr", "M_Cr", with which, usually, there are no problems with their physical modeling, $N_i = 50$ (determined by the capabilities of the test site: the more, the better). For class "H", the cluster center of which is not defined due to the large number of variants of defects of this type, it is sufficient to provide $N_i = 5 \ldots 10$ realizations of the defect, for example, with sizes 0.1×10^{-4} m^2 and 0.01×10^{-4} m^2. For the class of powerful events "G_R" one size is sufficient, for example, coinciding with the center of the corresponding cluster. The fact is that the events of the "H" and "G_R" classes are radically different from the classes in terms of the power of vibroacoustic emission, therefore, they are classified quite reliably when the parameter "power of vibroacoustic emission per unit of time" is included in the feature space. All implementations of imitation defects created in this way at a test site differ by at least the value of one parameter.

The values of the characteristic parameters of imitating defects, first of all it concerns the classes "S_Cr", "M_Cr", must uniformly cover the area of cluster scattering, which is determined by the dispersions of the scattering of its components.

That is, $\forall i \forall j, k : ||d_i^j - d_i^k|| \geq \varepsilon$, for some $\varepsilon > 0$. For each realization of a defect, it is necessary to carry out the agent being transported under various pressures corresponding to the actual modes of pipeline operation. For reasons of economy, a measurement cycle, the purpose of which is to form a training set, can be carried out for the smallest size of the pipeline. As in the case of solving a problem of the LD type, for solving the CL problem, observations $s(x|\Delta t_k)$ of the vibroacoustic field of the object are used. At the stage of training the system under the conditions of the test site, a training set of the following structure is formed:

$$S_L = \{([s(x,t), P_k]; d_i^j)|t \in \Delta t_L, P_k \in \mathbf{P} = \{P_1, P_2, \ldots, P_L\},$$
$$i \in \Theta_{PL}^{CL}; \; j = 1, \ldots N_i\}.$$

Here $\mathbf{P} = \{P_1, P_2, \ldots, P_L\}$ is the map of the model values of pressure in the pipeline, Δt_L is the training interval of the system. In the $([s(x,t), P_k]; d_i^j)$, the d_i^j element is marker and $f_p(t) = [s(x,t), P_p]$ element are features. Consider the following partitioning of the set S_L: $S_L = \bigcup_p S_L^{(p)} = \bigcup_p \{f_p(t); d_i^j)\}$. Each of the subsets of $S_L^{(k)}$ corresponds to a specific value of P_p. Each of these subsets is used to train the classifiers described in Sect. 5.2. To increase the generalizing ability, the methodology of cross-validation is used. Thus, we have $|\mathbf{P}|$ classifiers $C_p: f_p(t) \rightarrow \Theta_{PL}^{CL}$, each of which corresponds to a certain value of $P_p \in \mathbf{P}$. In real conditions, there is always information about the current value of pressure P in the pipeline. Therefore, from $\mathbf{P}$ we choose two adjacent values $P_p, P_{p+1} \in \mathbf{P}$ such that $P_p \leq P \leq P_{p+1}$, and solve the classification problem simultaneously for the classifiers C_k and C_{k+1}. Based on the hypothesis of the monotony of the effect of pressure in the pipeline on the parameters of the vibroacoustic field created by the movement of the transported agent, the solution of the CL problem is the class of Θ_{PL}^{CL} that has the highest probability value assigned to it by these classifiers.

6.3   Intelligent Diagnosis of Corrosion Degradation State

Detecting and tracking the dynamics of corrosion processes in pipeline systems is one of the most relevant in the PdM tasks group. For example, according to "Rostekhnadzor" reports, pipe metal corrosion is the main cause of accidents on Russian pipelines [18]. In this section, we will consider a new method for detecting and controlling the development of corrosion spots in pipelines, based on the use of fiber-optic monitoring using the single-photon counting technology in a receiving unit. The basis of the proposed method is the effect of focal metal loss due to corrosion on the dynamics of parameters $s(x|\Delta t_k)$. The problem is to restore the function of this influence as accurately as possible. To do this, it is necessary to estimate the parameters of the vibroacoustic field of the pipeline with a high degree of spatial resolution (2–3 cm along the linear coordinate), since the size of corrosion centers has a centimeter scale. This high resolution can be achieved using the C\F-OTDR system, the receiving unit of which is based on the principle of single photon counting. In addition, it is extremely important to create a training data sets that link models of corrosion foci of various shapes and depths with observations of $s(x|\Delta t_k)$. From practical studies it is known [19] that the maximum size of corrosion areas does not exceed 15–20 cm in linear size and has a shape similar to an ellipse. Under the conditions of the test site, artificially create a series $\Pi = \{\pi_i\}$ of mechanical depressions of ellipsoidal shape in the pipeline metal (with a maximum major axis size of 20 cm, with a certain proportion of the axle length ratio), with different depths $H = \{\eta_k\}$. Here, for predefined values $\varepsilon_\Pi > 0$, $\varepsilon_H > 0$, $\pi_i, \pi_i \in \Pi$: $|\pi_i - \pi_i| \geq \varepsilon_\Pi$, $\eta_i, \eta_i \in H$: $|\eta_i - \eta_i| \geq \varepsilon_H$. The smaller the values ε_Π, ε_H, the more accurately reproduced many possible corrosion defects, but the more expensive the full-scale experiment becomes. Thus, a set of markers $\Pi \otimes H$ is formed, each element $(\pi, \eta) \in \Pi \otimes H$ of which corresponds to a certain configuration of a model defect. The pressure in the pipeline is taken into account in the same way as in Sect. 6.2. As a result of such field experiments, the learning set S_L is formed. Here

$$S_L = \{([s(x, t), P_p]_j; (\pi, \eta)) | t \in \Delta t_L, P_p \in \mathbf{P}, (\pi, \eta) \in \Pi \otimes H, j = 1, \ldots N\},$$

$\mathbf{P} = \{P_1, P_2, \ldots, P_L\}$ is a map of the model values of pressure in the pipeline, Δt_L is the system learning interval, N is the sample size of measurements $[s(x, t), P_p]$ for each defect configuration (π, η). In the $([s(x, t), P_p]_j; (\pi, \eta))$ element, component (π, η) is the defect configuration marker, and $f_p(t) = [s(x, t), P_p]$ is the features vector. Consider the sets $\{f_p(t); (\pi, \eta)\}$, $p = 1, \ldots L$, each of which corresponds to a specific value of $P_p \in \mathbf{P}$. On each of these sets, using the cross-validation method, one of the classifiers described in Sect. 5.2 is trained. Thus, we have $|\mathbf{P}|$ classifiers $C_p: f_p(t) \to \Pi \otimes H$. The problem of the mismatch of the actual pressure in the pipeline P to the values of the map $\mathbf{P}$ is solved in the same way as this problem is solved in Sect. 6.2. Using this approach, in essence, creates the basis for the automatic solution of the "Predicting Repairs" and "Predictive Maintenance to the Rescue" tasks in relation to pipeline systems.

6.4 Optimizing the Sequence of Handling Incident Tickets in ML PdM Systems for Oil and Gas Pipeline

The features of the PdM processes for the pipeline infrastructure in the cryolithozone are due to both the influence of low ambient temperatures and possible soil shifts caused by the dynamics of thermokarst. The main tasks of PdM for pipeline infrastructure include monitoring the technical status of the equipment, forecasting the condition of the equipment, taking into account many factors, as well as planning timely maintenance and the maintenance procedures types. The following parameters are subject to control:

- pressure (hydrodynamic, static), measured at control points of the pipeline system;
- valves status (discrete parameter);
- temperature of the transported agent in the control points;
- density of the transported agent;
- the amount of transported agent passed through the pipeline for the period;
- technical parameters of pumping equipment (engine current, temperature, vector of vibration characteristics, working status, condition of fans, gas content parameter, heater condition, etc.);
- temperature and pressure in valves;
- vector of valves vibration characteristics;
- air temperature;
- precipitation in the area of infrastructure;
- shifts and temperature of the soil in the area of the foundations;
- angles of inclination of structural elements;
- vector of natural oscillations of structural elements;
- information on the condition of the anticorrosion coating of the pipeline;
- equipment failure statistics;
- emergency statistics (with technical details);
- statistics on the maintenance (with technical details);
- and others.

In addition, information about the state of the vibroacoustic field of the pipeline along its entire length, as well as information about automatically detected leaks and the places of origin of corrosion centers, are regularly received from the fiber-optic monitoring system. Thus, every moment of the pipeline system life is characterized by thousands of parameters, some of which are systematically stored in the **IHDB**. To collect these data, various types of sensors are used: pressure and temperature point sensors; fiber optic sensors on a single photon counting; point vibration sensors; underground inclinometric spits; inclinometers to control the angles of inclination of structures; infralow frequency sensors for monitoring of natural frequencies of supporting structures. Information about classified incidents, as well as maintenance activities data, comes from several of technological control systems. At the same time, on the basis of special predictive models (see Sect. 5) the equipment current status estimation problems are being solved. Basic values of the "equipment status" parameter are: "norm", "pre-fail" and "fail".

In this section, we will focus on the automatic classification procedures of the incident type, which is based on the analysis of its characteristics by AI methods, since these procedures play a significant role in the ML PdM process. As follows from the information in Sect. 5.1, the incident is not always associated with malfunctions. In pipeline systems, the cause of an incident can be maintenance procedures, urgent repair work on the pipeline or any other reasons that, at a purely technical level, induce the appearance of local signs of an incident in the sensor network. The pipeline control system receives information about several thousand incidents per day. The operators of these systems, in semi-automatic mode, are obliged to find out: is each specific incident related to a real pre-failure condition of the equipment? If the incident is not related to a pre-failure condition, it is transferred to the "closed" status. Crucially, a significant number of resources are spent on handling each incident. Due to limited resources, this is the reason for the inevitable omission of targeted incidents, which are really due to the pre-failure condition of the equipment. Thus, it is very important, first of all, to handle those incidents that are as close as possible to pre-incident ones. It is known that the importance of an incident is determined by the likelihood that the incident is due to a real pre-failure condition, as well as the consequences of the failure, for example: the economic effect.

In this section, we will focus only on the "likelihood of a pre-failure condition for a given incident" and build predictive models that predict the likelihood of a pre-failure state based on data from the IHDB and online information on pre-failure signs. So, we need to create such a classifier, which analyzing the group of incident features, in real time, gives an estimate: with what probability does this incident correspond to the real pre-fault condition? This problem is formulated as an ordinary binary classification problem, where:

- **class 1**: "the incident is associated with a pre-failure";
- **class 2**: "the incident is not associated with a pre-failure".

The highly efficient XGBoost algorithm was used to solve this classification problem [13].The predictive model is an ensemble of deep decision trees. In the process of learning, the XGBoost parameters were selected in such a way that 3400 decision trees of 12 levels were built. An example of a typical branch of a decision tree is shown in Fig. 3. When training was used IHDB base, collected in one of the oil companies. The system was trained on 750,000 incidents, of which only 45,800 corresponded to real pre-failure conditions. Размерность набора признаков инцидента равнялась 128. The dimension of the incident feature set was 128. As a result, the quality metrics [20] of the binary classification had the following values: precision: **0.98**; recall **0.98**: F1-score: **0.97**; Brier score: **0.018** (ideal value is 0), AUC: **0.92**. Each incident is assigned a rank equal to the estimated probability of belonging to class 1. The handling of incidents is carried out according to a decrease in their rank. The obtained results were recognized as practically effective, because of the **first 15% of processed incidents, more than 85% were indeed related to pre-failure conditions**.

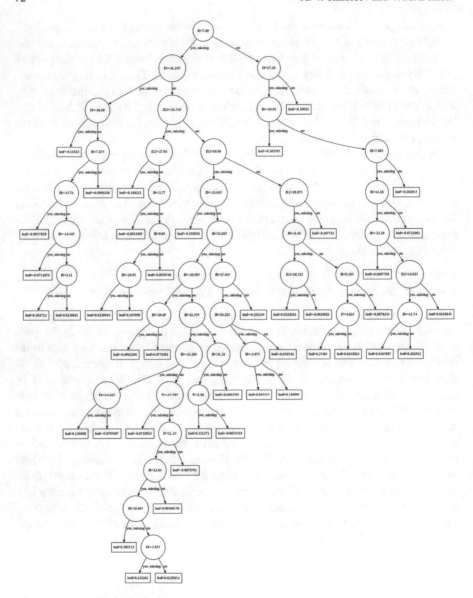

Fig. 3 Typical branch of a decision tree

7 Conclusions and Further Work

This chapter briefly were discussed the features of the use of Machine Learning-based Predictive Maintenance systems of infrastructure facilities in the cryolithozone. In particular, using the example of a pipeline system, solutions of a number of PdM problems are described, taking into account the influence of the cryolithozone factor. Due to the limited scope of the paragraph, some extremely important problems related to the operational control of the state of various types of structures (buildings, roads, bridges) with usage of intelligent monitoring methods set were beyond consideration. General approaches to solving these problems, in many respects, correspond to the approaches outlined in the chapter, but each of them has its own characteristics. In particular, when monitoring stress-strain structures in the cryolithozone, particular attention should be paid to the online assessment of the status of the stress-strain state of structural elements at ultra-low temperatures (below $-45\,°C$). This question, like many others, is the subject of future research.

References

1. Hashemian, H.M., Bean, W.C.: State-of-the-art predictive maintenance techniques. IEEE Trans. Instrum. Meas. **60**(10), 3480–3492 (2011). https://doi.org/10.1109/TIM.2009.2036347
2. Carnero, M.C.: Selection of diagnostic techniques and instrumentation in a predictive maintenance program. A case study. Decis. Support Syst. **38**(4), 539–555 (2005)
3. Swanson, D.C.: A general prognostic tracking algorithm for predictive maintenance. In: 2001 IEEE Aerospace Conference Proceedings (Cat. No.01TH8542), Big Sky, MT, USA, vol. 6, pp. 2971–2977. https://doi.org/10.1109/aero.2001.931317 (2001)
4. Zhou, X., Xi, L., Lee, J.: Reliability-centered predictive maintenance scheduling for a continuously monitored system subject to degradation. Reliab. Eng. Syst. Saf. **92**(4), 530–534 (2007)
5. Kaiser, K.A., Gebraeel, N.Z.: Predictive maintenance management using sensor-based degradation models. IEEE Trans. Syst. Man Cybern. Part A: Syst. Hum. **39**(4), 840–849 (2009)
6. Grall, L. Dieulle, C.B., Roussignol, M.: Continuous-time predictive-maintenance scheduling for a deteriorating system. IEEE Trans. Reliab. **51**(2), 141–150 (2002). https://doi.org/10.1109/tr.2002.1011518
7. PricewaterhouseCoopers: https://cdn-sv1.deepsense.ai/wp-content/uploads/2018/11/pwc-predictive-maintenance-4-0.pdf
8. Cline, B., Niculescu, R.S., Huffman, D., Deckel, B.: Predictive maintenance applications for machine learning. In: 2017 Annual Reliability and Maintainability Symposium (RAMS), Orlando, FL, pp. 1–7 (2017). https://doi.org/10.1109/ram.2017.7889679
9. Butte, S., Prashanth, A.R., Patil, S.: Machine learning based predictive maintenance strategy: a super learning approach with deep neural networks. In: 2018 IEEE Workshop on Microelectronics and Electron Devices (WMED), Boise, ID, pp. 1–5 (2018). https://doi.org/10.1109/wmed.2018.8360836
10. Timofeev, A.V., Denisov, V.M.: Multimodal heterogeneous monitoring of super-extended objects: modern view. recent advances in systems safety and security, 06/2016: chapter. In: Volume 62 of the series Studies in Systems, Decision and Control: pp. 97–116. Springer International Publishing, Berlin. ISBN: 978-3-319-32523-1. https://doi.org/10.1007/978-3-319-32525-5_6

11. Anger, C.: Hidden semi-Markov models for predictive maintenance of rotating elements. Technische Universität, Darmstadt (Ph.D. Thesis) (2018)
12. Bredensteiner, E., Bennett, K.: Multicategory classification by support vector machines. Comput. Optim. Appl. **12**, 53–79 (1999)
13. Chen, T., Guestrin, C.: XGBoost: a scalable tree boosting system. In: Krishnapuram, B., Shah, M., Smola, A.J., Aggarwal, C.C., Shen, D., Rastogi, R. (eds.) Proceedings of the 22nd ACM SIGKDD International Conference on Knowledge Discovery and Data Mining, San Francisco, CA, USA, August 13–17. ACM. pp. 785–794 (2016). arXiv:1603.02754. https://doi.org/10.1145/2939672.2939785
14. American Society for Testing and Materials E 1211-97
15. European Committee for Standardization E1211-97: Standard practice for leak detection and location using surface-mounted acoustic emission sensors
16. Savina, A.V.: Analysis of the risk of accidents when justifying safe distances from the main pipelines of liquefied petroleum gas to objects with the presence of people. Ph.D. Thesis: 05.26.03. Scientific-Technical Center of Research Industrial Problems Security, Moscow, vol. 121, p. il (2013). RSL OD, 61 14-5/120
17. Safety Guide: Methodical recommendations for the quality risk analysis of accidents in hazardous production facilities of main oil pipelines and main oil products. Approved by Order of the Federal Service for Environmental, Technological and Nuclear Supervision of June 17, 2016 n. 228: http://docs.cntd.ru/document/456007201 (2016)
18. Annual Report on the Activity of the Federal Service on Environmental, Technological and Atomic Supervision in 2014: Federal Service for Ecological, Technological and Nuclear Supervision of the Russian Federation, Moscow. http://www.gosnadzor.ru/public/annual_reports/ (2015)
19. Timashev, S.A., Bushinskaya, A.V.: Probabilistic methods for predictive maintenance of pipeline systems. In: Proceedings of the Samara Scientific Center of the Russian Academy of Sciences, vol. 12, no. 1–2, pp. 548-555 (2010)
20. Powers, D.M.W.: Evaluation: from precision, recall and F-measure to ROC, informedness, markedness & correlation). J. Mach. Learn. Technol. **2**(1), 37–63 (2011)

Cybersecurity Threats, Vulnerability and Analysis in Safety Critical Industrial Control System (ICS)

Xinxin Lou and Asmaa Tellabi

Abstract In this chapter, cybersecurity vulnerabilities and threats surrounding Industrial Control Systems (ICSs) are investigated. The main part is to investigate cybersecurity analysis approaches in safety critical Industrial Control Systems (ICSs). We taxonomy the selected representative cybersecurity analysis approaches, in a quantitative and qualitative way. The approaches are also classified based on the scope of the analysis, e.g. some approaches focus only on the system, while others also consider the organization structure. In the end, we propose a novel idea to perform the cybersecurity analysis logically by utilizing reverse engineering results. In addition, an idea of improving the system security by design, from the hardware level by considering Field Programmable Gate Arrays (FPGA) is proposed.

List of Abbreviation

BT	Bayesian Network
CFA	Casual Fault Analysis
CFG	Casual Fault Graph
CIA	Confidentiality, Integrity, and Availability
COBIT	Control Objectives for Information and Related Technology
CPU	Central Processing Unit
CPS	Cyber Physical System
CSRF	Cross-Site Request Forgery
CSRI	Cyber Security Risk Index

The original version of this chapter was revised: The figure 2 is updated with the correct figure. The correction to this chapter is available at https://doi.org/10.1007/978-3-030-31328-9_14

X. Lou (✉)
Bielefeld University, Bielefeld, Germany
e-mail: xLou@techfak.uni-bielefeld.de

A. Tellabi
University Siegen, Siegen, Germany
e-mail: asmaa.tellabi@uni-siegen.de

© Springer Nature Switzerland AG 2020, corrected publication 2021
E. Pricop et al. (eds.), *Recent Developments on Industrial Control Systems Resilience*,
Studies in Systems, Decision and Control 255,
https://doi.org/10.1007/978-3-030-31328-9_4

DoS	Denial of Service
DMZ	Demilitarized Zone
ET	Event Tree
HMI	Human Machine Interface
IACS	Industrial Automation Control System
ICS	Industrial Control System
IDS	Intrusion Detection system
IEC	International Electrotechnical Commission
ISO	International Organization for Standardization
IT	Information Technology
I&C	Instrumentation & Control
HTTP	HyperText Transfer Protocol
MSCs	Mixed Criticality Systems
NIST	National Institute of Standards and Technology
NPP(s)	Nuclear Power Plant(s)
OPC UA	OPC Unified Architecture
OT	Operational Technology
OS	Operating System
PG	(Programmiergerät in German)—Programming Device
SCADA	Supervisory Control and Data Acquisition
SIS	Safety Instrumented System
SMP	Semi Markov Process
STAMP	Systems-Theoretic Accident Model and Processes
STPA	Systems Theoretic Process Analysis
STPA-SafeSec	STPA and Security Analysis
XSS	Cross-Site Scripting

1 Introduction

The cybersecurity issue has increased significantly in recent years (e.g. the Stuxnet [1]), TRISIS Malware targets Schneider Electric's Triconex safety instrumented system (SIS) [2], attacks on Ukrainian critical infrastructure [3, 4]. A cyberattack can lead to huge losses, which are not only economical, but also can endanger the public´s safety and other aspects of human life. For example, a power outage can lead to inconvenience in public's daily life, and interrupts industrial continuous production. A cyberattack can also lead to safety related problems, e.g. the power outage in a hospital may occur during a surgery's operation. The attack on the public traffic control system may cause a chaos, even traffic incidents in a busy intersection [5]. Hence, it is necessary to deploy a cybersecurity analysis on these Industrial Control Systems (ICSs), to identify what kind of attacks might happen on an ICS before an unexpected situation occurs.

In this chapter, the core part is to investigate cybersecurity analysis approaches in safety critical ICS. The related cybersecurity vulnerabilities and threats will also be discussed. First, the difference of IT (Information Technology) security and cyber-security will be presented. The comparison of IT security and ICS security is based on security properties [6, 7] and the systems' differentiation.

Secondly, cybersecurity issues on current ICS researches, e.g. major threats, vulnerabilities and challenges, will be included. For example, threats from outside, inside and equipment vendors. Based on the mentioned cyber threats and vulnerabilities, the current challenges in the industrial domain will be discussed. This includes the complexity of ICS, the balance among information sharing, privacy protection, system architecture design. Besides technical skills difference, culture conflicts between OT (Operational Technology) engineers and IT engineers are also a key part of security issue in the real industrial domain, e.g. the OT team and IT team have different priorities when facing a problem/an accident (the priority of Confidentiality, Integrity, Availability), however, cybersecurity issues bring them together. When a cyberse-curity issue rises, knowing how to balance the priority of each side is a challenging task.

Thirdly, based on the identified cyber threats and vulnerabilities, a review of the current cybersecurity analysis approaches will be presented. It will cover how cyber-security engineers perform a cybersecurity analysis, and if it is possible to transfer the experience from typical IT security analysis to ICS security analysis. The analysis approaches are investigated and classified, e.g. some approaches evaluate the likelihood of an attack happening, while some approaches only perform the qualitative analysis. Some of quantitative approaches use semi-quantitative scales, to show the possibility of occurrence of an attack, or the consequence of an attack on the environment, to human safety or economics. For researches of qualitative analysis, some approaches focus on researching the offensive and defensive strategy, e.g. with the Diamond model. Some researches give more importance to the system features, e.g. the system architecture. Other researches cover a wider view when analyzing cybersecurity issues, e.g. STAMP/CAST (Systems-theoretic accident model and pro-cesses/Causal Analysis using system theory), they focus not only on the system but also cover the overall organization and operations.

Lastly, a cybersecurity analysis approach which focuses on the system´s functionality will be discussed in detail. The necessary background knowledge for this approach is presented, e.g. the concept of functional integrity and information integrity are referenced and introduced. This approach is based on the formal analysis method—CFA (Casual Fault Analysis) with the analysis result being represented as a CFG (Causal Fault Graph). In addition, an idea of dealing with cybersecurity issues at system architecture level is proposed. The structure of this chapter is shown in Fig. 1.

Fig. 1 The content
arrangement in this chapter

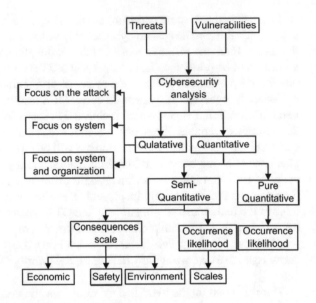

2 Security Characteristic of Industrial Control Systems and IT Systems

According to the IEC 62859 [8], cybersecurity is defined as:

– A set of activities and measures which have the objective of preventing, detecting, and reacting to: malicious disclosures of information (confidentiality) that could be used to perform malicious acts and could lead to an accident, an unsafe situation or plants performances degradation;
– Malicious modifications (integrity) of functions that may compromise the delivery or integrity of the required service by I&C programmable digital systems (incl. loss of control) which could lead to an accident, an unsafe situation or plants performances degradation;
– Malicious withholding or prevention of access to or communication of information, data or resources that could compromise the delivery of the required service by I&C systems (availability) which could lead to an accident, an unsafe situation or plant performance degradation.

According to ISO/IEC 27000:2018, the information security is the preservation of confidentiality, integrity and availability of information.

When talking about security objectives, one important difference between general IT system and ICS is the security objectives' priority. This is illustrated in Fig. 2.

The availability is a priority objective in ICS, while the confidentiality is the first consideration in IT system [6]. As for the safety critical system, we tend to consider the integrity as a first. It is due to that there are backup systems or channels in a safety critical system, we tend to consider the integrity as a first. So the availability of a safety critical system is assured normally e.g. by hot-standby redundancy, while the

Fig. 2 Comparison of security objectives priority among various systems

integrity will be the first consideration. As is shown in Fig. 2, the functional integrity and information integrity are considered separately. This is based on the concepts that are proposed by Prof. Ladkin in [9], we reference it here:

- **Functional integrity** is the property of a system or component that its system-relevant behavior remains the same. System-relevant behavior is the behavior of a system or a component of a system which contributes causally to the fulfillment of some part of the system requirements specification.
- **Information integrity** is the property that the meaning of the information held at any state St of the system Sys is conformant with:

  (a) either the real world (that is, the information corresponding to real-world parameters is veridical);
       or
  (b) veridical information held at other states $St1$ of the system, transformed by the functionally-correct transformations applied by Sys to $St1$ which result in St.

It is reasonable that the availability is the top priority in the ICS, as in the industrial domain, the system functions are different from those found in IT systems. Taking the power grid system as an example, if the system does not function well even for a short time, it means that a power outage will happen and it will affect many people's daily life, e.g. the Venezuelan outage [10], and Ukrainian electrical power outage attack [4]. Thus, Availability in an ICS is the first consideration. In addition, another difference between ICS and IT system is that, when the application software or the firmware/OS needs to be patched, the ICS cannot just execute it at any time. It must be scheduled at a suitable time as the factory manufacturing line or power plant cannot stop randomly. This is important to ensure the stability of operations of control systems, organization's economic interests, environment, and the public's safety. Since many differences exist between ICSs and typical IT systems, thus the analyses on them are different. However, referring to the experience of IT security analysis to analyze cybersecurity, is possible, as the analysis on IT security is somehow more mature than on cybersecurity.

Even though the CIA are important for ensuring the system's security, and the attacker mostly cannot exploit the system just by violating only one security property.

A successful attack should be due to a combination of threats that exploit vulnerabilities. Thus, the cybersecurity threats, vulnerabilities in ICS, and the current researches on cybersecurity analysis will be discussed in following parts of this chapter.

3   Cybersecurity Threats of Industrial Control Systems

A threat is any circumstance or event with the potential to adversely impact organizational operations (including mission, functions, image, or reputation), organizational assets, individuals, other organizations, or the Nation through an information system via unauthorized access, destruction, disclosure, modification of information, and/or denial of service [7]. Threats come from various sources, for example, adversary (insider or outsider), the environmental factor such as flood, earthquake, the system components failure, or a threat may even come from an operator's normal operation. When considering the cybersecurity, the adversary threats will be considered, for example, an adversary may sniff the network traffic to get the sensitive information to deploy further exploitation. While the force majeure, like earthquake, flood etc. unavoidable situations will not be included into the cybersecurity threat here.

Some technical threats and vulnerabilities on ICS are: DoS (Denial of Service), memory overflow, unnecessary ports and services, lack of encryption (using plaintext to transmit messages) and lack of authentication (Modbus) in communication protocols, viruses, Trojans, worms etc. Except for the technical threats, social engineering, insiders, unproper management policy, employees' security awareness, are also influenceable threats. Communication protocols may expose an ICS to a heavy DoS attack threat, for example. Lack of certification, encryption and authentication (e.g. Modbus [11]), leads to a situation where the attacker obtains sensitive information from the transmitted plaintext messages. If the communication protocol for ICS is not encrypted, it means that attackers can also get sensitive data (e.g. usernames and passwords) by sniffing the transmitted message between two devices. This poses a serious threat to the ICS. Threats against security properties such as, the threats on availability can be due to DoS, the threats on integrity can be due to Manipulation Masquerade etc. [12] (pp. 38–39).

Social engineering threats and insider threats tend to be neglected compared to typical security threats. Social engineering is a technique used to manipulate individuals into giving away private information, such as passwords [7]. It is "one of the simplest methods to gather information about a target through the process of exploiting human weakness that is inherit to every organization" [13]. Through the social engineering threats, hackers can collect the necessary information of devices layout, controller types and trace deeper.

Insider threat is a threat that is difficult to avoid and detect. Organizers are busy on defending outside hackers, by installing countermeasures such as using firewalls, IDS or by setting DMZ. They might forget that employees inside the organization (insiders) are also an important threat on security.

The improper management of the security policy including the daily behavior of employees is also a threat to ICS security. There are different kinds of situations, which are included in the insider threats category. First, employees tend to forget their passwords and hence store them on non-secure places which are accessible by others easily. Secondly, employees fail to distinguish between a compromised or malicious website and disclose their credentials via a phishing link. Thirdly, employees forget to change their passwords after sharing them with their colleagues to perform e.g. an emergency task, thus leading to a security breach or employees share their passwords with their co-workers due to naivety in response to a socially acceptable behavior [4]. There are also some other types of insider threats. For example, if an employee is not satisfied with the company´s management, or the employee holds a low salary, the employee is more susceptible to be bribed by a third party as the employee may have some specific demands (with or without the notice of the company's security policy). A research on insider threats modeling is developed in [4].

Employees' security awareness plays an essential role in reducing cybersecurity threats. Some people lack this kind of awareness, e.g. the username, and password used for logging into the HMI, controller, the Programming Devices (e.g. SIMATIC PG), and other systems are posted arbitrarily on the table or screen. Especially in a factory that needs people to be involved 24 h per day (e.g. at the power station) it is possible that the privacy information management and transmission are not consistent. One ignorable situation is that the operator charges his phone by connecting it to the industrial computer directly, this is reported in [14]. A situation like this is very risky for the overall ICS. The most notorious attack that resulted by inserting a mobile device and infecting the respective computer is Stuxnet. Even though in the Stuxnet accident, the inserted device is a USB, a smart phone tends to have more chances to be infected by an unnoticeable virus, as it has access to the internet normally.

Except for the above considerations, in the real world industrial domain, there are also some other challenges, for example, in the power plants and Nuclear Power Plant domain. According to the Chatham House report [14], there are communication/working culture conflicts between the OT and IT engineers. They frequently have difficulty on technically communicating with each side, which can lead to friction. The culture conflicts between OT engineers and IT engineers are also a key part of security issues in the real industrial domain, e.g. the OT team and IT team have different priority when facing a problem/an accident (the priority of Confidentiality, Integrity, Availability). However, cybersecurity issues bring them together. The potential conflict between working environments is a possible threat for dealing with security issues in real world plants.

4 Cybersecurity Vulnerabilities of Industrial Control Systems

A vulnerability is a weakness in an information system (including an ICS), in system security procedures, internal controls, or in the implementation that could be exploited or triggered by a threat source [7]. Vulnerabilities, no matter in the physical world (the hardware and the physical defense of the ICS), or the virtual world (software running, the communication network and the system operation), or even the management policy, are key parts for conceiving a successful attack on an ICS. There are always vulnerabilities existing in an ICS. How to identify and patch them before they are discovered and leveraged by the attacker is a challenging matter. A vulnerability may be located and found out from various aspects of an ICS, e.g. the origination level, the system operation level, the system architecture level, the software level, source code level etc.

To identify these vulnerabilities, there are various ways to categorize them, for example, in NIST SP 800-82, ICS vulnerabilities are categorized into different categories [7]. According to [15], zero-day vulnerabilities, non-prioritization of tasks in the OS and communication protocol issues are the major vulnerabilities in ICS. But in this chapter, we taxonomy the vulnerabilities according to their placements in the overall system structure, to assist the cybersecurity analysis on ICS from various levels. For example, a vulnerability may appear inside the hardware or software. It can also be found in the OS (Operating System) or firmware of the industrial controller. Vulnerabilities might exist in the communication protocol, or the imperfect organization management policy on physical access, remote access, and even the communication/cooperation policy among different departments. We represent the vulnerability taxonomy in Fig. 3.

For renowned industrial products, their vulnerabilities are exposed publicly to the worldwide online vulnerability databases after they are patched by vendors,

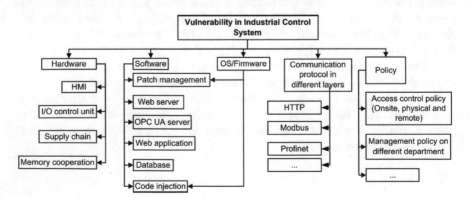

Fig. 3 Vulnerability classification in ICS

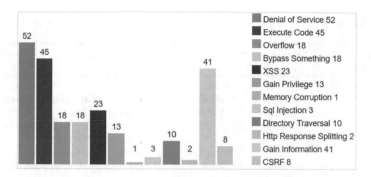

Fig. 4 The vulnerability statistics of Siemens products from 2000–2018 [64]

such as open vulnerabilities libraries/databases CVE (Common Vulnerabilities and Exposures), NVD (National Vulnerability Database) and ICS-CERT. Several most popular vulnerabilities in industrial products are: DoS, Memory Corruption, XSS (Cross Site Scripting), Execute code, Overflow, Bypass something etc. Figure 4 shows CVE statistics (2000–2018) on vulnerabilities found in Siemens industrial products. The reason behind taking Siemens as an example here is that, it is a worldwide industrial products vendor, and its products are widely used at different levels by industrial companies or production factories.

As Fig. 4 illustrates, there are various vulnerabilities on industrial products. Taking the worldwide ICS products vendor Siemens as an example, the top four vulnerabilities from 2000 to 2018 are: DoS, Execute Code, Gain information and XSS vulnerabilities. In fact, according to the statistics in CVE, the total number of vulnerabilities has increased significantly since 2011. To show the various vulnerabilities that exist in different components of an ICS, popular vulnerabilities in ICS are listed in Table 1 (according to the vulnerability classification in Fig. 4).

5 Cybersecurity Challenges in ICS

5.1 Cyber Threats Related to Information Sharing

Open source vulnerability databases are helpful for companies or users to know the vulnerable parts in their system. However, information about some incidents or even accidents may not be openly shared by all vendors. This might be due to some considerations, e.g. to maintain or protect their reputation, some industrial enterprises may choose to not expose the incidents and accidents that happened on their ICS(s). Even after the causes are found out, these kinds of information may be still be kept undisclosed. It is reasonable that industrial enterprises do not want such information to be leaked to public, and also, sometimes, there might be some private information inadvertently included. However, the advantage of sharing these kinds

Table 1 Common vulnerabilities in different components of ICS (considering only Siemens products)

System components	Vulnerabilities
HMI	Denial of service [16]; Execution of Arbitary Code [17], this vulnerability normally allows remote attackers to execute arbitrary code via crafted packets; Cross-Site Scripting (XSS) [18] may allow a remote authenticated user to inject arbitrary web script or HTML via unspecified data; CSRF (Cross-Site Request Forgery), it may allow an attacker to hijack the authentication of unspecified victims by leveraging improper configuration of HMI panels remotely [19]; Bypass a restriction or similar, this kind of vulnerability may allow an attacker to bypass the authentication process via a crafted cookie easier from remote side [20]; Directory traversal [21]
Server	A vulnerability in the OPC Foundation UA.NET Sample Code, may allow the attacker to access different kinds of sources by sending the crafted packets to an OPC UA server through a specific TCP port [22]; Execution Arbitrary Code vulnerability in an integrated web server, it may allow the attacker to execute the arbitrary code remotely [23]; XSS vulnerability in the integrated web server, it may allow the attacker to inject web script remotely [24]
Firmware/OS	DoS vulnerability may allow the attacker to bring the PLC into a "STOP" mode [25, 26]; Bypassing security controls vulnerability may allow the attacker to bypass a specific protection mechanism via some hacking knowledge [27]
Policy	Poor password policy, either hard-coded passwords are used, or the operators/engineers store passwords in a recoverable format (e.g., clear text); Remote Access Policies, the remote accessable server (e.g. the web server in an embedded device) provides a way for attackers to access the OT network; Policy of different department communication once an incident happens
Access physically	Attackers accessing to the network, a way to allow further malicious behaviors, e.g. sniffing, listening
Vulnerability on communication protocol	Attacks by leveraging privileges vulnerabilities on communication protocol in various layers, for example, the detailed vulnerabilities on various layers are reported in [28]

of information is that, it allows others to learn from such experience, thus, similar disasters can be avoided. How to balance between the industrial enterprises' privacy and the advantage of sharing these types of information is a challenging topic.

5.2   The Complexity of ICS

ICSs are different to IT systems as we have mentioned before. An ICS tends to be more complex than an IT system, due to the interaction among different hardware components, between hardware components and software. It is difficult to analyze the overall system completely. Challenges such as securing access to devices, data transmissions, securing data storage, securing actuation are discussed in [29]. Cybersecurity challenges of ICS like access control, IDS and firewalls, vulnerability assessment, security management, Cryptography, key management, device and OS security are reported in [30].

5.3   The Management Policy and Implementation

After the Stuxnet attack, many organizations started paying more attention to cybersecurity issues, e.g. to formulate security policies and improve the employee's awareness by training. However, these policies are not always followed by all employees. For example, many corporations would forbid the introduction of personal mobile devices like smartphones in secure environments. But the problem is they (the company) don't enforce it [14].

In addition, once a cybersecurity problem occurs, how to balance/handle the priority of OT engineers work and IT engineers work, and the communication between them becomes a challenging issue, e.g. the OT engineers may care more about the system´s availability and safety, however, the IT engineers may tend to solve the security confidentiality problems as priority.

6   Cybersecurity Analysis Approaches of ICS

Cybersecurity issues can be studied from various perspectives. For example, in order to improve the robustness of an ICS when facing cyberattacks, we can deal with it from two aspects. On one side, preparing before an accident happens, e.g. by improving the system's defense ability to cyberattacks. This should be considered during the system design, as part of the "Security by Design" approach. On the other side, study what to do if an attack happens, especially the response of the system to the attack. From this angle, using remedies to support the system running as usual once the attack happened is a reasonable approach, e.g. by activating a backup system. No matter which approach is used to improve the system resilience, knowing the possible attacks on our target system is a prerequisite. Because once we know what kind of attacks might happen on our system, or even the probability of the occurrence (if possible), we can take specific measures to defend against them or decide whether we can accept the risk level, either before or after the attacks happen, e.g. by improving the system design, or by using remedies after the cyberattacks happened.

Identifying possible attacks can be achieved by conducting cybersecurity analysis. It includes potential threats, vulnerability and risk analysis. ICSs may have similar parts from one system to another. However, when considering cyberattack aspects, we need to analyze our target system specifically, e.g. which specific controller is being used, its working environment, whether operators always stay in front of the actuator/sensor/controller. All the specific details determine the possible attacks on the system. Once we have a thorough cybersecurity analysis on the ICS, we can do work properly on improving system resilience before the attack happens. The cybersecurity assessment is described by Kaplan and Garrick [31] as [32]:

What can go wrong? (the hazard identification)
What is the likelihood that it would go wrong? (the probability evaluation)
What are the consequences? (the event analysis)

Researches on each question are investigated in this chapter, beyond above questions, some more concerns like, what will be included in the cybersecurity analysis? Will it be only the system, or only the attack approach, or the analysis will cover the overall organization structure and operations. In this part, we will represent some representative researches or approaches on cybersecurity analysis. The previous cybersecurity risk analysis approaches review made before 2015 is presented in [32]. A review of IT risk analysis until 2017 is developed in [33]. After reviewing researches, we present these analysis approaches in a taxonomy that includes various vectors as Fig. 5 shows.

The investigated cybersecurity analysis mainly includes two categories: the qualitative and quantitative analysis. However, it is difficult to deploy a pure quantitative analysis, as the attacker's technical skills are unpredictable. Thus, most of the quantitative researches are pursuing a semi-quantitative way, which means, they are using various metrics to scale the probability instead of an accurate number. For the qualitative cybersecurity analysis, researches focus on various aspects, for example, some researchers give more attention to some specific attacks on a system, e.g. the attack modeling, offensive and defensive modeling. The representative approaches are cyber kill chain and diamond model. Selected approaches which only focus on the analysis of system are: STPA-SafeSec, the analysis approach based on CFA (Casual Fault Analysis), and an approach considering cybersecurity in the system hardware design. Instead of focusing on the attack or the system, there are also researches that consider the cybersecurity issue from the overall organization´s perspectives, e.g. the approach based on STAMP. An approach is used for IT systems is COBIT, which is considered to be referenced into ICS, and is represented with a dashed rectangle.

7 Quantitative Analysis

Researches on cybersecurity analysis from the quantitative part include assessing the likelihood of the occurrence of cyberattacks, and the impact of various attacks. Modeling techniques are used in this domain, such as modeling an attack by describing and dividing it into discrete steps, and evaluating the probability of each step. The pure quantitative analysis on ICS, e.g. the possibility of one attack is implemented

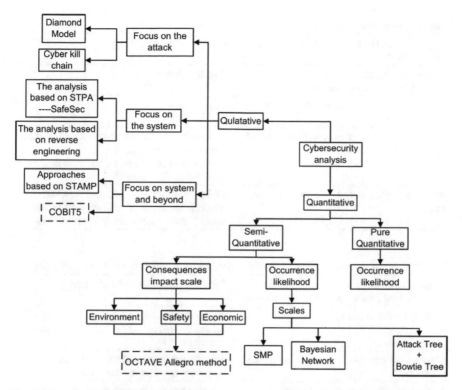

Fig. 5 The cybersecurity analysis approach taxonomy. SMP: Semi Markov Process, CFA: Casual Fault Analysis, STAMP: Systems-Theoretic Accident Model and Processes; the dotted box means this method has not applied into cybersecurity analysis yet

successfully, is difficult to calculate, as there are too many unsure factors, e.g. attackers' skills, the system vulnerabilities. Instead of the pure quantitative cybersecurity analysis, using the matrix level to represent the qualitative likelihood level (semiquantitative) [34] is a popular approach.

The semi-quantitative assessment approach is not being used to calculate the precise probability of the occurrence of an attack, but a risk matrix is used to categorize the loss/consequence of an hazard into various levels, e.g. the research in [35]. Some impactors are used in the security assessment, e.g. the average time-to-compromise [36] is utilized to reflect the difficulty level of an attack. The severity [36] level of a consequence is used to represent the scale of the loss (human life or economic). Two representative researches of semi-quantitative analysis methods are: using BN (Bayesian Network) and combining the attack tree with the Bowtie tree, the detailed descriptions are illustrated in Tables 2 and 3.

In the later research of Shin et al. [38], they use the BT (Bayesian Network) and ET (Event Tree) to do the cybersecurity risk analysis, and provide a model to evaluate the cybersecurity risk, which includes the exact probability of cybersecurity risk. The cybersecurity risk model is integrated by the activity-quality model and the

Table 2 Cybersecurity risk analysis by using BN (Bayesian Network)

Steps of using BN to do cybersecurity risk analysis [37]	How to achieve each step
Identify the threats Identify the countermeasures	The literature review
Classify the relation between threats and countermeasures	N/A
Scale the expected value of each stage of the cyber security life cycle	The probability of the occurrence of an attack (the architecture model) The level of compliance with implementation (activity-quality model)
Assign the cyber security risk index (CSRI) value	Using the formula to calculate the CSRI value, however, there is description on how to specify some parameters in the formulate

Table 3 Cybersecurity analysis approach that combines the attack tree and Bowtie Tree (BT)

Steps of safety/security risk analysis on ICS by combining attack tree and bowtie analysis	Detailed description of each step
Identifying risk scenarios	By combining AT (Attack Tree) and adjust BT (Bowtie Tree) to identify safety incidents and security threats
Likelihood evaluation scenarios of safety and security risk	Specify the likelihood scales of input events, one for safety and another for security
Severity of consequences evaluation	This step quantifies system assets loss, human life and environment damage loss, but it is not presented in the paper

structure model. The activity-quality model is developed based on the checklist in the guidance document from KINS/RG-N 08.22. They group the checklist and map the impact of each group to the mitigation measures.

The analysis approach in Table 3 only considers the cyber-security breaches that can lead to major hazards that impact human life and the environment. The security properties like CIA are not included specifically.

Semi-Markov Process (SMP) is one of the popular methods to model the cyber-attacks. When the SMP is considered in an analysis approach, researchers [39] tend to use some impactors, for example, taking the time (attack time interval [40], or the mean time-to-compromise [39]) as a factor to assess and evaluate the reliability of the system.

The pure quantitative analysis approach considers how to calculate a successful attack rate. For example, research in [41] proposes the successful-attack-probability index and attack-impact index to reflect the potential risk of a system. This method is based on the idea that a successful attack is decided by the likelihood that the vulnerabilities are successfully exploited. But how to identify the complete vulnerability set of the overall system is not mentioned.

8 Qualitative Analysis

8.1 Analysis Focus on the Attacks

Researches with a focus on the attacks tend to use the defender and intrusion roles to simulate the adversary model, e.g. diamond model, using SMP (Semi-Markov Process) to model random attack process. The diamond model [42] is based on analysis from the adversary's intrusion perspective. The cyber kill chain [43] focuses on the attack process. It models the exploitation on an ICS into various staged, e.g. the attacker collects the necessary information of the target, then produces the specific weapons (virus, worms etc.) based on the attack goal, followed by delivery of the weapon to the system. The weapons execute its exploitation and install the Remote Access Trojan (RAT) to access the corporate network. They establish the connection with their CC (Command and Control) server, and eventually execute the exploitation finally [43].

Researches that focus on the attack may also consider the consequences of the attack on a specific system, e.g. the analysis in [44] focus on a critical infrastructure, with unspecific potential attack and the qualitative evaluation of impact. Research in [40] focuses on the analysis of influence of cyberattacks on Cyber Physical Systems (CPS). It does not focus on where or on which component the attack may happen, but focus on the time and the interval time of an attack on the system (e.g. when the attack happens, how long does it last) as well as the impact of these factors on the system. The Semi-Markov Process (SMP) is considered to model various stochastic intrusion processes, as it is commonly used [45]. In research [46], an approach using a tool to generate the attack scenario and validate the scenario (in the practical system) is developed, but the analysis is actually not detailed into the CPU part.

8.2 Analysis Focus on Systems

Analysis focus on the system mainly considers the target system only, e.g. the system architecture, specific system vulnerabilities, and system application background. The proposed analysis approach in [47], considers to involve the security risk analysis during the design phase of I&C systems in NPPs, the assessment even down to the subparts of the system, and to evaluate the possible vulnerabilities on each subparts qualitatively. The work in [48] proposes to use the fault tree analysis to determine the cybersecurity vulnerability and identify the potential consequences of cyberattacks. The fuzzy theory decision is used for the quantitative evaluation of the risks related to financial losses restoration time.

A representative research is the work in [29], which focus on the cyber physical system itself, to analyze the system from different layers in the cyber physical system architecture, such as the perception layer, transmission layer and application layer. The system structure model is based on the system´s architecture (including the sub-systems/components). The well-known vulnerabilities (virus, Trojan, worm, denial of service) and mitigation measures (firewall, online monitoring after regular patching and testing, intrusion prevention system, intrusion detection system) are mapped to the specific components; this is the architecture model. The mitigation measures, as a bridge to connect the previous activity-quality model, and the architecture model, after connecting both models, they denote the final model as a cybersecurity risk model.

8.3   Analysis Focus on Systems and Beyond

According to NIST SP 800-82, a cybersecurity risk should be addressed in three tiers: (i) the organization level; (ii) mission/business process level; and (iii) information system level (IT and ICS). Leveson [49] developed an accident model—STAMP (System-Theoretic Accident Model and Processes). It can be used to do hazard and accident/incident analysis. Applying the STAMP to do hazard analysis is STPA, applying STAMP to do accidents and incidents analysis is CAST (Causal Analysis based on STAMP). A following work in [50], uses this model to manage cybersecurity risk. Another work uses it to analyze the cyberattack on cyber-physical systems [51]. They concern more about the management, structure (more about the company structure), instead of the detailed parts inside of an industrial control system. The general process of using STAMP to do accident analysis is presented in [49]. This approach has an obvious advantage, as compared to typical analysis approaches. It considers the organization structure, the operational structure and system structure. These are all considered at the same time. While the technical part of the control system is not analyzed deeply. And as far as we know, by the time of writing, applying the STAMP into security/cybersecurity analysis is limited to cyberattacks that had already happened but not before they occur. Another analysis approach is STAP-SafeSec [52], which is based on the STAP, the STPA is proposed by Leveson [49]. The STPA-SafeSec is proposed to do the safety and security analysis before the attack happens. Main steps and how they can be achieved are described in Table 4 (Note: not all steps are represented here).

Even this approach is based on the STPA, it does not focus on the organization, but only on the system, safety and security together, and only the general threats and general vulnerabilities are provided.

Some IT risk analysis approaches can be considered for conducting a cybersecurity analysis on ICS, such as the OCTAVE Allegro method [54] and COBIT 5 for risk [55]. The OCTAVE Allegro assess to what extent the computed threat influences the organization. The worksheet of risk measuring criteria is provided to support the risk analysis process. In the COBIT 5 for risk, the COBIT 5 framework is proposed to guide the IT related risk analysis and management on various aspects, e.g. organizational structures, cultures and services.

Table 4 The STAP-SafeSec analysis approach

STAP-SafeSec [52] main steps	The way to achieve each step
Identify system losses and system hazards	Consulting experts
Derive high level safety and security constraints	By negation of identified hazards
Identify the hazardous control action	Using a semi-automated manner which is defined in [53]
Map safety and security constraints between control and component layer	Manually
Describe the hazardous scenario according to the safety or security flaw of the system	The scenario and sub-scenario are illustrated in a tree structure

8.4 Cybersecurity Analysis of ICS Functionality Based on CFA (Casual Fault Analysis)

We propose to do cybersecurity analysis from the perspective of the system functionality. This analysis approach will focus on the qualitative analysis and only on the system. It does not include the management, or the human decision-making part, but only considers potential hazards that might be caused by attacks of the system. In this analysis approach, questions like the following will be answered:

(1) What can go wrong (unexpected situations)?
(2) What are the consequences of these unexpected situations?
(3) How can the "wrong" happen?

This analysis process is based on the CFA (Casual Fault Analysis), which is a formal analysis approach based on the WBA (Why-Because Analysis). We deduce the reasons of a hazard in a formal way. The Fault Tree or the Attack Tree has a similar analysis idea, however, there is a mechanism to check if the casual relations that we deduced are correct, thus we say that this approach is stricter on semantic than Fault and Attack Tree (the checking mechanism is called the Counterfactual Test [56]). This is the reason that we use the CFA.

As we all know, the system is used to realize various functions, according to the designers' intention. Hence, if there is a successful attack on the system, which means that one or more related/specific function(s) does not work as what they are expected to be. Looking for which function is compromised, and how that may happen (the attack path), is the core idea of this approach. The steps are illustrated in Fig. 6 briefly.

We first perform a reverse engineering of an existing safety critical ICS, which is to produces the formal functional specification. The way of conducting reverse engineering on current ICS cybersecurity analysis domain is novel. Then, based on the examined formal functional specification, we identify the hazards which have a potential to lead to an accident or a disaster on human life safety and environment.

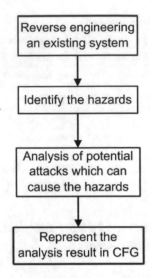

Fig. 6 Cybersecurity analysis of ICS functionality based on CFA (Casual Fault Analysis)

The system vulnerabilities are identified based on knowing of the system, the system functional specification and some online open source vulnerability databases, e.g. CVE. After the hazards are identified, we look for which function might be compromised (the function's functional integrity is violated), and how that could happen. The question of how to look for the compromised function is proposed to consider the formal functional specification. The functional specification describes the preconditions of executing this function and what are the postconditions after executing the specific function. Thus, when we look for the compromised function, we should check not only the preconditions but also the postconditions.

The compromised function checking process takes the formal functional specification as a reference. Following this way, we then deduce the potential attacks, including: where, when and how these attacks might happen. Finally, the analysis results will be presented in CFG (Casual Fault Graph). The reason that we choose CFG is discussed in [57]. The deducing of when, where, and how the attacks might happen and lead to the hazard is investigated around the overall system, located on various layers, e.g. the system architecture, the component, inside a CPU and even down to the code level. Potential attacks are modeled based on the specific system architecture and system vulnerabilities. We call this analysis process "cybersecurity analysis of system functionality". In [57], the cybersecurity analysis approach from the system functionality perspective is illustrated. It uses a safety critical Industrial Control System (ICS) which is commonly used inside NPPs as an example to show how to perform a holistic cybersecurity analysis.

As for how to check the functional specification in our analysis process, we proposed to use the artificial intelligence planner in previous work [58].

8.5 Considerations on Dealing with Cybersecurity Issues at the System Architecture Level

Securing the ICSs at the architecture level is one way to reduce the cybersecurity risk, for example, in a safety critical industrial domain (e.g. Nuclear Power Plants), inter-actions between safety-critical systems and non-safety systems have to be restricted. In case a failure happens in a non-critical (low-critical) application it must not disturb safety-critical (high-critical) applications and mixed criticality helps in solving this issue. The Field-Programmable Gate Arrays (FPGAs) is proposed as a bridge to implement this intention for ICS in NPPs [59], as FPGAs can enhance the reliability and also reduce the complexity of hardware features [59].

The Mixed Criticality Systems (MSCs) have gained much attention in research and also in ICS design. As the computational power is available with current hardware, including FPGAs, this offers multiple advantages by allowing the implementation of several different applications within one system. Thus, using the MCS permits numerous applications of different criticality levels to run in a single system is possible. Virtualization is a technology that is frequently implemented in computers and servers to isolate execution environments, and to support the execution of multiple OS on the same entity [60]. For embedded systems, virtualization offers improved performance, better transparency, portability and interoperability by integrating hardware and software resources, and networking services into one hardware platform. It makes the integration of MCS become easier. System virtualization is seen as an abstraction and management layer of system resources like CPU, memory, and peripherals.

Thus, the security by design idea is proposed to consider in the early stages of design. Using a Xilinx Soc 7000 ZC 706 Board, with self-test partitions on the software layer and hardware layer, also some basic partitions containing C applications are running on the board, to implement a test platform is a perspective to realize this security by design. And port a General Operating System (OS) such as Linux on top of Xtratum might be a feasible way to make the platform more suitable under the idea of security by design.

9 Security by Design, Secure Interoperability

Current industrial solutions deploy legacy network communication protocols, like MODBUS over TCP/IP. With newer solutions these inherent risks can be avoided, by consistently enforcing secure interoperability methods. For industrial automation systems the interoperability does not only consider the point-to-point, multicast or broadcast exchange via one network. Interoperability considers the whole path from a client software application (e.g. on a tablet computer used in the plant shop floor) to an embedded system reached via switches and intermediated automation systems. New approaches like Industrie 4.0 in Germany, Industry du Futur in France and IIoT in the US are starting to address these topics by the use of OPC Unified Architecture (defined via 14 standard parts in IEC 62541) and Time Sensitive Networks (TSN) [61]. The secure design of these new protocols has been evaluated by the German BSI [62] and the whole protocol stack is considered as securely designed. Additional standards related to these "Security by Design" approaches are introduced in [63]. As a future work the qualitative and quantitative approaches described in this chapter will need to consider to be applied to these "Security by Design" approaches with regard to their consistent implementation, their impact on functional safety and the real-time assumptions. This is a considerable effort, as some of these IIoT approaches, like the use of OPC UA introduce more sophisticated concepts with regard to object oriented information models, typing and sub-typing constraints, negotiation of security properties to be implemented by clients and servers, role based access control etc. Our work on these topics already started and publications will follow.

10 Conclusion

In this chapter, we started by investigating typical threats to Industrial Control Systems (ICSs). We introduced the classification the common vulnerabilities of ICSs. The core part is to investigate the cybersecurity analysis approaches and to taxonomy them according to quantitative and qualitative vectors. For each type of analysis approach, more detailed classification is developed, e.g. the pure quantitative analysis, semi-quantitative analysis, analysis focus from the offensive and defensive perspective, analysis focus on the system and even the overall organization. At last, a cybersecurity analysis of system functionality approach, and an idea to deal with cybersecurity issue from system architecture level based on virtualization are proposed.

Acknowledgements We appreciate the guidance of Prof. Ladkin, Dr, Karl Waedt, and Prof. Ruland. We also thank all reviewers.

References

1. Langner, R.: To kill a centrifuge. Langer. https://www.langner.com/wp-content/uploads/2017/03/to-kill-a-centrifuge.pdf (2013). Accessed 4 July 2019
2. Dragos Inc.: TRISIS malware-analysis of safety system targeted malware. https://dragos.com/wp-content/uploads/TRISIS-01.pdf(2017). Accessed 4 July 2019
3. Hamilton, B.A.: When the lights went out. https://www.boozallen.com/content/dam/boozallen/documents/2016/09/ukraine-report-when-the-lights-went-out.pdf (2016). Accessed 4 July 2019
4. Lee, R.M., Assante, M.J., Conway, T.: SNAS ICS. Analysis of the cyber attack on the Ukrainian power grid. https://ics.sans.org/media/E-ISAC_SANS_Ukraine_DUC_5.pdf (2016). Accessed 4 July 2019

5. Kaspersky: Traffic Lights are Easy to Exploit. https://www.kaspersky.com/blog/traffic-light-attacks/5830/ (2014). Accessed on 31 Mar 2019
6. IEC Technical Specification: IEC 62443 Industrial communication networks. In: Network and System Security—Part 1-1: Terminology, Concepts and Models (2014)
7. Stouffer, K., Stouffer, K., Abrams, M.: NIST SP 800-82 r2: Guide to Industrial Control Systems (ICS) Security. https://nvlpubs.nist.gov/nistpubs/SpecialPublications/NIST.SP.800-82r2.pdf (2015). Accessed 4 July 2019
8. Walter, T.: IEC62859: Nuclear power plants instrumentation and control systems. In: Requirements for Coordinating Safety and Cybersecurity, p. 10 (2016)
9. Ladkin, P.B.: Chapter 5: Integrity. In: A Critical-System Assurance Manifesto: Issues Arising from IEC 61508. RVS Group, Bielefeld. https://rvs-bi.de/publications/RVS-Bk-17-01.html (2017). Accessed 7 July 2019
10. Venezuelanalysis: Venezuela: New Widespread Power Outage as Gov't Denounces Alleged Attacks. https://venezuelanalysis.com/news/14404 (2019). Accessed on 31 Mar 2019
11. Irmak, E., Erkek, I.: An overview of cyber-attack vectors on SCADA systems. In: 2018 6th International Symposium on Digital Forensic and Security (ISDFS), Antalya, Turkey, March 2018. pp. 1–5, IEEE (2018)
12. European Telecommunications Standards Institute: ETSI-TR 102 893-V1.1.1—ETSI TR 102. Intelligent Transport Systems (ITS); Security; Threat, Vulnerability and Risk Analysis (TVRA). https://www.etsi.org/deliver/etsi_TR/102800_102899/102893/01.01.01_60/tr_102893v010101p.pdf (2010). Accessed 4 July 2019
13. Conteh, N.Y., Schmick, P.J.: Cybersecurity: risks, vulnerabilities and countermeasures to prevent social engineering attacks. Int. J. Adv. Comput. Res. **6**(23), 31–38 (2016)
14. Baylon, C., Brunt, R., Livingstone, D.: Cyber security at civil nuclear facilities understanding the risks. Chatham House Report. https://www.chathamhouse.org/sites/default/files/field/field_document/20151005CyberSecurityNuclearBaylonBruntLivingstone.pdf (2015). Accessed 7 July 2019
15. Babu, B. et al.: Security issues in SCADA based industrial control systems. In: 2nd International Conference on Anti-cyber Crimes (ICACC), Abha, Saudi Arabia, March 2017. pp. 47–51, IEEE (2017)
16. CVE Online Vulnerability Database: CVE-2015-2822. https://www.cvedetails.com/cve/CVE-2015-2822/ (2015). Accessed 31 Mar 2019
17. CVE Online Vulnerability Database: CVE-2016-5743. https://www.cvedetails.com/cve/CVE-2016-5743/ (2016). Accessed 31 Mar 2019
18. CVE Online Vulnerability Database: CVE-2013-0672. https://www.cvedetails.com/cve/CVE-2013-0672/ (2013). Accessed 31 Mar 2019
19. CVE Online Vulnerability Database: CVE-2013-4911. https://www.cvedetails.com/cve/CVE-2013-4911/ (2013). Accessed 31 Mar 2019
20. CVE Online Vulnerability Database: CVE-2011-4508. https://www.cvedetails.com/cve/CVE-2011-4508/ (2011). Accessed 31 Mar 2019
21. CVE Online Vulnerability Database: CVE-2013-0671. https://www.cvedetails.com/cve/CVE-2013-0671/ (2013). Accessed 31 Mar 2019
22. NIST NVD Online Vulnerability Database. CVE-2017-12069. https://nvd.nist.gov/vuln/detail/CVE-2017-12069#vulnCurrentDescriptionTitle (2017). Accessed 31 Mar 2019
23. CVE Online Vulnerability Database: CVE-2014-1697. https://www.cvedetails.com/cve/CVE-2014-1697/ (2014). Accessed 31 Mar 2019
24. CVE Online Vulnerability Database. CVE-2014-2246. https://www.cvedetails.com/cve/CVE-2014-2246/ (2014). Accessed 31 Mar 2019
25. CVE Online Vulnerability Database: CVE-2016-2200. https://www.cvedetails.com/cve/CVE-2016-2200/ (2016), Accessed 31 Mar 2019
26. CVE Online Vulnerability Database: CVE-2014-2256. https://www.cvedetails.com/cve/CVE-2014-2256/ (2014). Accessed 31 Mar 2019
27. CVE Online Vulnerability Database: CVE-2016-2846. https://www.cvedetails.com/cve/CVE-2016-2846/(2016). Accessed 31 Mar 2019

28. National Cybersecurity and Communications Integration Center: US-CERT. Attack Possibilities by OSI Layer. https://www.us-cert.gov/sites/default/files/publications/DDoS%20Quick%20Guide.pdf (2014). Accessed 4 July 2019
29. Ashibani, Y., Mahmoud, Q.H.: Cyber physical systems security: analysis, challenges and solutions. Comput. Secur. **68**, 81–97 (2017)
30. Igure, V.M., Laughter, S.A., Williams, R.D.: Security issues in SCADA networks. Comput. Secur. **25**(7), 498–506 (2006)
31. Kaplan, S., Garrick, B.J.: On the quantitative definition of risk. Soc. Risk Anal. **1**(1) (1981)
32. Cherdantseva, Y. et al.: A review of cyber security risk assessment methods for SCADA systems. Comput. Secur. **56**, 1–27 (2016)
33. Gritzalis, D., Stavrou, V.: Exiting the risk assessment maze: a meta-survey. ACM Comput. Surv. **51**(11), 1–30 (2018)
34. Abdo, H., Kaouk, M., Flaus, J.M., Masse, F.: A safety/security risk analysis approach of industrial control systems: a cyber bowtie-combining new version of attack tree with bowtie analysis. Comput. Secur. **72**, 175–195 (2017)
35. Zheng, Y., Zheng, S. (2015) Cyber security risk assessment for industrial automation platform. In: International Conference on Intelligent Information Hiding and Multimedia Signal Processing, Adelaide, SA, Australia, July 2015. pp. 341–344, IEEE
36. Ledwaba, L., Venter, H.S.: A threat-vulnerability based risk analysis model for cyber physical system security. In: 50th Hawaii International Conference on System Sciences. Hawaii, USA, January 2017. pp. 6021–6030, IEEE (2017)
37. Shin, J., Son, H., Heo, G.: Cyber security risk evaluation of a nuclear I&C using BN and ET. Nucl. Eng. Technol. **49**(3), 517–524 (2017)
38. Shin, J., Son, H., Khalil Ur, R., Heo, G.: Development of a cyber security risk model using Bayesian networks. Reliab. Eng. Syst. Saf. **134**, 208–217 (2015)
39. Zhang, Y., Wang, L., Xiang, Y., Ten, C.W.: Inclusion of SCADA cyber vulnerability in power system reliability assessment considering optimal resources allocation. IEEE Trans. Power Syst. **31**(6), 4379–4394 (2016)
40. Fang, Z.H., Mo, H.D., Wang, Y.: Reliability analysis of cyber-physical systems considering cyber-attacks. In: IEEE International Conference on Industrial Engineering and Engineering Management, Singapore. pp. 364–368, IEEE (2017)
41. Wu, W., Kang, R., Li, Z.: Risk assessment method for cybersecurity of cyber-physical systems based on inter-dependency of vulnerabilities. In: IEEE International Conference on Industrial Engineering and Engineering Management, Indonesia, December 2016. pp. 1618–1622, IEEE (2016)
42. Caltagirone, S., Pendergast, A.: The diamond model of intrusion analysis. https://apps.dtic.mil/dtic/tr/fulltext/u2/a586960.pdf (2013). Accessed 7 July 2019
43. Hutchins, E.M., Cloppert, M.J., Amin, R.M.: Intelligence-driven computer network defense informed by analysis of adversary campaigns and intrusion kill chains. In: 6th International Conference on Information Warfare and Security, George Washington University, March 2011. pp. 113–125 (2011)
44. Hansen, A., Staggs, J., Shenoi, S.: Security analysis of an advanced metering infrastructure. Int. J. Crit. Infrastruct. Prot. **18** (2017)
45. Xiang, Y., Ding, Z., Zhang, Y., Wang, L.: Power system reliability evaluation considering load redistribution attacks. IEEE Trans. Smart Grid **8**(2), 889–901 (2017)
46. Kang, E., Adepu, S., Jackson, D., Mathur, A.P.: Model-based security analysis of a water treatment system. In: IEEE/ACM 2nd International Workshop on Software Engineering for Smart Cyber-Physical Systems, Austin, Texas, May 2016. pp. 22–28, IEEE/ACM (2016)
47. Song, J.-G., Lee, J.-W., Lee, C.-K., et al.: A cyber security risk assessment for the design of I&C systems in nuclear power plants. Nucl. Eng. Technol. **44**(8), 919–992 (2012)
48. de Gusmão, A.P.H., et al.: Cybersecurity risk analysis model using fault tree analysis and fuzzy decision theory. Int. J. Inf. Manage. **43**, 248–260 (2018)
49. Leveson, N.G.: Engineering a Safer World-System Thinking Applied To Safety (Draft). MIT Press, Cambridge. http://sunnyday.mit.edu/safer-world.pdf (2011). Accessed 04 July 2019

50. Salim, H.M.: Cyber Safety: A Systems Thinking and Systems Theory Approach to Managing Cyber Security Risks. MIT Press, Cambridge. http://web.mit.edu/smadnick/www/wp/2014-07.pdf (2014). Accessed 04 July 2019
51. Whyte, D.: Using a Systems-Theoretic Approach to Analyze Cyber Attacks on Cyber-Physical Systems. MIT Press, Cambridge. https://dspace.mit.edu/handle/1721.1/110143(2017). Accessed 04 July 2019
52. Friedberg, I., McLaughlin, K., Smith, P., et al.: STPA-SafeSec: safety and security analysis for cyber-physical systems. J. Inf. Secur. Appl. **34**, 183–196 (2017)
53. Thomas, J.: Extending and automating a systems-theoretic hazard analysis for requirements generation and analysis. Ph.D. thesis, MIT Press, Cambridge. http://sunnyday.mit.edu/JThomas-Thesis.pdf (2013). Accessed 04 July 2019
54. Caralli, R.A., Stevens, J.F., Young, L.R., Wilson, W.R.: Introducing OCTAVE Allegro: Improving the Information Security Risk Assessment Process. https://resources.sei.cmu.edu/asset_files/TechnicalReport/2007_005_001_14885.pdf (2007). Accessed 7 July 2019
55. ISACA: COBIT5 for risk. https://www.isaca.org/COBIT/Documents/COBIT-5-for-Risk-Preview_res_eng_0913.pdf(2013). Accessed 7 July 2019
56. Ladkin, P.B.: An Example of Why-Because Analysis in Digital System Safety. RVS Group, Bielefeld. https://rvs-bi.de/publications/RVS-Bk-17-02.html (2017). Accessed 4 July 2019
57. Lou, X., Waedt, K., Schürmann, T., et al.: Cybersecurity analysis of industrial control system towards system function view. In: International Conference on Industrial Cyber-Physical Systems, Taipei, May 2019, IEEE (2019)
58. Lou, X., Waedt, K. et. al.: Combining AI planning advantages to assist preliminary formal analysis on ICS cybersecurity vulnerabilities. In: 10th Edition Electronics, Computers and Artificial Intelligence, Iasi, Romania, June 2018. IEEE (2018)
59. Jung, J., Ahmed, I.: Development of field programmable gate array-based reactor trip functions using systems engineering approach. Nucl. Eng. Technol. **48**(4) (2016)
60. Tellabi. A., Peters, L., Ruland, C., et al.: Security Aspects of Hardware Virtualization Technologies for Industrial Automation and Control Systems. In: GIACM WS on I4.0/IACS Standardization, Berlin (2018)
61. DIN/DKE/VDE. DEUTSCH ENORMUNGSROADMAP Industrie 4.0-Version 3. https://www.din.de/blob/95954/97b71e1907b0176494b67d8d6d392c54/aktualisierte-roadmap-i40-data.pdf (2018). Accessed 17 July 2019
62. OPC Foundation: The Industrial Interoperability Standard. https://opcfoundation.org/. Accessed 17 July 2019
63. Sino-German Industrie 4.0: Intelligent Manufacturing Standardisation Sub-Working Group. Security Standards White Paper for Sino-German Industrie 4.0/Intelligent Manufacturing. https://www.dke.de/resource/blob/1711300/9e7add87021790df6d2dc57312e05302/security-standards-white-paper-for-sino-german-industrie-40-data.pdf (2018). Accessed 17 July 2019
64. CVE Online Vulnerability Database. Siemens: Vulnerability Statistics. https://www.cvedetails.com/vendor/109/Siemens.html (2019). Accessed 31 Mar 2019

Automatic Attack Graph Generation for Industrial Controlled Systems

Mariam Ibrahim, Ahmad Alsheikh and Qays Al-Hindawi

Abstract Industrial Controlled Systems (*ICSs*) are prone to cyber-attacks exploiting weaknesses in their units. This chapter illustrates through an example of a pressurized water Nuclear Power Plant (*NPP*) control system how these vulnerabilities may be exploited by an attacker compromising the system. The control system is described by Architecture Analysis and Design Language (*AADL*), and then checked with a security property via *JKind* checker. The resulted *Lustre* file is later fed to an Attack Scenarios Generation and Filtration Tool (*ASGFT*). This tool automatically generates all possible attack scenarios resulting in overall plant disruption. The union of these attack scenarios is visualized by a *Visualizer.exe*, a Graphical User Interface (*GUI*) built employing C#.

Keywords Nuclear power plant · Attack scenario · Assets · Vulnerabilities · *ASGFT*

1 Introduction

Industrial Control Systems (*ICSs*) are computer-based systems that are controlled by Supervisory Control and Data Acquisition (*SCADA*) [1]. The security and safety of these systems are critical issues to satisfy an acceptable state of operation, and prevent them from malfunctioning, causing a delay of production, eavesdropping of potential and sensitive data, and automated mechanical disruption as a result of miscommunication or cutoff in transmitted data. Hence, a powerful tool is needed

M. Ibrahim (✉) · A. Alsheikh · Q. Al-Hindawi
Department of Mechatronics Eng, Faculty of Applied Technical Sciences, German Jordanian University, Amman, Jordan
e-mail: mariam.wajdi@gju.edu.jo

A. Alsheikh
e-mail: a.alsheikh@gju.edu.jo

Q. Al-Hindawi
e-mail: q.alhindawi@gju.edu.jo

© Springer Nature Switzerland AG 2020
E. Pricop et al. (eds.), *Recent Developments on Industrial Control Systems Resilience*,
Studies in Systems, Decision and Control 255,
https://doi.org/10.1007/978-3-030-31328-9_5

Table 1 Cyber-attacks on *ICN*s and *NPP*s

Date	Country	Name	Description
February 1992	Lithuania	Ignalia NPP	Disruption attempt [2]
June 1999	United Kingdom	Bradwell NPP	Data destroyed by employee [3]
January 2003	United States	Davis-Besse *NPP*	Access to core information prevented by virus [4]
June 2009	Iran	Uranium enrichment facility at the Natanz plant	Series of attacks to damage the centrifuges [5]
September 2011	France	Areva	Network intrusions [6]
2014	Germany	Steel mill	Malware caused furnace malfunction [7]
December 2014	Korea	Korea Hydro and Nuclear Power Company	Release of stolen data [8]
February 2015	Japan	Nuclear Material Control Center (NMCC)	File-sharing software causing leakage of data [9]
April 2016	Germany	Gundremmingen NPP	Viruses affecting fuel rod monitoring system [10]
June 2016	Japan	University of Toyama, Hydrogen Isotope Research Center	Stealing research data via spear-phishing [11]
May 2017	United States	Companies operates NPPs e.g., Wolf Creek	Compromise enterprise networks [12]

to evaluate these systems side by side with their existed vulnerabilities, analyzing and presenting all potential attacks that might compromise the system as given by its Attack graph.

As a case study, we consider Nuclear Power Plants (*NPPs*) which are prone to cyber-attacks exploiting vulnerabilities in their structural units. Hence, it is vital to determine these vulnerabilities and the potential attacks corrupting them (their Attack graphs). Table 1 shows a summary of the surveyed cyber-attacks against Industrial Control Networks and Nuclear Power Plants.

1.1   Related Work

Various methods and tools for generating Attack graphs have been explored in literature. As an example, in [13] a graph-based algorithm is utilized to encode a model-based Attack graph for multi-host. An approach to use extensive scenario graphs to develop cyber-scenarios that may reflect the fundamental attributes and internal structure of *NPP*s facilities networks is proposed by Ahn et al. [14]. The paper modeled attacks through determining the paths, which are initiated at an attacker node and terminated at a goal node in overall scenario.

One of early papers that linked pre/post condition of attack actions and multiple attacker goals to generate Attack graph is [15]. In their work, attacks are showed as a set of capabilities rather than a series of events. A flexible extensible model is used in security analysis to eliminate the lack of expressing complex scenarios which is inherited in attack single sequence approach.

A tool Presented by Noel and Jajodia [16] generates a reduced Attack graph for real networks adapting the reduction of graph complexity utilizing interactive visualization. An Attack graph is developed by Tao et al. [17] utilizing forward-*search*, *breadth-first* and *depth-limit* algorithms for a network. In order to find a general technique to generate complex Attack graphs for heterogeneous networks, [18] developed a tool that proved to be simpler than [17], and can immediately find all minimal attack paths.

A logical Attack graph generation tool is proposed by Xinming et al. [19]. Their tool is designed using *MulVAL* based on *XSB* [20]. *MulVAL* evaluates the Data log interaction of the "security advisories", "network configuration", and "Machine configuration", and then sends the trace of the evaluation to a graph builder resulting in the final graph.

An easy-to-go toolkit interface designed by Sheyner and Wing [21] to automatically generate Attack graph which helps researchers to analyze networks. The toolkit is used to perform systematic experiments with different network configurations. The result is aimed to help system administrator to test network connectivity and services running on hosts. The open source and the commercial tools, are compared and investigated by Yi et al. [22] in terms of Attack graph types, scalability, complexity, and the degree of visualization.

The commercial products *Amenaza*, (a secure tree software for attack graph analysis) [23] and *Skybox*, (an automated tool for networks security analyses) [24] are used to perform a cyber-security analysis on communication and control networks to implement Attack graphs. The generated graphs are utilized by researchers to determine all possibilities, probabilities, and vulnerabilities that can be used by intruders to compromise the system facilities. A review is done by Lippmann and Ingols [25] comparing previous studies on Attack graphs in terms of their goals, construction of scenarios and vulnerabilities, as well as their scaling.

An Attack graph consists of nodes, defining potential attacks, which may happen. Edges illustrate the dynamic variation in system state resulted by the attacker's actions [26]. A survey managed by IIs [27] to examine security and safety levels in various power stations globally. The collected data illustrated a disparity in these levels, and the need to implement more attempt to increase them.

An approach is presented by Sklyar [28] to protect *NPP* against cyber-attacks. This included employing Programmable controllers that possess a small level of weaknesses, and doing further cyber-security analysis for such attacks. In order to keep facilities computer machines and measuring devices secure, and prevent any cyber-attack from affecting its performance, the computers OS and devices firmware have to be continuously updated [29]. In [30], the authors stated that the capabilities in control systems' facilities create vulnerabilities. An equipment can be remotely reprogrammed, or traffic can be monitored by a suspicious user behaving unexpectedly. This causes a physical damage to mechanical equipment, infrastructure, and life. The paper discussed in details ways to avoid these damages.

In this work, we present a model-based approach to generate the Attack graph causing system compromise. The system architecture and the security property are encoded with *AADL* [31], and then verified by *JKind* [32]. To illustrate the issue of Attack graph implementation, an illustrative example *NPP* is considered, and the developed Attack graph is generated automatically through our *Java* based tool and visualized using C# in Microsoft Visual Studio (*VS*) [33].

2 Illustrative Example: NPP Control System

This Section describes an illustrative example of *NPP* control system. This includes its architecture, possible attack instances/actions, and system formal description.

2.1 NPP Architecture

Figure 1 illustrates *NPP* adapted from [10] with the following levels:

Field Level, F
It consists of three water loops, primary, secondary and cooling loops. The pressurizer [34] maintains the water pressure at 155 bar (2235 psig [35]) by heating the water using electrical heaters, or cooling it using water sprays [36]. The steam generator

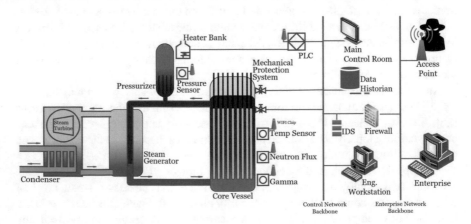

Fig. 1 A pressurized water reactor

[37] transfers the heat of the pressurized water with the water in the secondary loop. Thus, changing water state to steam that will flow up till it approaches steam turbine.

The safety and protection system [38], smart sensors *(Sen)* (e.g., temperature sensor, Gamma sensor), Actuators *(Act)*, and Heater Bank are field components as well.

Control Level, C
It includes Programmable Logic Controller *(PLC)* and Remote Terminal Unit *(RTU)*.

Supervisory Level, S
It includes:

- Main Control Room *(CR)*: it has Supervisory Control and Data Acquisition *(SCADA)* software.
- Data Historian *(DH)*: a database that contains information on the network.
- Engineering Workstation *(EW)*: it is utilized for inspection and maintenance of F components.

Enterprise Level, E
Enterprise Site Management computer *(SM)* collects data, sends results and reports to decision makers. While Wireless Access point *(WA)* [10] used for external internet communication.

Network Backbones

- Control network *(CB)*: links C with S.
- Enterprise network *(EB)*: links S with E.

In addition, there is a firewall isolating the system from the remaining of the enterprise network. The firewall does not apply any access control restrictions on the flow of network traffic. An Intrusion Detection System *(IDS)* monitors the network traffic flow among (E; S), whereas the flow among (S; C) is not monitored. For an attack instance/action that is detectable, the *IDS* triggers an alarm upon its detection, while a stealthy attack remains undetected. *SM*, *CR*, *DH*, and *EW* have Commercial off-the-shelf *(COTS)* OS vulnerability, while *PLC*, *Sen*, and *Act* have *firmware* vulnerability.

2.2 Possible Attacks

1. *Intelligent Gathering (IG)*: this attack is used when the attacker aims to get more information. This attack can be either stealthy or detectable.
2. *Malware Injection (MI)*: the attacker can edit, copy, or install the code at the host Operating Software. It can give the attacker a *root* access. This is a stealthy attack.

3. *Bypass Security Mechanism* (*BSM*): this attack is used inside a firmware update to attack a specific Hardware (*PLC, Sen*) to get control over it. This attack is detectable by *IDS*.
4. *Denial-of-Service* (*DoS*): this attack is employed by an attacker to flood the server with requests for a specific service. Thus, restraining the service provided to other users. This attack is detectable by *IDS* (e.g., using *Outliers Techniques* [39].)
5. *Man-in-the-Middle* (*MiM*): this attack can be conducted when the attacker accesses a connection between two users, giving the attacker a *root* access on the communication transmitter. This attack is detectable by *IDS*.
6. *Alteration-of-Data* (*AoD*): this attack can occur when the attacker has access to a software. It causes data processing *latency* and data corruption.

2.3   Formal Description for NPP System

1. The attacker is assumed to be located at *WA* and has a *root* privilege (static).
2. Set of field level components F; variable $f \in \{Sen, Act\}$ (static).
3. Set of control level components C; variable $c \in \{PLC, RTU\}$ (static).
4. Set of supervisory level components S; variable $s \in \{CR, DH, CB, EW\}$ (static).
5. Set of enterprise level components E; variable $e \in \{WA, SM\}$ (static).
6. System connectivity, $Z \subseteq E \times E, E \times S, S \times S, S \times C, C \times F$; $z_{ij} = 1$ iff component i connected to component j (static).
7. System vulnerabilities V; Boolean $v_i = 1$ iff vulnerability $v \in \{COTS, firmware\}$ exists on host i (static).
8. Set of possible attacks A; variable $a \in \{IG, MI, BSM, DoS, MiM, AoD\}$ (static).
9. Attack instances, $AI \subseteq A \times (E \times E, E \times S, S \times S, S \times C, C \times F)$; labeled $a_{ij} \equiv$ attack a from source i to target j, $a \in A$ (static).
10. Attacker level of privilege P on machine/host $i \in \{S, E\}$; variable $p_i \in \{none, user, root\}$ (dynamic).
11. Hardware control H on device $i \in \{F, C\}$; Boolean $h_i = 1$ iff attacker gains control over the firmware of device i (dynamic).
12. Latency L from Component i; Boolean $l_i = 1$ iff communication from i is delayed (dynamic).
13. Input data N into host i; Boolean $n_i = 1$ iff input of component i is corrupted (dynamic).
14. Output data O from host i; Boolean $o_i = 1$ iff output of component i is corrupted (dynamic).
15. Data knowledge K; Boolean $k_j = 1$ iff attacker gets knowledge from j (dynamic).
16. Intrusion detection system *IDS*: $A \times E \times S \to \{0,1\}$; Boolean $ids(a_{ij}) = 1$ iff attack a from source i to target j is detectable (static).
17. A global Boolean dg tracks whether an IDS alarm has been triggered for any previously executed atomic attack (dynamic).

18. Attack instances/actions pre-conditions:

- $Pre(IG_{ij}) \equiv (z_{ij} = 1) \bigwedge (p_i \geq user)$.
- $Pre(MI_{ij}) \equiv (z_{ij} = 1) \bigwedge (p_i \geq user) \bigwedge (COTS_j = 1) \bigwedge (\exists y \in \{S, E\}: k_y = 1)$.
- $Pre(BSM_{ij}) \equiv (z_{ij} = 1) \bigwedge (p_i \geq user \bigvee (h_i = 1)) \bigwedge (firmware_j = 1) \bigwedge (\exists y \in \{F, C\}: k_y = 1)$.
- $Pre(DoS_{ij}) \equiv (z_{ij} = 1) \bigwedge (p_i = root \bigvee (h_i = 1)) \bigwedge (COTS_j = 1 \bigvee (firmware_j = 1))$.
- $Pre(MiM_{ij}) \equiv (z_{ij} = 1) \bigwedge (p_i = root \bigvee (h_i = 1))$.
- $Pre(AoD_{ij}) \equiv (z_{ij} = 1) \bigwedge (p_i = root \bigvee (h_i = 1))$.

19. Attack instances/actions post-conditions:

- $Post(IG_{ij}) \equiv k_j = 1 \bigwedge ((i = SM \bigvee WA) \bigwedge (dg = 0) \Rightarrow (dg = 0) \bigvee (dg = 1))$.
- $Post(MI_{ij}) \equiv p_j = root$.
- $Post(BSM_{ij}) \equiv h_j = 1 \bigwedge ((i = SM \bigvee WA) \bigwedge (dg = 0) \Rightarrow dg = 1)$.
- $Post(DoS_{ij}) \equiv l_j = 1 \bigwedge k_j = 1 \bigwedge ((i = SM \bigvee WA) \bigwedge (dg = 0) \Rightarrow dg = 1)$.
- $Post(MiM_{ij}) \equiv o_j = 1 \bigwedge k_j = 1 \bigwedge ((i = SM \bigvee WA) \bigwedge (dg = 0) \Rightarrow dg = 1)$.
- $Post(AoD_{ij}) \equiv l_j = 1 \bigwedge n_j = 1$.

20. Initial state: $p_{WA} = root \bigwedge (l_j \in \{F, C, S, E\}: P_j = none \bigwedge (l_j = h_j = k_j = n_j = o_j = 0)) \bigwedge dg = 0$. (The attacker has root privilege on *WA* and no privilege, no latency, no hardware control, no data knowledge, no input/output data on all system's machines/hosts, and *IDS* has not detected security violation).

21. *Security property* φ is that attacker cannot disrupt the system (i.e., no *latency* in *PLC*, *PLC* is not controlled by attacker) or attack gets detected. This property can then be described by a Computational Tree Logic (*CTL*):

$$\varphi \equiv AG((l_{PLC}=0) \bigwedge (h_{PLC}=0) \bigvee (dg =1)) \equiv AG(((l_{PLC} = 1) \bigvee (h_{PLC} = 1) \bigwedge (dg = 0))).$$

3 Attack Graph Generation

In this section, several software tools are employed to conduct Attack graph generation and visualization as illustrated through *NPP* control system.

Attack Scenarios Examples For NPP

JKind is an infinite state model checker for checking safety properties of synchronous systems [40], which are expressed in *Lustre*, a programming language for reactive systems [41]. A falsified property is reported with an explicit Counter-Example (*CE*), demonstrating the property breaching that is given here as an attack scenario (a sequence of attack instances causing system disruption). The *AADL* architecture level model is enclosed by *AGREE Annex* plug-in [42].

To conduct attack scenarios, the following *security property* φ is considered, where an attacker aims to cause system disruption by either making a denial of service from *PLC* or gaining control on *PLC* without being detected. This target can be accomplished by gaining a *root* privilege on both *CR*, and *EW* machines. When verifying the system model against φ, the *JKind* generated the following counterexample/attack-scenario, $CE1: = IG_{WADH} \rightarrow MI_{WAEW} \rightarrow MI_{WACR} \rightarrow BSM_{CRPLC}$ as a spreadsheet which was then saved as a separate *CSV* file for later use.

By encoding this counterexample in disjunct with φ, that is $\varphi \bigvee CE1$, a new counterexample *CE2* is generated, where $CE2: = IG_{WADH} \rightarrow MI_{WACR} \rightarrow MI_{WAEW} \rightarrow DoS_{CRPLC}$. In this scenario, after gaining a *root* privilege on both *EW* and *CR* machines, DoS_{CRPLC} attack is lunched to cause latency from *PLC* and deny other machines from requesting any information from it. *DoS* attack also aims to gather information about the measured data delivered from F Level Components through *PLC* to S Level machines (e.g., temperature values, Gamma radiation, Neutron flux, and Pressurizer Pressure).

Figure 2 shows the workflow of the Attack graph automatic generation. It consist of two main stages. In the first stage, the attack scenarios are generated as .csv files using our Attack Scenarios Generation and Filtration Tool (*ASGFT*). In the second stage, we visualize the Attack graph using our *Visualizer.exe* tool. In the following sections, we explain these tools.

3.1 Attack Scenarios Generation and Filtration Tool

So far, the attack scenarios were generated by repeatedly running *AGREE*, and updating the security property in every run to exclude the previously generated attack scenarios. In this part, the generated *NPP* model and the security property of interest are translated to *Lustre* model which is later fed to *ASGFT* that generates all possible attack scenarios automatically.

ASGFT is developed through *NetBeans* which is an integrated development environment (*IDE*) for Java [43]. *Maven* was selected as the main project for *ASGFT* tool in *NetBeans* [44, 45]. *Maven* is a build automation tool used primarily for *Java* projects [46].

When running *AGREE* based *JKind* model-checker for *Lustre* models within *Osate2*, it can only generate one counterexample at a time. Yet, when running it a couple of times, it may regenerate the same counterexample. *ASGFT* ensures that a explicit counterexample is generated every time the *JKind* is executed and automatically all the possible attack scenarios are visualized using MS Visual Studio software.

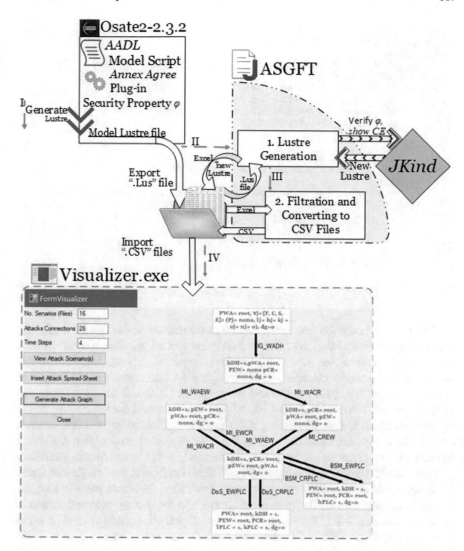

Fig. 2 Attack graph generation workflow

```
 1 INPUT:  System Model (Lustre 0.lus), Attack instances A[], maximum length (n)
 2 OUTPUT: All possible attack scenarios in (.csv)
 3 procedure
 4    Insert attack instances in the Single-Dimensional Array (A)
 5    Define maximum length (n)
 6    loop 1: # for (A^n), All possible attack combinations
 7        CE_1 = possible comnination from A[] of length n
 8        New Lustre= do new (.lus) copy (Lustre 0.lus)
 9        New Lustre= New Lustre + CE_1
10        goto loop 1
11    loop 2:
12        call JKIND through cmd
13        New Lustre.xlsx= do result in (.xlsx) format
14    loop 3:
15        New Lustre.csv= replace (New Lustre.xlsx) format to (.csv) format
16    loop 4: # delete non-violating attack scenarios
17        if  New Lustre.csv contians CE_1 = False :
18            delete(New Lustre.csv)
19            goto loop 4
20
21        generate violting attack scenarios
```

Fig. 3 *ASGFT* Algorithm

ASGFT takes only the first *Lustre* model (a translation of the system model and the security property from *AGREE*), then it generates all possible combinations of potential attack scenarios (i.e., *CEs*) as (*.lus*) format files. Next, *ASGFT* communicates with the model checker *JKind* through the Command Prompt Commands (*CMD*) to check the system model and the potential *CE* iteratively against security property within *lustre* files. Doing so, *ASGFT* generates all corresponding results as separate Excel sheets, illustrating if the potential *CE* is truly an attack scenario or not. Having generated all excel files, a special project is used within *ASGFT* to convert the (.xlsx) format to (.csv) format to be used later in Visual Studio Visualizer.

The *ASGFT* algorithm is shown in Fig. 3. The user inserts the first *Lustre* Model as the input and defines the attack instances constituting the attack paths in a single-dimensional array. Also, the user must choose the maximum expected length of an attack scenario. *ASGFT* in turns, generates all potential combinations of attack scenarios that is A^n, where A is number of attack instances, and n is the maximum length of an attack scenario. Each new potential attack scenario is stored as a variable *CE1* within (.lus) generated files. Next *JKind* is called iteratively to check these files against security property φ, If *CE1* violates the security property, then it is a potential attack scenario and the result is given as an excel spreadsheet (.xlsx), otherwise it will reject *CE1*. Afterward, since it is easier for our visualizer tool to read (.csv) files, *ASGFT* converts .xlsx to .csv files and feed them to the *Visualizer.exe* for generating the Attack graph graphically.

For the modeled *NPP* system, *ASGFT* generated 6561 .lus files containing all possible combinations of attack scenarios, where each scenario is a distinct *CE*. Yet, only 16 .lus files violate the security property φ and hence are truly attack scenarios.

Step	0	1	2	3
Inputs				
CE_1	FALSE	FALSE	FALSE	FALSE
thr_AoD_SenSen.val	FALSE	FALSE	FALSE	FALSE
thr_BSM_CRPLC.val	FALSE	FALSE	FALSE	TRUE
thr_BSM_EWPLC.val	FALSE	FALSE	FALSE	FALSE
thr_BSM_PLCSen.val	FALSE	FALSE	FALSE	FALSE
thr_DoS_CRDH.val	FALSE	FALSE	FALSE	FALSE
thr_DoS_CRPLC.val	FALSE	FALSE	FALSE	FALSE
thr_DoS_EWDH.val	FALSE	FALSE	FALSE	FALSE
thr_DoS_EWPLC.val	FALSE	FALSE	FALSE	FALSE
thr_DoS_SMDH.val	FALSE	FALSE	FALSE	FALSE
thr_DoS_WADH.val	FALSE	FALSE	FALSE	FALSE
thr_IG_CRDH.val	FALSE	FALSE	FALSE	FALSE
thr_IG_EWDH.val	FALSE	FALSE	FALSE	FALSE
thr_IG_SMDH.val	FALSE	FALSE	FALSE	FALSE
thr_IG_WADH.val	TRUE	FALSE	FALSE	FALSE
thr_MI_CREW.val	FALSE	FALSE	FALSE	FALSE
thr_MI_EWCR.val	FALSE	FALSE	FALSE	FALSE
thr_MI_SMCR.val	FALSE	FALSE	FALSE	FALSE
thr_MI_SMEW.val	FALSE	FALSE	FALSE	FALSE
thr_MI_WACR.val	FALSE	FALSE	TRUE	FALSE
thr_MI_WAEW.val	FALSE	TRUE	FALSE	FALSE
CR_sub_thr_MiM_CRCB.val	FALSE	FALSE	FALSE	FALSE
control_PLC.val	FALSE	FALSE	FALSE	TRUE

Fig. 4 A spreadsheet result for a false *CE1*

It can be seen that *ASGFT* iteratively calls *JKind* and generates 16 attack scenarios as .csv files that are fed to a visualizer tool. To illustrate how *ASGFT* deals with counterexamples, Fig. 4 shows a spreadsheet file for $IG_{WADH} \rightarrow MI_{WACR} \rightarrow MI_{EWCR} \rightarrow BSM_{EWPLC}$ potential attack, where *CE1* value is false (i.e., it doesn't violate the security property), *JKind* generates a random true *CE1* ($IG_{WADH} \rightarrow MI_{WAEW} \rightarrow MI_{WACR} \rightarrow BSM_{EWPLC}$). However, this file will be rejected by *ASGFT* as it is not a true attack scenario. For a true attack scenario $IG_{WADH} \rightarrow MI_{WAEW} \rightarrow MI_{EWCR} \rightarrow BSM_{EWPLC}$, the *CE1* value is true (i.e., it violates the security property) as shown in Fig. 5, and the .csv file will be fed to the visualizer *GUI*.

3.2 Visualization of Attack Graph

Once *ASGFT* has generated and saved all *CEs* (attack scenarios) as Comma-separated values (*csv*) files, these files can be imported into a *Visualizer.exe*, a windows application that generates a Graphical User Interface (*GUI*) which was programmed using C# within Microsoft Visual Studio [47]. The hierarchical architecture of *visualizer.exe* is illustrated in the Appendix.

Figure 6 illustrates the main screen of the programmable *GUI*. This screen has three main components; *CEs*' attributes, actions, and results. In *CEs*' attributes, the number of *csv* files (*CEs*) to be read is entered in "*No. Scenarios (Files)*". The "*Attack*

	Step	0	1	2	3
1	Step	0	1	2	3
2	Inputs				
3	CE_1	FALSE	FALSE	FALSE	TRUE
4	thr_BSM_CRPLC.val	FALSE	FALSE	FALSE	FALSE
5	thr_BSM_EWPLC.val	FALSE	FALSE	FALSE	TRUE
6	thr_BSM_PLCSen.val	FALSE	FALSE	FALSE	FALSE
7	thr_DoS_CRDH.val	FALSE	FALSE	FALSE	FALSE
8	thr_DoS_CRPLC.val	FALSE	FALSE	FALSE	FALSE
9	thr_DoS_EWDH.val	FALSE	FALSE	FALSE	FALSE
10	thr_DoS_EWPLC.val	FALSE	FALSE	FALSE	FALSE
11	thr_DoS_SMDH.val	FALSE	FALSE	FALSE	FALSE
12	thr_DoS_WADH.val	FALSE	FALSE	FALSE	FALSE
13	thr_IG_CRDH.val	FALSE	FALSE	FALSE	FALSE
14	thr_IG_EWDH.val	FALSE	FALSE	FALSE	FALSE
15	thr_IG_SMDH.val	FALSE	FALSE	FALSE	FALSE
16	thr_IG_WADH.val	TRUE	FALSE	FALSE	FALSE
17	thr_MI_CREW.val	FALSE	FALSE	FALSE	FALSE
18	thr_MI_EWCR.val	FALSE	FALSE	TRUE	FALSE
19	thr_MI_SMCR.val	FALSE	FALSE	FALSE	FALSE
20	thr_MI_SMEW.val	FALSE	FALSE	FALSE	FALSE
21	thr_MI_WACR.val	FALSE	FALSE	FALSE	FALSE
22	thr_MI_WAEW.val	FALSE	TRUE	FALSE	FALSE
23	thr_MI_WASM.val	FALSE	FALSE	FALSE	FALSE
24	control_PLC.val	FALSE	FALSE	FALSE	TRUE

Fig. 5 A csv result for a true *CE1*

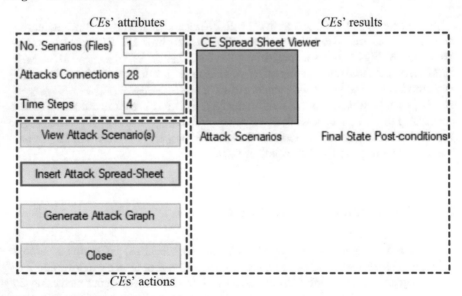

Fig. 6 GUI generated by Visulizer.exe

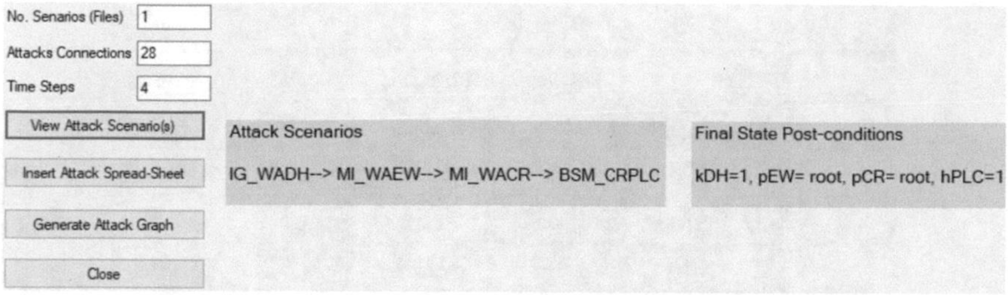

No. Senarios (Files)	1		
Attacks Connections	28		
Time Steps	4		
View Attack Scenario(s)		**Attack Scenarios**	**Final State Post-conditions**
Insert Attack Spread-Sheet		IG_WADH--> MI_WAEW--> MI_WACR--> BSM_CRPLC	kDH=1, pEW= root, pCR= root, hPLC=1
Generate Attack Graph			
Close			

Fig. 7 1st Counter-Example, *CE1* and its final state post-condition

Connections" field indicates all possible attack instances between *NPP* model's components. "Time Steps" field indicates the maximum number of time steps (transitions between attack instances in the attack scenarios) to be presented. Once these fields are determined, attack scenario(s) can be represented by choosing "*View Attack Scenario(s)*", "*Insert Attack Spread-Sheet*", and "*Generate Attack Graph*" actions. The results part shows the *CE Spread-Sheet viewer*, *Attack Scenarios*, and *Final State Post-conditions*.

To run this *GUI*, we consider the following *counter-example/attack-scenario*, $CE1: = IG_{WADH} \rightarrow MI_{WAEW} \rightarrow MI_{EWCR} \rightarrow BSM_{CRPLC}$, which was generated by *JKind* as a spreadsheet and was then saved as a separate *csv* file.

Having generated all 16 *CEs* csv files using *ASGFT*, the number of *CEs* to be presented is entered in "*No. Scenarios (Files)*" field, also in these *CEs*, the number of attack instances/actions is 28 (these instances are also defined in the *AADL* model), and number of time steps as 4. Figure 7shows a simplified representation of the first attack scenario, *CE1* given by choosing "*View Attack Scenario(s)*".

To present the desired *CE* as a spreadsheet, the "*Insert Attack Spread-Sheet*" action can be used, and the sheet can be presented in the *Spread-Sheet viewer*. Figure 8 shows a spreadsheet illustrating a potential generated attack scenario *CE1*.

By choosing "*Generate Attack Graph*", the union of attack scenarios causing system disruption as given by its Attack graph is shown in Fig. 9. Each arrow illustrates a possible occurrence of an attack instance/action. Each Node describes the system state at that step. A path from the initial node to the final node represents a sequence of attack instances (i.e., an attack scenario) that the attacker can conduct to achieve its goal without getting detected. This Attack graph has 16 attack scenarios that lead to two final states.

4 Conclusion

An Attack graph implementation and visualization for *NPP* control system was demonstrated in this chapter. The proposed *VS* program is capable of reading all scenarios spreadsheets and automatically representing the possible attack paths, their

CE Spread Sheet Viewer

threats	time0	time1	time2	time3
AoD_CRCR	FALSE	FALSE	FALSE	FALSE
AoD_PLCPLC	FALSE	FALSE	FALSE	FALSE
AoD_SenSen	FALSE	FALSE	FALSE	FALSE
BSM_CRPLC	FALSE	FALSE	FALSE	TRUE
BSM_EWPLC	FALSE	FALSE	FALSE	FALSE
BSM_PLCSen	FALSE	FALSE	FALSE	FALSE
DoS_CRDH	FALSE	FALSE	FALSE	FALSE
DoS_CRPLC	FALSE	FALSE	FALSE	FALSE
DoS_EWDH	FALSE	FALSE	FALSE	FALSE
DoS_EWPLC	FALSE	FALSE	FALSE	FALSE
DoS_SMDH	FALSE	FALSE	FALSE	FALSE
DoS_WADH	FALSE	FALSE	FALSE	FALSE
IG_CRDH	FALSE	FALSE	FALSE	FALSE
IG_EWDH	FALSE	FALSE	FALSE	FALSE
IG_SMDH	FALSE	FALSE	FALSE	FALSE
IG_WADH	TRUE	FALSE	FALSE	FALSE
MI_CREW	FALSE	FALSE	FALSE	FALSE
MI_EWCR	FALSE	FALSE	FALSE	FALSE
MI_SMCR	FALSE	FALSE	FALSE	FALSE
MI_SMEW	FALSE	FALSE	FALSE	FALSE
MI_WACR	FALSE	FALSE	TRUE	FALSE
MI_WAEW	FALSE	TRUE	FALSE	FALSE
MI_WASM	FALSE	FALSE	FALSE	FALSE
MiM_CRCB	FALSE	FALSE	FALSE	FALSE
MiM_EWCB	FALSE	FALSE	FALSE	FALSE
MiM_PLCSen	FALSE	FALSE	FALSE	FALSE
MiM_SMCB	FALSE	FALSE	FALSE	FALSE
MiM_WACB	FALSE	FALSE	FALSE	FALSE

Fig. 8 *CE1* spreadsheet

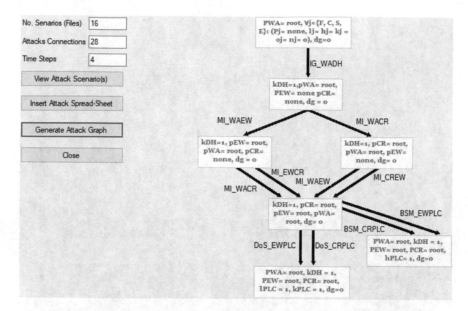

Fig. 9 *NPP* developed Attack graph

final state post-conditions, and *CEs* spreadsheet viewer. The generated Attack Graph can aid system designers choosing the best placement of countermeasures of such attacks. Further security analysis such as risk and resiliency assessment, and the deployment of Markov chain to determine the probability of reaching undesirable states where the security is breached are left as future directions.

Acknowledgements The authors would like to acknowledge Deanship of Graduation Studies and Scientific Research at the German Jordanian University for the Seed fund SATS 02/2018.

Appendices

In this Appendix, Fig. 10 illustrates the *"Visuazer.exe* hierarchal architecture.

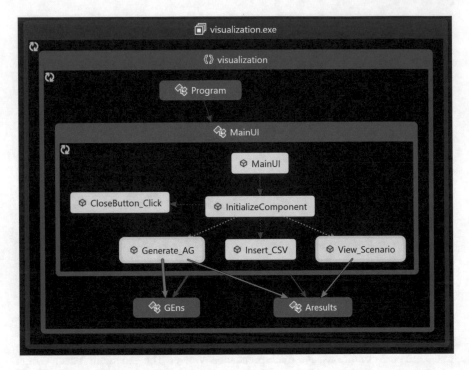

Fig. 10 The *Visualizer.exe* architecture

References

1. Morris, T.H., Gao, W.: 1st International Symposium on ICS & SCADA Cyber Security Research 2013, pp. 22–29. Leicester, UK (2013)
2. Potter, W.C.: Less Well Known Cases of Nuclear Terrorism and Nuclear Diversion in Russia. (NIT). http://nti.org/analysis/articles/less-well-known-cases-nuclear-terrorism-and-nuclear-diversion-russia/. (Aug, 1997)
3. Maguire, K.: Guard tries to sabotage nuclear reactor. The Guardian. http://theguardian.com/uk/2001/jan/09/nuclear.world (Jan, 2001, Ed.)
4. Kesler, B.: The vulnerability of nuclear facilities to cyber attack. Strat. Insights **10**(1), 15–25 (2011)
5. Farwell, J.P., Rohozinski, R.: Stuxnet and the future of cyber war. Surv. Global Politics Strategy, **53**(1), 23–40. (28 Jan 2011). https://doi.org/10.1080/00396338.2011.555586
6. Arène, V.: Le réseau informatique d'Areva piraté. Le Monde Informatique. Retrieved from https://lemondeinformatique.fr/actualites/lire-le-reseau-informatique-d-areva-pirate-42092.html. (30 Sept 2011)
7. The 4 most iconic industrial cyberattacks. (Sentryo). https://www.sentryo.net/4-most-iconic-industrial-cyberattacks/. (28 Sept 2016)
8. Meeyoung, C., Park, J.-m.: South Korea blames North Korea for December hack on nuclear operator. Infosec Institute. http://resources.infosecinstitute.com/cyber-attacks-against-nuclear-plants-a-disconcerting-threat/#gref. (17 Mar 2015)
9. Nuclear Material Control Center data leak report says sensitive info unlikely affected. The Mainichi. Retrieved from https://mainichi.jp/english/articles/20160519/p2a/00m/0na/010000c. (19 May 2016)

10. Varuttamaseni, A., Bari, R.A., Youngbl, R.: Construction of a Cyber Attack Model for Nuclear Power Plants. 2017 ANS Annual Conference, pp. 1–10. San Francisco. https://bnl.gov/isd/documents/94595.pdf (2017)

11. Cimpanu, C.: Hackers steal research and user data from Japanese Nuclear Research Lab. Softpedia Niews. http://securitynewspaper.com/2016/10/18/hackers-steal-research-user-data-japanese-nuclear-research-lab/ (17 Oct 2016)

12. Perlroth, N.: Hackers are targeting nuclear facilities, Homeland Security Dept. and F.B.I. Say. The Newyork Times. https://nytimes.com/2017/07/06/technology/nuclear-plant-hack-report.html (6 July 2017)

13. Ammann, P., Wijesekera, D., Kaushik, S.: Scalable, Graph-Based Network Vulnerability Analysis. 9th ACM Conference on Computer and Communications Security, pp. 217–224. ACM, Washington. https://doi.org/10.1145/586110.586140 (2002)

14. Ahn, W., Chung, M., Min, B.-G., Seo, J.: Development of cyber-attack scenarios for nuclear power plants using scenario graphs. Int. J. Distrib. Sens. Netw. **2015**, 12 (2015). https://doi.org/10.1155/2015/836258

15. Templeton, S.J., Levitt, K.N.: A requires/provides model for computer attacks. 2000 Workshop on New security Paradigms, pp. 31–38. ACM, Ballycotton, County Cork, Ireland. https://doi.org/10.1145/366173.366187 (2000)

16. Noel, S., Jajodia, S.: Managing attack graph complexity through visual hierarchical aggregation. The 2004 ACM Workshop on Visualization and Data Mining for Computer Security, pp. 109–118. Washington. https://doi.org/10.1145/1029208.1029225 (2004)

17. Zhang, T., Hu, M.-Z., Li, D., Sun, L.: An effective method to generate attack graph. In: The Fourth International Conference on Machine Learning and Cybernetics, pp. 3926–3931. IEEE, Guangzhou. https://doi.org/10.1109/ICMLC.2005.1527624. (Aug, 2005)

18. Sheyner, O., Haines, J., Jha, S., Lippmann, R., Wing, J.: Automated generation and analysis of attack graphs. In: IEEE Symposium on Security and Privacy, pp. 273–284. IEEE, Oakland, California. https://doi.org/10.1109/SECPRI.2002.1004377 (May 2002)

19. Xinming, O., Boyer, W., McQueen, M.: A scalable approach to attack graph generation. In: The 13th ACM conference on Computer and communications security, pp. 336–345. ACM, Alexandria, Virginia, USA. https://doi.org/10.1145/1180405.1180446 (2006)

20. Rao, P., Sagonas, K., Swift, T., Warren, D.S., Freire, J.: XSB: A system for efficiently computing. In: International Conference on Logic Programming and Nonmonotonic Reasoning (LPNMR'97), pp. 2–17. Springer, Dagstuhl, Germany. https://doi.org/10.1007/3-540-63255-7_33 (1997)

21. Sheyner, O., Wing, J.: Tools for generating and analyzing attack graphs. In: International Symposium on Formal Methods for Components and Objects, pp. 344–371. Leiden, The Netherlands. https://doi.org/10.1007/978-3-540-30101-1_17 (2003)

22. Yi, S., Peng, Y., Xiong, Y. (eds.): Overview on attack graph generation and visualization technology. In: 2013 International Conference on Anti-Counterfeiting, Security and Identification (ASID), pp. 1–6. IEEE, Shanghai, China. https://doi.org/10.1109/ICASID.2013.6825274 (2013)

23. Amenaza.: Secur/Tree for Attack Tree analysis. (Amenaza Technologies). https://amenaza.com (2001). Accessed on 10 May 2018

24. Skybox.: Scybox Security. (Skybox Inc) Retrieved 2018 from https://skyboxsecurity.com (2002). Accessed on 10 May 2018

25. Lippmann, R.P., Ingols, K.W.: An Annotated Review of Past Papers on Attack Graphs. Massachusetts Inst of Tech Lexington Lincoln Lab (2015)

26. Swiler, L.P., Phillips, C., Ellis, D., Chakerian, S.: Computer-attack graph generation tool. In: DARPA Information Survivability Conference and Exposition II, 2001, pp. 307–321. IEEE, Anaheim, CA, USA (2001)

27. Institute for Security and Safety (ISS) in Cooperation with the Nuclear Threat Initiative (NTI).: Cyber Security at Nuclear Facilities: National Approache. Research Paper, Institute for Security and Safety (ISS) at the Brandenburg University of Applied Sciences. www.nti.org/media/pdfs/Cyber_Security_in_Nuclear_FINAL_UZNMggd.pdf?_=1466705014 (2015)

28. Sklyar, V.: Cyber security of safety-critical infrastructures: a case study for nuclear facilities. Inf. Secur. Int. J. **28**(1), 98–107 (2012)
29. Stoutland, P.: Cyberattacks on Nuclear Power Plants: How Worried Should We Be?. Nuclear Threat Initiative. https://nti.org/analysis/atomic-pulse/cyberattacks-nuclear-power-plants-how-worried-should-we-be/ (Mar, 2018). Accessed on 16 May 2018
30. Angle, M.G., Madnick, S., Kirtley, J.L., Khan, S.: Identifying and anticipating cyber attacks that could cause physical damage to industrial control systems. IEEE Power Energy Technol. Syst. J. https://doi.org/10.1109/JPETS.2019.2923970 (June, 2019)
31. Carnegie-Mellon-University.: Open Source AADL Tool Environment for the SAE Architecture. http://osate.github.io/index.html (2018)
32. Gacek, A., Backes, J., Whalen, M., Wagner, L., Ghassabani, E.: The JKind model checker. Computer Aided Verification 2018. Oxford, UK. https://doi.org/10.1007/978-3-319-96142-2_3 (2018)
33. Microsoft.: Visual Studio. https://visualstudio.com/vs/ (2018). Accessed on 20 May 2018
34. Li, Y., Ma, J., Chan, A., Huang, Y., Wang, B.: Mechanism model of pressurizer in the pressurized water reactor nuclear. In: 2012 24th Chinese Control and Decision Conference (CCDC), pp. 178–182. IEEE, Taiyuan, China. https://doi.org/10.1109/CCDC.2012.6244026 (2012)
35. USNRC HRTD.: Westinghouse Technology Systems Manual. U.S. Nuclear Regulatory Commission. https://nrc.gov/docs/ML1122/ML11223A287.pdf. Accessed on 21 May 2018 (n.d.)
36. USNRC Technical Training Center.: Pressurized Water Reactor Systems. https://nrc.gov/reading-rm/basic-ref/students/for-educators/04.pdf. Accessed on 21 May 2018 (n.d.)
37. Green, S.J., Hetsroni, G.: PWR steam generators. Int. J. Multiph. Flow **21**(null), 1–97, (1995). https://doi.org/10.1016/0301-9322(95)00016-q
38. Nuclear Power Plant Safety Systems.: Canadian Nuclear Safety Commission. https://cnsc-ccsn.gc.ca/eng/reactors/power-plants/nuclear-power-plant-safety-systems/ (2016). Accessed on 23 May 2018
39. Razak, T.A., Ibrahim Salim, M.: A study on IDS for preventing Denial of Service attack using outliers techniques. In: 2nd IEEE International Conference on Engineering and Technology (ICETECH), pp. 768–775. IEEE, Coimbatore, India. https://doi.org/10.1109/ICETECH.2016.7569352 (Mar 2016)
40. JKind, An infinite-state model checker for safety properties.: Loonwerks. Available http://loonwerks.com/tools/jkind.html (n.d.). Accessed on 11 Nov 2018
41. Halbwachs, N., Caspi, P., Raymond, P., Pilaud, D.: The synchronous data flow programming language LUSTRE. IEEE **79**(9), 1305–1320 (Sept, 1991). https://doi.org/10.1109/5.97300
42. Uof-Minnesota, R.-C. a.: The Assume Guarantee Reasoning Environment. Pittsburgh, Pennsylvania, USA. http://standards.sae.org/as5506/ (2016). Accessed on 11 Jan 2018
43. HTML5 Web Development Support NetBeans: https://netbeans.org/features/html5/index.html. (2014). Accessed on 17 Feb 2018
44. Tulac.: Happy Birthday NetBeans. J, Interviewer. (17 May 2008)
45. Maven, Using NetBeans with Apache Maven.: NetBeans. http://wiki.netbeans.org/Maven (2014). Accessed on 10 Oct 2018
46. Böck, H.: The definitive guide to NetBeans™ Platform 7, Apress. Apress (2011). https://doi.org/10.1007/978-1-4302-4102-7
47. David, M.: Visual Studio IDE Offers Many Advantages For Developers. (SearchSoftware Quality) from https://searchsoftwarequality.techtarget.com (9 Sept 2015). Accessed on 20 May 2018

Determining Resiliency Using Attack Graphs

Mariam Ibrahim and Ahmad Alsheikh

Abstract System Resiliency is concerned with "its capability to cope with adverse events". Its quantification allows system designers to assess the level of system security and adopt the best scheme and countermeasures utilizing the available resources. This chapter introduces a novel approach for determining the Level-of-Resilience (*LoR*) of a system employing its Attack graph constituting distinct attack scenarios corrupting the system. This requires an overall system characterization, defining its architecture and connectivity, components and performance, assets, mitigation, vulnerabilities, and attacks. Two communication networks and two power systems are evaluated, and formally characterized using Architecture Analysis & Design Language (*AADL*). The developed system designs are then checked with a security property using *JKind* verifier. The union of the resulted attack sequences/scenarios causing a system breach is the Attack Graph. Each attack sequence has a correlated *LoR*. Then, by identifying all potential attack scenarios, their worst case *LoR* can be evaluated as illustrated through the communication networks and power system examples.

Keywords Resilience · Attack graph · Power system · Performance

1 Introduction

Industrial Controlled Systems (*ICSs*) are often a sitting target for cybercriminals. The majority of these systems monitor complex industrial process and critical infrastructures such as electric, water, transportation, manufacturing, healthcare, chemical and other essential services. Disruptive events can significantly affect the *ICS*, threatening human life safety, and causing severe destruction to

M. Ibrahim (✉) · A. Alsheikh
Department of Mechatronics Eng., Faculty of Applied Technical Sciences, German Jordanian University, Amman, Jordan
e-mail: mariam.wajdi@gju.edu.jo

A. Alsheikh
e-mail: a.alsheikh@gju.edu.jo

© Springer Nature Switzerland AG 2020   117
E. Pricop et al. (eds.), *Recent Developments on Industrial Control Systems Resilience*,
Studies in Systems, Decision and Control 255,
https://doi.org/10.1007/978-3-030-31328-9_6

the environment and other financial concerns. As a result, when scheming a system, it is vital to select the most secure and resilient scheme under various attack scenarios as depicted by the Attack graph.

The term "resilience" is not well defined as stated by [1]. For example, in the power systems due to the lack of interpretability standards, limited real-time data, and poor observability in power distribution systems. Holling [2] Defined resilience as the ability of a system to maintain its functionality and behavior after a disturbance. Holling et al. [3] updated the definition by including the buffer capacity for absorbing perturbations in a timely fashion. Walker et al. [4] Extended the definition to include the ability to self-heal during disturbances. A crucial aspect of resilience defined by Kendra and Wachtendorf [5] as "bouncing back from a disturbance". Brown et al. [6] defined resilience as the ability to reduce the magnitude and duration of disturbances. It depends upon the system's ability to predict, absorb and adapt to disturbances and recover rapidly. Haimes [7] represented resiliency by the ability of the system to recover within an acceptable time and composite costs and risks.

Various papers examined the Industrial Controlled Systems (*ICSs*) resiliency such as [8], where the control system resiliency was described in terms of the system Quality of Control (*QoC*). They introduced new index called Resiliency Index (*RI*), which indicates how resilient the control system is. The *RI* was used on a Wireless Networked Control Systems (*WNCSs*) to keep operational normalcy against wireless interference incidents, like Radio Frequency (*RF*) jamming and signal blocking. However, this work only focused on sensor data delay and loss with wireless sensors in the feedback loop. Garcia et al. [9] proposed a Resilient Condition Assessment Monitoring (*ReCAM*) system, which was comprised of information, assessment, and sensor selection layers. *ReCAM* was applied on a simplified power plant model to detect and measure natural or malicious disturbances that may occur at each operational unit of the monitored system. While sensor data may not be trustworthy due to cyber-attacks. Their goal was to dynamically collect and interpret sensor data, and correctly assess the physical condition or health of the monitored system within desired timeliness requirements.

An intelligent Resilient Control Strategy (*RCS*) for Model-based building control is proposed by Ji et al. [10] to improve the Building Automation Systems (*BASs*) performance against unanticipated adverse conditions or incidents. Dinh et al. [11] described a three system states an industrial process operation could be in, which are under faults/attacks state, normal state, and upset/catastrophic state. They also proposed six principles (Flexibility, Controllability, Early Detection, Minimization of Failure, Limitation of Effects, and Administrative Controls/Procedures), and five factors (Design, Detection Potential, Emergency Response Plan, Human Factor, and Safety Management) that contribute to the resilience of a process. These principles and factors were applied on a reactor case, where a release of flammable materials led to an explosion following a runaway reaction and the rupture of the reactor as a result of an increase in temperature.

Wei and Kun [12] proposed resiliency metrics to estimate the resiliency of an industrial control system, these metrics are (protection time, degrading time, identification time, recovery time, performance degradation, performance loss, total loss, and overall potential critical loss). A review was done by Hosseini et al. [13] on recent research articles defining and quantifying resilience in various disciplines, with a focus on engineering systems. This review provided a classification scheme that focused on qualitative and quantitative resiliency approaches and their subcategories.

The resilience of interdependent electric power and natural gas infrastructure systems were assessed by Ouyang and Wang [14] under multiple hazards, noting how interdependent network performance could be measured in physical engineering terms, or in terms of societal impact. Three stochastic resilience metrics were introduced by Pant et al. [15], and applied to a supply chain network. The metrics were Time to Total System Restoration, Time to Full System Service Resilience, and Time to α%-Resilience. Francis and Behailu [16] proposed an alternative metric for measuring system resilience that incorporated knowledge uncertainty associated with the nature of an event as an integral input into evaluating system resilience. The basic idea of resilience was expressed as a resilience factor, which included the performance level immediately post-disruption, the performance level after an initial post-disruption equilibrium state is reached, the performance at the new stable level after recovery efforts were exhausted, and the speed recovery factor.

The concepts of time-dependent operational resilience and infrastructure resilience metrics were introduced by Panteli et al. [17]. These metrics were slope of the resilience degradation during the event, the resilience degradation level, and the time that the network remains in the post-disturbance degraded state. MacKenzie and Kash [18] employed a Dynamic Inoperability Input-Output Model to quantify the resilience of a critical infrastructure sector. Then, applied it on a regression model and a mixed effects model to anticipate rates of post-disruption recovery, and system resiliency. Using data from 2005 Katrina hurricane in the U.S. Gulf of Mexico, they developed a "spatial-temporal non-stationary random process to model large-scale disruptions of power distribution induced by severe weather." Their stochastic approach yielded resiliency estimates for components, and identified dynamical neighborhoods within the network. Panteli [19] introduced a novel sequential Monte-Carlo-based time-series simulation model to assess power system resilience for estimating the impact of weather on Electrical Power Infrastructure, in terms of the frequency and duration of power interruptions, the loss of load frequency, and loss of load expectation.

In this chapter, the Levels-of-Resilience were assessed for two Communication Networks (*CNs*), and two Power Systems (*PSs*) examples employing their Attack graphs.

2 Illustrative Examples

Here, an illustrative examples of two communication networks, and two power systems are evaluated, and formally characterized using Architecture Analysis and Design Language (*AADL*) [20].

2.1 Communication Networks Examples

Figure 1a and b illustrate a pair of communication networks with identical users and services (Email, FTP, and Video), but with distinct topologies. The 1st communication network, CN_1, has three routers; R_1, R_2, and R_3 which are using Routing Information Protocol (*RIP*). R_1 is linked to three Local Area Networks (*LANs*). R_2 is linked to LAN_4. Router R_3 has three more connections, which link it respectively, to an Email server through the Internet, an FTP server, and a Video workstation. The 2nd communication network, CN_2 acquires a different topology, in which LAN_1 is linked to R_3. The application configurations, the node models in use, and link models in use for the two networks are tabulated in the Appendix of [21].

Formal Characterization of Communication Networks CN_1 And CN_2

(1) *Set of Routers R = 1, 2, 3, IP Cloud; Variable I* $\in$ *{1, 2, 3, 4} (static).*
(2) *Set of LANs N = 1, 2, 3, 4; Variable k* $\in$ *{1, 2, 3, 4} (static).*

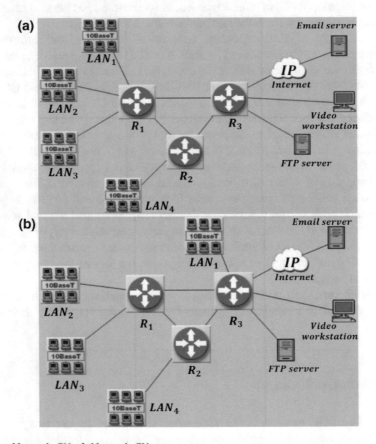

Fig. 1 **a** Network CN_1, **b** Network CN_2

(3) *Set of Service Providers S; Variable $s \in \{Ftp, Email, Video\}$ (static).*

(4) *Set of Connection Links $L \subseteq R \times R$, $R \times N$, $R \times S$; Labeled $l_{ij} \equiv$ Link is placed between component i and component j (dynamic).*

(5) *System Connectivity $C = L$; Boolean $c_{ij} = 1$ if there is a connection between component i and component j (dynamic).*

(6) *System Stability T; Boolean $t = 1$ if system is stable (dynamic).*

(7) *System Performance $P \subseteq S$; Boolean $f_k = 1$ if ftp service is provided to LAN k, Boolean $e_k = 1$ if Email service is provided on LAN k and Boolean $v_k = 1$ if Video service is provided on LAN k (dynamic).*

(8) *System recovery Action R; Variable $r \in \{p, a, d\}$, in case of normal operation $r = p$, in case of recovery action $r = a$, and in case of no action can be done, $r = d$ (dynamic).*

(9) *Number of faulted Links that occur sequentially N; Variable $n \in \{0, 1, 2\}$, in case of no fault $n = 0$, in case of first fault $n = 1$, and in case of second fault $n = 2$ (dynamic).*

(10) *Attack Instance $AI \subseteq A \times R \times R$, $A \times R \times N$, $A \times R \times S$, Labeled a_{ij}^m $\equiv$ Attack a on the Link between component i and component j, where $m \subseteq L$ is a sequence of the previous faulted link(s) if exists.*

(11) *Pre-Attack conditions for CN_1:*

- $Pre(a_{13}) \equiv (c_{13} = 1) \wedge (t = 1) \wedge (r = p) \wedge (n = 0)$
- $Pre(a_{12}) \equiv (c_{12} = 1) \wedge (t = 1) \wedge (r = p) \wedge (n = 0)$
- $Pre(a_{23}) \equiv (c_{23} = 1) \wedge (t = 1) \wedge (r = p) \wedge (n = 0)$
- $Pre(a_{23}^{13}) \equiv (c_{23} = 1) \wedge (r = a) \wedge (n = 1)$
- $Pre(a_{23}^{12}) \equiv (c_{23} = 1) \wedge (r = p) \wedge (n = 1)$
- $Pre(a_{13}^{23}) \equiv (c_{13} = 1) \wedge (r = a) \wedge (n = 1)$
- $Pre(a_{13}^{12}) \equiv (c_{13} = 1) \wedge (r = p) \wedge (n = 1)$
- $Pre(a_{12}^{13}) \equiv (c_{12} = 1) \wedge (r = a) \wedge (n = 1)$
- $Pre(a_{12}^{23}) \equiv (c_{12} = 1) \wedge (r = a) \wedge (n = 1)$

(12) *Post-Attack conditions for CN_1:*

- $Post(a_{13}) \equiv (c_{13} = 0) \wedge (r = a) \wedge (n = 1)$
- $Post(a_{12}) \equiv (c_{12} = 0) \wedge (r = p) \wedge (n = 1)$
- $Post(a_{23}) \equiv (c_{23} = 0) \wedge (r = a) \wedge (n = 1)$
- $Post(a_{23}^{13}) \equiv (t = 0) \wedge (c_{23} = 0) \wedge (f_1 = f_2 = f_3 = f_4 = e_1 = e_3 = e_4 = v_2 = v_3 = 0) \wedge (r = d) \wedge (n = 2)$
- $Post(a_{23}^{12}) \equiv (t = 0) \wedge (c_{23} = 0) \wedge (f_4 = e_4 = 0) \wedge (r = d) \wedge (n = 2)$
- $Post(a_{13}^{23}) \equiv (t = 0) \wedge (c_{13} = 0) \wedge (f_1 = f_2 = f_3 = f_4 = e_1 = e_3 = e_4 = v_2 = v_3 = 0) \wedge (r = d) \wedge (n = 2)$
- $Post(a_{13}^{12}) \equiv (t = 0) \wedge (c_{13} = 0) \wedge (f_1 = f_2 = f_3 = e_1 = e_3 = v_2 = v_3 = 0) \wedge (r = d) \wedge (n = 2)$
- $Post(a_{12}^{13}) \equiv (t = 0) \wedge (c_{12} = 0) \wedge (f_1 = f_2 = f_3 = e_1 = e_3 = v_2 = v_3 = 0) \wedge (r = d) \wedge (n = 2)$
- $Post(a_{12}^{23}) \equiv (t = 0) \wedge (c_{12} = 0) \wedge (f_4 = e_4 = 0) \wedge (r = d) \wedge (n = 2)$

(13) *Pre-Attack conditions for CN_2*:

- $\text{Pre}(a_{13}) \equiv (c_{13} = 1) \wedge (t = 1) \wedge (r = p) \wedge (n = 0)$
- $\text{Pre}(a_{12}) \equiv (c_{12} = 1) \wedge (t = 1) \wedge (r = p) \wedge (n = 0)$
- $\text{Pre}(a_{23}) \equiv (c_{23} = 1) \wedge (t = 1) \wedge (r = p) \wedge (n = 0)$
- $\text{Pre}(a_{23}^{13}) \equiv (c_{23} = 1) \wedge (r = a) \wedge (n = 1)$
- $\text{Pre}(a_{23}^{12}) \equiv (c_{23} = 1) \wedge (r = p) \wedge (n = 1)$
- $\text{Pre}(a_{13}^{23}) \equiv (c_{13} = 1) \wedge (r = a) \wedge (n = 1)$
- $\text{Pre}(a_{13}^{12}) \equiv (c_{13} = 1) \wedge (r = p) \wedge (n = 1)$
- $\text{Pre}(a_{12}^{13}) \equiv (c_{12} = 1) \wedge (r = a) \wedge (n = 1)$
- $\text{Pre}(a_{12}^{23}) \equiv (c_{12} = 1) \wedge (r = a) \wedge (n = 1)$

(14) *Post-Attack conditions for CN_2*:

- $\text{Post}(a_{13}) \equiv (c_{13} = 0) \wedge (r = a) \wedge (n = 1)$
- $\text{Post}(a_{12}) \equiv (c_{12} = 0) \wedge (r = p) \wedge (n = 1)$
- $\text{Post}(a_{23}) \equiv (c_{23} = 0) \wedge (r = a) \wedge (n = 1)$
- $\text{Post}(a_{23}^{13}) \equiv (t = 0) \wedge (c_{23} = 0) \wedge (f_2 = f_3 = f_4 = e_3 = e_4 = v_2 = v_3 = 0) \wedge (r = d) \wedge (n = 2)$
- $\text{Post}(a_{23}^{12}) \equiv (t = 0) \wedge (c_{23} = 0) \wedge (f_4 = e_4 = 0) \wedge (r = d) \wedge (n = 2)$
- $\text{Post}(a_{13}^{23}) \equiv (t = 0) \wedge (c_{13} = 0) \wedge (f_2 = f_3 = f_4 = e_3 = e_4 = v_2 = v_3 = 0) \wedge (r = d) \wedge (n = 2)$
- $\text{Post}(a_{13}^{12}) \equiv (t = 0) \wedge (c_{13} = 0) \wedge (f_2 = f_3 = e_3 = v_2 = v_3 = 0) \wedge (r = d) \wedge (n = 2)$
- $\text{Post}(a_{12}^{13}) \equiv (t = 0) \wedge (c_{12} = 0) \wedge (f_2 = f_3 = e_3 = v_2 = v_3 = 0) \wedge (r = d) \wedge (n = 2)$
- $\text{Post}(a_{12}^{23}) \equiv (t = 0) \wedge (c_{12} = 0) \wedge (f_4 = e_4 = 0) \wedge (r = d) \wedge (n = 2)$

(15) *Initial state*: $(t = 1) \wedge (c23, c12, c13 = 1) \wedge (f1 = f2 = f3 = f4 = e1 = e3 = e4 = v2 = v3 = 1) \wedge (r = p) \wedge (n = 0)$. *(Initially, the system is stable, normally operated, and no service outages).*

(16) *The security/resiliency property ϕ is that both CN_1 and CN_2 are always stable under the given attacks/faults. This can then be given by a Computational Tree Logic (CTL) formula:*

$$\phi \equiv AG(t = 1) \equiv AG(\neg(t = 0)).$$

2.2 Power Systems Examples

Figure 2a and b show two power systems with exact buses, generators and loads but with distinct topologies. For the 1st power system, PS_1, bus 1 is the slack bus, buses 2 and 3 are the generator buses, and buses 4, 5, and 6 are the load buses. The 2nd power

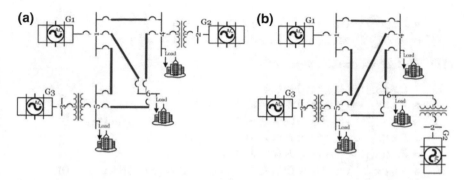

Fig. 2 **a** System PS_1, **b** System PS_2

system, PS_2 has different topology, where line L_{16} looks like line L_{45}. The machine, load and line data, generation schedule, and reactive power limits for the regulated buses, along with power flow solution data for PS_1 are given in the Appendix of [22]. The attacks/faults on these systems can include transmission line or node failures.

Formal Characterization of Power Systems PS_1 And PS_2

(1) *Set of Buses B = 1, 2, 3, 4, 5, 6; Variable $i \in \{1, 2, 3, 4, 5, 6\}$ (static).*
(2) *Set of Generators G = 1, 2, 3; Variable $k \in \{1, 2, 3\}$ (static).*
(3) *Set of transmission Lines $L \subseteq B \times B$; Labeled $l_{ij} \equiv$ Link l is placed between bus i and bus j (static).*
(4) *Power Flow $F = \subseteq B \times B$; Boolean $f_{ij} = 1$ if there is a Power Flow between bus i and bus j (dynamic).*
(5) *System Stability T; Boolean t = 1 if system is stable (dynamic).*
(6) *System Performance $P \subseteq B \times G$; Boolean $p_{ik} = 1$ if load of bus i is served by generator k. (dynamic).*
(7) *System recovery Action, R; Variable $r \in \{p, a, d\}$, in case of normal operation r = p, in case of recovery action r = a, and in case of no action can be done r = d (dynamic).*
(8) *Number of faulted Lines that occur sequentially N; Variable $n \in \{0, 1, 2, 3\}$, in case of no fault n = 0, in case of first fault n = 1, in case of second fault n = 2, and in case of third fault n = 3 (dynamic).*
(9) *Attack Instance $AI \subseteq A \times B \times B$; Labeled $a_{ij}^m \equiv$ Attack a on the Line between bus i and bus j, where $m \subseteq L$, is a sequence of the previous faulted line(s) if exists.*
(10) *Pre-Attack conditions for PS_1:*

   - *$\text{Pre}(a_{15}) \equiv (f_{15} = 1) \wedge (t = 1) \wedge (r = p) \wedge (n = 0)$*
   - *$\text{Pre}(a_{46}^{15}) \equiv (f_{46} = 1) \wedge (t = 1) \wedge (r = a) \wedge (n = 1)$*
   - *$\text{Pre}(a_{56}^{46, 15}) \equiv (f_{56} = 1) \wedge (t = 1) \wedge (r = a) \wedge (n = 2)$*
   - *$\text{Pre}(a_{56}) \equiv (f_{56} = 1) \wedge (t = 1) \wedge (r = p) \wedge (n = 0)$*

- $\text{Pre}(a_{14}^{56}) \equiv (f_{14} = 1) \wedge (t = 1) \wedge (r = a) \wedge (n = 1)$
- $\text{Pre}(a_{46}^{14,\,56}) \equiv (f_{46} = 1) \wedge (t = 1) \wedge (r = a) \wedge (n = 2)$

(11) *Post-Attack conditions for PS$_1$*:

- $\text{Post}(a_{15}) \equiv (f_{15} = 0) \wedge (r = a) \wedge (n = 1)$
- $\text{Post}(a_{46}^{15}) \equiv (f_{46} = 0) \wedge (r = a) \wedge (n = 2)$
- $\text{Post}(a_{56}^{46,\,15}) \equiv (f_{56} = 0) \wedge (t = 0) \wedge (r = d) \wedge (n = 3) \wedge (p_{53} = 0)$
- $\text{Post}(a_{56}) \equiv (f_{56} = 0) \wedge (r = a) \wedge (n = 1)$
- $\text{Post}(a_{14}^{56}) \equiv (f_{14} = 0) \wedge (r = a) \wedge (n = 2)$
- $\text{Post}(a_{46}^{14,\,56}) \equiv (f_{46} = 0) \wedge (t = 0) \wedge (r = d) \wedge (n = 3) \wedge (p_{62} = 0)$

(12) *Pre-Attack conditions for PS$_2$*:

- $\text{Pre}(a_{15}) \equiv (f_{15} = 1) \wedge (t = 1) \wedge (r = p) \wedge (n = 0)$
- $\text{Pre}(a_{46}^{15}) \equiv (f_{46} = 1) \wedge (t = 1) \wedge (r = a) \wedge (n = 1)$
- $\text{Pre}(a_{56}^{46,\,15}) \equiv (f_{56} = 1) \wedge (t = 1) \wedge (r = a) \wedge (n = 2)$
- $\text{Pre}(a_{56}) \equiv (f_{56} = 1) \wedge (t = 1) \wedge (r = p) \wedge (n = 0)$
- $\text{Pre}(a_{14}^{56}) \equiv (f_{14} = 1) \wedge (t = 1) \wedge (r = a) \wedge (n = 1)$
- $\text{Pre}(a_{46}^{14,\,56}) \equiv (f_{46} = 1) \wedge (t = 1) \wedge (r = a) \wedge (n = 2)$

(13) *Post-Attack conditions for PS$_2$*:

- $\text{Post}(a_{15}) \equiv (f_{15} = 0) \wedge (r = a) \wedge (n = 1)$
- $\text{Post}(a_{46}^{15}) \equiv (f_{46} = 0) \wedge (r = a) \wedge (n = 2)$
- $\text{Post}(a_{56}^{46,\,15}) \equiv (f_{56} = 0) \wedge (t = 0) \wedge (r = d) \wedge (n = 3) \wedge (p_{53} = 0)$
- $\text{Post}(a_{56}) \equiv (f_{56} = 0) \wedge (r = a) \wedge (n = 1)$
- $\text{Post}(a_{14}^{56}) \equiv (f_{14} = 0) \wedge (r = a) \wedge (n = 2)$
- $\text{Post}(a_{46}^{14,\,56}) \equiv (f_{46} = 0) \wedge (t = 0) \wedge (r = d) \wedge (n = 3) \wedge (p_{62} = 0)$

(14) Initial state: $(t = 1) \wedge (f_{56}, f_{15}, f_{14}, f_{46}, f_{16}, f_{42}, f_{35} = 1) \wedge (\forall\ i, k \in B \times G,$ $p_{ik} = 1) \wedge (r = p) \wedge (n = 0)$. *(Initially, the system is stable, normally operated, and no service outages).*

(15) *The security/resiliency property ϕ is that both PS$_1$ and PS$_2$ are always stable against the given attacks/faults. This can then be given as CTL formula*:

$$\phi \equiv AG(t = 1) \equiv AG(\neg(t = 0)).$$

3 Attack Scenarios Implementation

Here, the model-based application of the attack scenarios constituting Attack graphs for the given systems are presented. The generated Attack graphs will aid determining the associated *LoR,* and worst case *LoR.*

3.1 Communication Network

Here, we show the outcomes of executing *JKind* for the encoded models of CN_1 and CN_2, respectively, against ϕ. *JKind* is an infinite state model checker for verifying safety properties of synchronous systems [23]. The verification is based on k-induction and property directed reachability using a back-end *SMT* solver. A verified property is approved to be true for all runs of the system. A falsified property is reported with an explicit Counter-Example (*CE*) demonstrating the property violation, which is given here as an attack scenario represented by a sequence of attack instances causing system disruption.

Figure 3 presents an EXCEL-sheet illustrating a potential attack scenario, $CE1: = a_{13} \to a_{23}^{13}$. In this scenario, the attacker's aim is to compromise CN_1 through driving it unstable. In such a scenario, two links are corrupted in the sequel. Initially, a fault is made on link L_{13} enclosed by routers R_1 and R_3, (CN_1 remains stable). Next, another fault is made on link L_{23} enclosed by routers R_2 and R_3 resulting in CN_1 loss of stability. (Where network instability is given by the unlimited loss of traffic with time [24].) When encoding this generated $CE1$ in disjunct with ϕ, that is $\phi \bigvee CE1$, a new counter example $CE2: = a_{13\bigvee} \to a_{12}^{13}$ is produced that complies

83	CE_1	FALSE	FALSE
84	a12.val	FALSE	FALSE
85	a13.val	TRUE	FALSE
86	a23.val	FALSE	FALSE
87	a1213.val	FALSE	FALSE
88	a1223.val	FALSE	FALSE
89	a1312.val	FALSE	FALSE
90	a1323.val	FALSE	FALSE
91	a2312.val	FALSE	FALSE
92	a2313.val	FALSE	TRUE
93	e1.val	1	0
94	e3.val	1	0
95	e4.val	1	0
96	f1.val	1	0
97	f2.val	1	0
98	f3.val	1	0
99	f4.val	1	0
100	v2.val	1	0
101	v3.val	1	0
102	security property	TRUE	FALSE
103	r.val	1	2
104	t.val	TRUE	FALSE

Fig. 3 A possible attack scenario, *CE1* for CN_1

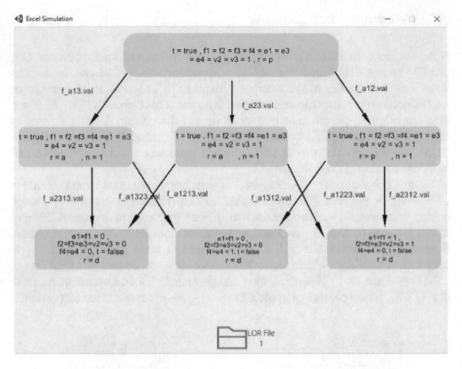

Fig. 4 Network CN_1 Attack graph

with: $\neg\ (\phi\ CE1) = \neg\ \phi \bigwedge \neg\ CE1$. By imitating this procedure many (but limited) *CEs* can be generated, resulting the "Attack graphs" for CN_1 and CN_2, as illustrated in Figs. 4 and 5, respectively. These graphs are visualized using *Unity* tool [25].

In the developed attack graphs, CN_1 and CN_2 have six attack scenarios which are: S_1: $a_{13} \rightarrow a_{23}^{13}$, S_2: $a_{13} \rightarrow a_{12}^{13}$, S_3: $a_{12} \rightarrow a_{23}^{12}$, S_4: $a_{23} \rightarrow a_{13}^{23}$, S_5: $a_{23} \rightarrow a_{12}^{23}$, S_6: $a_{12} \rightarrow a_{13}^{12}$.

3.2 Power Systems

When executing *JKind* for the encoded models of PS_1 and PS_2, against ϕ. Figure 6 illustrates an EXCEL-sheet showing a potential attack scenario, $CE1$: $= a_{56} \rightarrow a_{14}^{56} \rightarrow a_{46}^{14,\ 56}$. In this scenario, the attacker's aim is to disrupt PS_1 through driving it unstable. For this scenario, three lines are corrupted in the sequel. Initially, a fault is made on line L_{56} enclosed by bus 5 and bus 6, (PS_1 remains stable). Next, another fault is made on line L_{14} enclosed by bus 1 and bus 4, (PS_1 remains also stable). Eventually, a third fault is made on line L_{46} enclosed by bus 4 and bus 6, resulting in PS_1 loss of stability. (Where system stability is given by the rotor angle transient stability [22].)

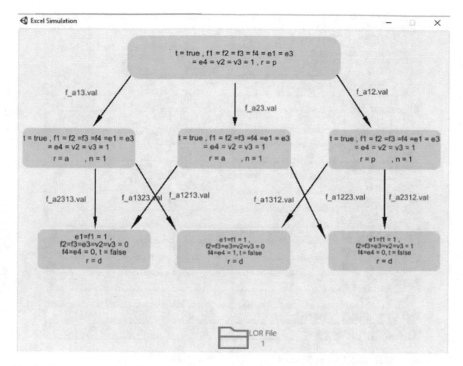

Fig. 5 Network CN_2 Attack graph

When encoding this generated counterexample $CE1$ in disjunct with ϕ, that is ϕ $\vee CE1$, a new counterexample $CE2 := a_{15} \rightarrow a_{46}^{15} \rightarrow a_{56}^{46,\,15}$ is produced that complies with: $\neg (\phi \vee CE1) = \neg \phi \wedge \neg CE1$. By imitating this procedure many (but limited) CEs can be generated, resulting of the "Attack graphs" for PS_1 and PS_2, as illustrated in Figs. 7 and 8, respectively.

In the developed attack graphs, PS_1 and PS_2, have two attack scenarios which are: S_1: $a_{15} \rightarrow a_{46}^{15} \rightarrow a_{56}^{46,15}$, S_2: $a_{56} \rightarrow a_{14}^{56} \rightarrow a_{46}^{14,56}$.

4 Level-of-Resilience Assessment

In the developed Attack graphs, every path is an attack sequence causing system corruption, and has a correlated LoR. Then, by identifying all potential attack scenarios, their worst case LoR can be evaluated.

To evaluate the worst case Level-of-Resilience of a system employing its Attack graph, the consecutive definition can be utilized. The system is the worst resilient to an attack scenario for the given graph if this attack acquires the maximum loss of stability, or otherwise, the maximum loss of performance, or otherwise the maximum time to recover.

68	_TOP.R_1..ASSUME.HIST	TRUE	TRUE	TRUE
69	_TOP.R_2..ASSUME.HIST	TRUE	TRUE	TRUE
70	_TOP.R_3..ASSUME.HIST	TRUE	TRUE	TRUE
71	_TOP.R_4..ASSUME.HIST	TRUE	TRUE	TRUE
72	_TOP.T..ASSUME.HIST	TRUE	TRUE	TRUE
73	f_a15.val	FALSE	FALSE	FALSE
74	f_a46_1456.val	FALSE	FALSE	TRUE
75	f_a56.val	TRUE	FALSE	FALSE
76	f_a56_4615.val	FALSE	FALSE	FALSE
77	f_a1456.val	FALSE	TRUE	FALSE
78	f_a4615.val	FALSE	FALSE	FALSE
79	s_P53.val	1	1	1
80	s_P62.val	1	1	0
81	s_f14.val	1	0	0
82	s_f15.val	1	1	1
83	s_f46.val	1	1	0
84	s_f56.val	0	0	0
85	s_r.val	1	2	3
86	security property	TRUE	TRUE	FALSE
87	test_t_value.val	TRUE	TRUE	FALSE

Fig. 6 A possible attack scenario for PS_1

Definition 1 *Given a system M and an Attack-Graph A_G comprising a set of attack scenarios $S \equiv \cup S_i$, $i \in \{1,..., z\}$, where z is the number of attack scenarios, the $LoR(M, S_i)$ is the worst if:*

$[LoS_R(M, S_i) > LoS_R(M, S-S_i)]$
$\vee [[LoS_R(M, S_i) = LoS_R(M, S- S_i)$

$\quad \wedge [LoP_R(M, S_i) > LoP_R(M, S-S_i)]]$

$\vee [[LoS_R(M, S_i) = LoS_R(M, S-S_i)]$

$\quad \wedge [LoP_R(M, S_i) = LoP_R(M, S-S_i)]$
$\quad \wedge [RT(M, S_i) > RT(M, S-S_i)]].$

To evaluate the *LoR* of several systems against an Attack scenario, the following definition can be utilized. A system is more resilient (as compared to other systems) if it acquires a minimum loss of stability, or otherwise, a minimum loss of performance, or otherwise a minimum time to recover.

Definition 2 *Given a set of systems $M \equiv \cup M_j$, $j \in \{1,..., y\}$, where y is the number of systems, and an attack scenario $S_i \in S$, the $LoR(M_i, S_i) > LoR(M-M_i, S_i)$ if:*

$[LoS_R(M_i, S_i) < LoS_R(M-M_i, S_i)]$
$\vee [[LoS_R(M_i, S_i) = LoS_R(M-M_i, S_i)]$

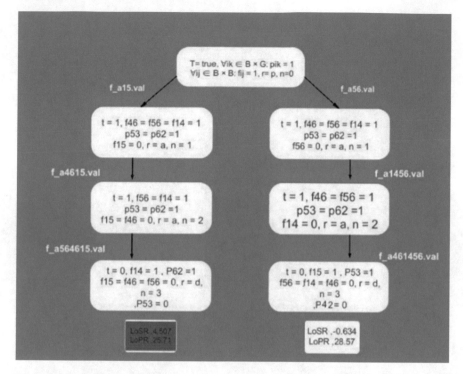

Fig. 7 System PS_1 Attack graph

$$\wedge [LoP_R(M_i, S_i) < LoP_R(M\text{-}M_i, S_i)]]$$

$$\vee [[LoS_R(M_i, S_i) = LoS_R(M\text{-}M_i, S_i)]$$

$$\wedge [LoP_R(M_i, S_i) = LoP_R(M\text{-}M_i, S_i)]$$

$$\wedge [RT(M_i, S_i) < RT(M\text{-}M_i, S_i)]].$$

4.1   LoR Assessment for Communication Networks Example

Tables 1, 2, and 3 show the Level-of-Stability-Reduction (LoS_R), the final Level-of-Performance-Reduction (LoP_R) in the application Quality-of-Service (QoS), and the Recovery-Time (RT), respectively, of CN_1 and CN_2 providing their A_{GS}. The given data are obtained using Optimized Network Engineering Tools (*Opnet Modeler*) [26].

Using **Definition** 1. For CN_1, the maximum LoS_R is under S_4: $a_{23} \rightarrow a_{13}^{23}$. Thus, $LoR(CN_1, S_4)$ is the worst. In the same way, For CN_2, the maximum LoS_R is under S_4. Thus, $LoR(CN_2, S_4)$ is the worst.

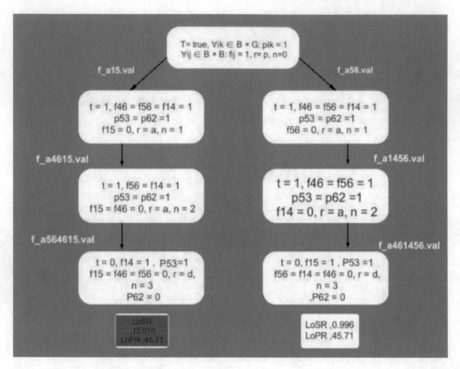

Fig. 8 System PS_2 Attack graph

Table 1 CN_1 and CN_2 LoS_R
(%) under A_G scenarios

S	CN_1 LoS_R	CN_2 LoS_R
S_1	7×10^{-4}	5×10^{-4}
S_2	7×10^{-4}	7×10^{-4}
S_3	3.84×10^{-5}	4.55×10^{-5}
S_4	0.0266	0.0291
S_5	7.31×10^{-5}	7.09×10^{-5}
S_6	5.03×10^{-5}	5.6×10^{-5}

Table 2 CN_1 and CN_2 LoP_R (%) under A_G scenarios

S	CN_1			CN_2		
	Email LoP_R	FTP LoP_R	Vedio LoP_R	Email LoP_R	FTP LoP_R	Vedio LoP_R
S_1	∞	∞	∞	34.63	13.20	∞
S_2	36.58	23.82	∞	35.57	15.61	∞
S_3	26.11	4.78	14.83	7.5	1.29	3.45
S_4	∞	∞	∞	9.71	11.92	∞
S_5	34.76	11.79	-0.75	45.27	15.71	-2.38
S_6	-2.8	1.02	∞	0.56	3.46	∞

Table 3 CN_1 and CN_2 RT (SEC) under A_G scenarios

S	CN_1 RT	CN_2 RT
S_1	7.6939	7.5340
S_2	7.7017	7.4404
S_3	7.9426	7.5819
S_4	5.9360	5.9360
S_5	7.3811	7.3879
S_6	8.0867	7.5905

These outcomes comply with the developed Attack graphs for both networks, where the sequences constituting S_4 end in states where CN_1 and CN_2 are no longer stable, and have the worst performance, and without recovery made.

It can also be shown that against S_4, CN_1 has higher LoP_R than CN_2 over all applications. Moreover, both CN_1 and CN_2 have almost the same LoS_R. Then, using **Definition** 2. $LoR(CN_2, S_4) > LoR(CN_1, S_4)$. These outcomes comply with the developed Attack graphs of the networks, where the sequence constituting S_4 for CN_1 ends in a state where the network is no longer stable, and has no services available, as compared to CN_2, the state shows that the network is unstable, but some services are yet available.

4.2 LoR Assessment for Power Systems Example

Tables 4 and 5 show the LoS_R and eventual LoP_R, respectively, of systems PS_1 and PS_2 under A_G scenarios. The RT for power systems does not change adequately from one system topology to another, and thus not considered for this example. The given data are gathered using *MATLAB*.

Using **Definition** 1. For PS_1, the maximum LoS_R is 4.507%, that is under S_1: $a_{15} \rightarrow a_{46}^{15} \rightarrow a_{56}^{46,15}$. In this attack scenario, the load connected to bus 5 is no longer served by generator 3. Hence, $LoR(PS_1, S_1)$ is the worst. In the same way, For PS_2,

Table 4 PS_1 and PS_2 LoS_R (%) under A_G scenarios

S	PS_1 LoS_R	PS_2 LoS_R
S_1	4.507	15.018
S_2	−0.634	0.996

Table 5 PS_1 and PS_2 LoP_R (%) under A_G scenarios

S	PS_1 LoP_R	PS_2 LoP_R
S_1	25.71	45.71
S_2	28.57	45.71

the maximum LoS_R is 15.018%, which is under S_1 where the load connected to bus 6 is no longer served by generator 2. Hence, $LoR(PS_2, S_1)$ is the worst.

We also note that for PS_1 under S_1, the $LoP_R = 25.71\%$, while for PS_2, the $LoP_R = 45.71\%$. Hence, PS_2 has higher LoP_R than PS_1. Moreover, PS_2 has higher LoS_R (15.018%) as compared to PS_1 (4.507%), by using **Definition** 2. $LoR(PS_1, S_1) > LoR(PS_2, S_1)$. Thus, topology PS_1 is more resilient in comparison to PS_2, against S_1.

# 5	Conclusion

In this work, the levels of resilience were assessed for two communication networks, and two power systems using their associated Attack graphs. The Attack graphs forming the attack scenarios compromising the systems (as given by their loss of stability) were generated using *JKind* checker tool, and graphically presented using Unity software. A future work is to investigate risk analysis and mitigation techniques utilizing the developed graphs.

Acknowledgements The authors would like to acknowledge Deanship of Graduation Studies and Scientific Research at the German Jordanian University for the Seed fund SATS 02/2018, and Suha Tayseer Sabha, for her development support in C#.

References

1. Reza, A., Von Meier, A., Mehrmanesh, L., Mili, L.: On the definition of cyber-physical resilience in power systems. Renew. Sustain. Energy Rev. **58**, 1060–1069 (2016)
2. Holling, C.S.: Resilience and stability of ecological systems. Ann. Rev. Ecol. Syst. 4 **1**, 1–23 (1973
3. Holling, C.S., Gunderson, L.H., Light, S.: Barriers and Bridges to the Renewal of Ecosystems. Columbia University Press, New York (1995)
4. Walker, B., Carpenter, S., Anderies, J., Abel, N., Cumming, G., Janssen, M., Lebel, L., Norberg, J., Peterson, G.: Resilience management in social-ecological systems: a working hypothesis for a participatory approach. Conserv. Ecol. vol. 6 (2002)
5. Kendra, J., Wachtendorf, T.: Elements of resilience after the world trade center disaster: reconstituting New York City's Emergency Operations Centre. Disasters **27**, 37–53 (2003)
6. Brown, G., Matthew, C., Javier, S., Kevin, W.: Defending critical infrastructure. Interfaces **36**, 530–544 (2006)
7. Haimes, Y.Y.: On the definition of resilience in systems. Risk Anal. Int. J. **29**, 498–501 (2009)
8. Ji, K., Dong, W.: Resilient control for wireless networked control systems. Int. J. Control Autom. Syst. **9**, 285–293 (2011)
9. Garcia, H.E., Wen-Chiao, L., Semyon, M.M.: A resilient condition assessment monitoring system. In: 5th International Symposium on Resilient Control Systems (ISRCS), IEEE, pp. 98–105 (2012)
10. Ji, K., Yan, L., Linxia, L., Zhen, S., Dong, W.: Prognostics enabled resilient control for model-based building automation systems. In: Proceedings of the 12th Conference of International Building Performance Simulation Association, pp. 286–293 (2011)

11. Dinh, L.T., Hans, P., Xiaodan, G., Sam, M.M.: esilience engineering of industrial processes: principles and contributing factors. J. Loss Prev. Process Ind. **25**, 233–241 (2012)
12. Wei, D., Kun, J.: Resilient industrial control system (RICS): concepts, formulation, metrics, and insights. In 3rd International Symposium on Resilient Control Systems (ISRCS), IEEE, pp. 15–22 (2010)
13. Hosseini, S., Kash, B., Jose, E.R.-M.: A review of definitions and measures of system resilience. Reliab. Eng. Syst. Saf. **145**, 47–61 (2016)
14. Ouyang, M., Wang, Z.: Resilience assessment of interdependent infrastructure systems: with a focus on joint restoration modeling and analysis. Reliab. Eng. Syst. Saf. **141**, 74–82 (2015)
15. Pant, R., Barker, K., Ramirez-Marquez, J., Rocco, C.: Stochastic measures of resilience and their application to container terminals. Comput. Ind. Eng. **70**, 183–194 (2014)
16. Francis, R., Behailu, B.: A metric and frameworks for resilience analysis of engineered and infrastructure systems. Reliab. Eng. Syst. Saf. **121**, 90–103 (2014)
17. Panteli, M., Pierluigi, M., Dimitris, N.T., Elias, K., Nikos, D.H.: Metrics and quantification of operational and infrastructure resilience in power systems. IEEE Trans. Power Syst. **32**, 4732–4742 (2017)
18. MacKenzie, C.A., Kash, B.: Empirical data and regression analysis for estimation of infrastructure resilience with application to electric power outages. J. Infrastruct. Syst. **19**, 25–35 (2012)
19. Panteli, M., Pierluigi, M.: Modeling and evaluating the resilience of critical electrical power infrastructure to extreme weather events. IEEE Syst. J. **11**, 1733–1742 (2017)
20. SEI: Architecture analysis and design language, Carnegie-Mellon University, Pittsburgh, Pennsylvania, USA, [Online]. Available http://standards.sae.org/as5506/
21. Ibrahim, M.: A Resiliency Measure for Communication Networks, Amman. IEEE, JORDAN (2017)
22. Ibrahim, M., Jun, C., Ratnesh, K.: A resiliency measure for electrical power systems. In: 13th International Workshop on Discrete Event Systems (WODES), pp. 385–390 (2016)
23. Singh, S., Stålmarck, G., Sheeran, M.: Checking safety properties using induction and a SAT-Solver. In: International Conference on Formal Methods in Computer-Aided Design, Berlin, Heidelberg (2000)
24. Alvarez, C., Blesa, M., Serna, M.: The robustness of stability under link and node failures. Theor. Comput. Sci., Elsevier (2011)
25. Craighead, J., Burke, J.: Using the unity game engine to develop sarge: a case study. In: International Conference on Intelligent Robots and Systems (2008)
26. Sethi, A.S., Hnatyshin, V.Y.: The Practical OPNET User Guide For Computer Network Simulation. CRC Press, Tayler and Frances Group (2013)

Modern Methods for Analyzing Malware Targeting Control Systems

Nitul Dutta, Kajal Tanchak and Krishna Delvadia

Abstract Industrial control systems are critical infrastructure of nation. ICSs are sensor-actuator networks that control physical systems. The core components are Programmable Logic Controllers (PLCs), Supervisory Control and Data Acquisition (SCADA), distributed control systems (DCS). Traditional ICS had specialized hardware without Internet connection. Nowadays ICS are commodity computers comes with high configuration and internet connection which makes it defenseless for most common attacks. Defensive mechanism are limited because ICSs are not using typical solutions like anti-viruses. They developed a malware-tolerant ICS network architecture that operate in secure manner even if attacker can attack on some of components. They provide ProVerif proofs to show the correctness of the network protocol. They added self-healing mechanism they implemented it on top of FreeRTOS and ARM TrustZone. The architecture automatically repair ordinary and malicious faults is known as self-healing. Governmental organizations recommend a strategy called "defense in depth" which tries to deploy defenses at every layer of the network. But author of paper use new approach. They distribute trust over each component on the network so malware cannot break the security policies. This approach is called malware tolerant.

Keywords Industrial control system · Distributed control system · Programmable logic control · SCADA systems · Malware analysis · Dynamic taint analysis

N. Dutta (✉) · K. Tanchak
Computer Engineering Department, MEF Group of Institutions, Rajkot, Gujarat, India
e-mail: nituldutta@ieee.org

K. Tanchak
e-mail: tanchakkajal123@gmail.com

K. Delvadia
Chhotubhai Gopalbhai Patel Institute of Technology, Bardoli, Gujarat, India
e-mail: krishnadalsaniya10@gmail.com

© Springer Nature Switzerland AG 2020
E. Pricop et al. (eds.), *Recent Developments on Industrial Control Systems Resilience*,
Studies in Systems, Decision and Control 255,
https://doi.org/10.1007/978-3-030-31328-9_7

1 Introduction of Industrial Control System

ICS is the heart of the industrial process, operation, automation, and control system applications. A term used to include the numerous applications and uses of industrial control systems. ICS refers to "A collection of hardware and software that affect or impact the safe, secure, and reliable operation of an industrial process." ICS consists of different control systems like supervisory control and data acquisition (SCADA) systems, distributed control systems (DCS), and other control system configurations such as Programmable Logic Controllers (PLC), Remote terminal unit (RTU). Control systems are used in many critical infrastructures such as power plant, nuclear power plant, oil, electrical, water and wastewater, oil and natural gas, chemical, transportation, pharmaceutical, pulp and paper, food and beverage, and discrete manufacturing industries electrical, water and wastewater, oil and natural gas, chemical, transportation, pharmaceutical, pulp and paper, food and beverage, and discrete manufacturing industries [1].

2 History of Evolution of ICS System

In 1620, Cornelis Drebbel designed a feedback loop, or closed loop control system, to operate a furnace, effectively designing the first thermostat. René-Antoine Ferchault de Réamur (1683–1757) designed automatic devices to control the temperature of the incubator. A U-tube containing mercury which measures temperature by extension of the liquid container. The draft to a furnace is controlled by a float in the mercury via a mechanical linkage. The rate of combustion and heat output was changed as the draft was opened or closed. This concept was called closed-loop feedback as the feedback was provided to liquid by incubator temperature and goes back to draft control [2]. A Steam engine governor was developed during the 18th century to control the gap between the grinding-stones in both wind and water mills via the lift-tenter mechanism. The first steam governor was produced by James watt which provides some control without exact speed control. Many efforts were found to improve the watt governor. In 1868 James Clerk Maxwell (1831–1879) developed a well-known Maxwell formula for governors. From 18th to early 19th-century control system focused on basic process activities of controlling temperatures, pressures, liquid levels, water levels and the speed of rotating machines [3]. During the 2nd industrial revolution (1900–1970), the concept of "negative feedback" was incorporated into new control theory concepts and design of control systems (Bell laboratories). Richard moreley developed a programmable controller which has 125 words of memory. The first programmable controller (PC) implemented was Modicon 084 in 1969. The name Modicon stood for MOdular DIgital CONtroller. The 3rd revolution (1970–2000), is all about Automation of production by electronics. Allen-Bradley designed 1774 PLC and coined the term "Programmable Logic Controller (PLC)" in 1971. New inventions like Modbus allow to talk to PLCs and remote IO was introduced. Later on, PLCs were connected to PC and various field bus protocol include ControlNet, DeviceNet, Profibus, Fieldbus Foundation were developed. In 1992 Ethernet and TCP/IP were connected to PLCs. All network devices of ICS have embedded web server which provides a control panel for configuring the devices were introduced in 2003 [2].

3 ICS Operations and Components

The basic operation of the ICS system is shown in Fig. 1 [4]. The key components of ICS systems are a human-machine interface (HMI), controller, actuators, sensors, remote diagnostics and maintenance tools, and controlled processes include various security mechanism.

Controlled process is manipulated by control loops which consist of sensors actuators and controller. Sensors are devices that can detect and measures physical phenomena then send this information to the controller as a controlled variable as a signal. This signal interpreted by the controller and generate output variable based on the control algorithm and target set points which transmits to the actuators. As per the commands are given from controllers, actuators like control valves, breakers, switches, and motors manipulate the controlled process. To monitor and configure

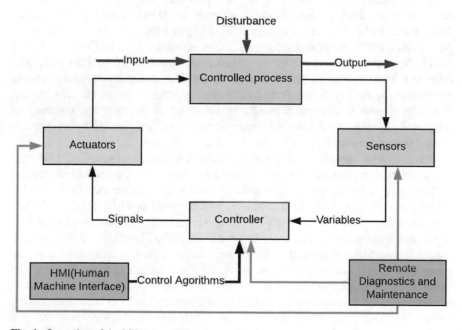

Fig. 1 Operation of the ICS system [1]

set points, algorithms and to adjust parameters in controllers, operators and engineers use human-machine interface, it also displays the status information of process
and historical information. Control loops can be nested and/or cascading, by which
the set point for one loop is determined by another loop based on the process variable. Supervisory-level loops and lower-level loops operate continuously in order of
milliseconds to minutes [1].

4 Types of ICS

Larger systems are usually implemented by Supervisory Control and Data Acquisition (SCADA), distributed control systems (DCS), programmable logic controllers
(PLCs), Remote terminal unit (RTU). SCADA and PLCs can be used by small systems with few control loops.

4.1 SCADA System

SCADA stands for Supervisory Control and Data Acquisition. These systems are
used for controlling and monitoring at the supervisory level as well as centralized
data acquisition are important. SCADA software is placed on the top of hardware
through an interface like a PLC system. SCADA systems are used in steel industries,
power plant, and the nuclear plant also. SCADA systems are run on DOS, VMS, and
UNIX [5]. Data transmission system and data acquisition are combined with each
other in a SCADA system. Monitoring and controlling of input and output process
are handled by HMI software. Field information is gathered by the SCADA system
that is transferred to the central computer, the operator can view this information
graphically or textually. An entire system can be control and monitor by an operator
from a central location [6].

Figure 2 shows the configuration of the SCADA system [1]. Typical SCADA system consists of various hardware like communication routers, control server placed
in the control center, engineering workstations, human-machine interface, and different geographically distributed field sites which consist of PLCs, RTUs and IEDs
connected through modem or WAN card. Control center is responsible for alarming,
report and trend analysis. The local process is controlled by PLC and RTU whose
output is stored and processed in the control server. Data are transferred between
the control server and local process are handled by communication systems like
radio, satellite, telephone line, cellular system. Standard communication protocols
are used for communication. System was guided by software to monitor, ranging the
parameters that are acceptable for work. Intelligent electronic devices may directly
communicate to a local server or local RTU may poll the IEDs to collect the data and
pass to the server. SCADA uses various communication topology like point-to-point,
series, series-star, and multi-drop.

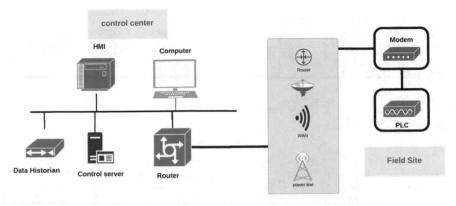

Fig. 2 SCADA general layout [1]

Advantages of SCADA System [7]

- Redundancy and increased reliability.
- SCADA system have control over the remote equipment, that leads to reduced cost and increased equipment life and maintenance.
- Easily extendable system, new equipment can be added to existing ones hence scalability is improved.
- Availability of process history by data storage.
- Reduced cost, improve performance and system efficiency.

Disadvantages of SCADA System [7]

- Communication system issues.
- Dependability and security issues.
- Installation cost is high.
- System upgrade and update issues.

4.2 Distributed Control System

To control large, complex and geographically distributed application in industrial control process distributed control system is designed. As shown in Fig. 3 [8], field devices and network elements consist of various layers such as a supervisory layer, intermediary supervisory level, and field level. At supervisory level, HMI is used to send a control request to control the server through a network. This request supplies the processing data to the receiving device and then transmit to set points at a lower level [8].

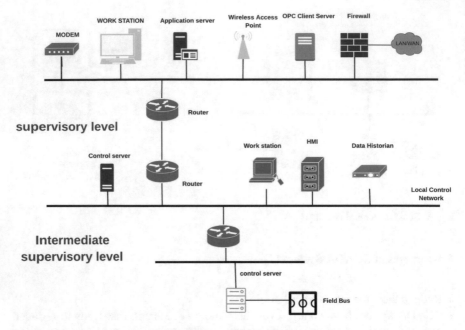

Fig. 3 Architecture of a distributed control system [1]

Control server sends this process set point or request data from set points to a lower level that is one or more intermediatory supervisor level. Control server at this level poll for data and send data set point to the lower level which consists of PLCs and other field devices. PLCs receives data from the sensor and sends output in terms of signals to actuators. Sensors and actuators communicate to control server through the network are called fieldbus [9]. Various communication protocols such as Mod-Bus [10], Fieldbus [11], Distributed Network Protocol v3 (DNP3) [12] and for fieldbus communication, Device Net [13] or Mod-Bus are used.

Advantages of Distributed Control System [9]

- Redundancy can be achieved at every level.
- With more number of I/O's DCS can handle the complex process.
- DCS structure is more scalable.
- DCS provides a secure system to handle complex functions.

Disadvantages of Distributed Control System [14]

- Failure of one controller cause failure in more than one loop.
- It requires a skilled operator to understand all the information and data.

4.3   *Programmable Logic Controllers*

Programmable logic control (PLC) used in large automation industry. SCADA and DCS both systems are using PLCs widely to provide local management of process and function through control. As shown in Fig. 4 [1], PLCs consists of Memory, I/O, CPU, power supply and programming devices. PLCs are accessed using a workstation. All the components are connected over the field bus using LAN. Data are stored in a data historian [15].

Advantages of Programmable Logic Control [16]

- Easy to operate by the user.
- Remove hardwire logic.
- Easily programmed.
- Speedy and performance are high.

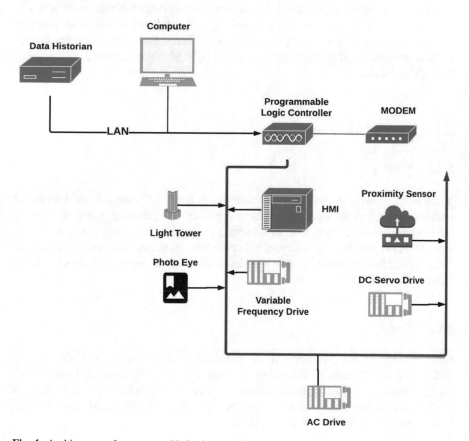

Fig. 4 Architecture of programmable logic controllers [1]

Disadvantages of Programmable Logic Control [16]

- Difficult to find errors, it requires skilled operators.
- Difficulties with change and replacements.

5   Introduction of Malware

Most of the large industries are using automation for controlling the process. With increased developments of internet and communication, various malicious attacks were performed on ICS. After the successful attack of 'Stuxnet' security system of ICS become more and more attentive on securing the industrial control system. ICS consists of various hardware and software manufactured by various companies. Many software does not use standardized protocols for particular devices. The operator of this industry does not know about the implementation details of computer systems. Various types of malware can be attacked on ICS. Malware is a type of program which gains the access of a computer, steals the data and harms the computer system without knowledge of operator or user of ICS [17]. So it's necessary to know about types of malware.

5.1   Types of Malware

5.1.1   Virus

Virus is the most common type of malware. It steals information and harms the host computer and network. It is having capabilities of copying itself and spread the infection to other computers by attaching themselves to a programme, scripts or executable. When the user executes this program, virus starts spreading infection. There are different kinds of viruses like Email virus, Macro virus, Resident virus and Boot sector virus [17].

5.1.2   Worm

It is similar to a virus. Worms have capabilities to self-replication and self-copying and spread its infection to computers through a network by sending a mass mail. Main difference between virus and worm is a virus require human intervention to execute the program while worm spread by its own. ILOVEYOU, Melissa, Morris, Code Red are the example of worms [17].

5.1.3 Trojan Horse

This type of malware appears as legitimate software and make people think that they download legitimate software but actually downloading the malware. Once get access to the system it will collect information from the system and installs more malware. Example of this type of malware is Trojan DDoS, Trojan dropper, Trojan gamer [18].

5.1.4 Ransomware

It is the most dangerous malware which encrypts the user's file and demand ransom for decrypting malware. First, it spread like a normal worm after gaining access to the system it can freeze the system and displays a text message to pay money in the form of bitcoin to decrypt the file. Examples of ransomware are crypto jacker, wanna cry, locky [17].

5.1.5 Rootkit

It allows attackers to remotely access or control system without being noticed by users. Once gaining access to the system, install more malware and steals information. It hard to detect this type of malware and removal is even harder. Example of this type of malware is knark, Rkit and Adore [18].

5.1.6 Spyware

This type of malware can spy on your internet activity to send adware. This is actually not harmful. They send popups that cannot be blocked by windows system [17].

5.1.7 Adware

The only purpose of this malware type is displaying advertisements on the computer. Often adware can be seen as a subclass of spyware and it will degrade the processing power of the computer system [18].

5.2 Types of Malware Attacks on ICS

5.2.1 Stuxnet

It is a computer worm that developed and launched by the United States and Israel in 2010 that targets programmable logic controllers which handles the automatic electrochemical process like centrifuges. It is the first malware designed to attack industrial control system. Stuxnet enters into a system via USB and infects all the machines, by gaining digital certificate that shows it comes from a genuine company. Worm cannot be detected by the detection system. Stuxnet checks whether a machine is part of ICS made by Simense. To enrich the uranium Iran deployed such kind of system to run high-speed centrifuge. If the machine is not part of ICS then it does nothing otherwise worm connected to the internet for downloading its updated version and target the PLC, exploit a zero-day vulnerability that hasn't been detected by the security system. Worm spies on targeted system and use the information that takes control of centrifuge and make them fail. During this, it provides false feedback to the outside controller so that they cannot do anything [19].

5.2.2 Havex

It is a remote access Trojan discovered in 2013 by espionage campaign targets ICS mainly energy grid operators, major electricity generation firms, petroleum pipeline operators, and industrial equipment providers. It involves three phases of delivery. It uses sphere phishing to infect victims computer and collect information on targets in their first phase. Havex was delivered via watering hole attacks in which users of legitimate websites are redirected to havex infected server. In the third phase, victim downloaded this application. Once installed it infects SCADA system [19].

5.2.3 BlackEnergy

It was first appeared as a DDoS in 2007. BE3 was involved in Ukraine attack that results in power outages. Its main role is not about cutting lights off but in collecting information about the ICS system. BE3 was delivered to the Ukrainian energy companies via spear-phishing emails and weaponized Microsoft word documents [19].

5.2.4 Irongate

ICS—focused malware running within simulated Siemens control system environment. This malware family named as Irongate. Two samples of the malware payload were uploaded by different sources in 2014, but none of the antivirus vendors flagged them as malicious [20].

5.2.5 TRITON

It is not publically known malware for ICS which interact with Triconex Safety Instrumented System (SIS) controllers. Attacker gained remote access to SIS workstation and placed TRITON to reprogram SIS controllers. Amid this attack, a few SIS controllers entered a failed safe state, which naturally shut down the industrial procedure and incited the benefit proprietor to start an examination. The examination found that the SIS controllers started a safe shutdown when application code between repetitive handling units failed an approval check—bringing about an MP demonstrative disappointment message [19].

5.2.6 Industroyer

On December 17th, 2016, the Ukrainian capital Kiev was hit by a power outage for around 75 min. Nearby examiners later affirmed that the vitality blackout was brought about by a cyber-attack. This malware is named Industroyer. This malware targeting industrial control system and communication-protocol which are widely used. At that time this protocol did not have any security measures so it is easy to attack and that malware spread the infection to one local sub-station to another which leads to energy blackout [21].

5.3 Malware Analysis

Antivirus vendors generate signature manually. Before writing a signature, an analyst must know if an unknown sample is malicious sample or not. Different malware analysis techniques can be used by malware analyst to understand the risk and intention of threat. Analyst understand the behavior of malware and gave a fight to disinfect the malicious code. Different malware analysis technique and tools are available, to prevent malware from analysis attackers came up with evasion techniques. This technique modifying the malicious code or conceal its code in legitimate software, so anti-virus cannot detect it. In industrial control system, there are different techniques for malware analysis to provide security against malware attacks. Techniques like fuzz test, static taint analysis and dynamic taint analysis are most commonly used in ICS. In control based system a taint label is associated with program counter, condition on branch decision is tainted a program counter is also tainted. Because of assignment, target variable is also tainted. Label of program counter is restored at a merge point.

5.3.1 Static Taint Analysis

In static taint analysis, for a set of variables assigned in untaken branch it calculates the upper bound for those variables. If a branch is taken, all those variables are marked as sensitive and program counter is also determined to be sensitive. It preserves the non-interference in simple language that doesn't support array. Information flow techniques of a static analysis component run for different languages. There are some disadvantages of static analysis like it is not accurate as the dynamic analysis. Some of the methods like code obfuscation employed by malware make them intractable for analysis. For this dynamic analysis is suitable.

5.3.2 Dynamic Taint Analysis

Dynamic taint analysis technique is used during program execution for propagating taint propagation across memory locations to track information flow within a software application it is used to detect 0 day attack and information leakage from software. It is based on dynamic binary instrumentation (DBI) frameworks or whole-system emulators/virtual machine monitors. With DBI frameworks the boundary of information tracking is a single process while with system emulators it is a single system. At machine level instruction this analysis tracks data flow dependencies. DTA tools that uses a BDI framework can be attached to any software without recompilation and it doesn't require any source code. It can also support colour tainting so it is possible to track multiple information flow. Inter-process communication (IPC) provides facilities in each monitored process is given to track information flow between processes.

Single Process Dynamic Taint Analysis

To track information flow in single program they have used dynamic taint analysis tool [22]. A running program is consist of a sequential execution of machine instructions. When instruction is executed data is involved as instruction takes one or more source and destination operands. In case of single program execution data are copied from only one location to another or some performs translation of data. In these both case dynamic dependencies exists among memory location and data. We can identify information flow from this transformation (Fig. 5).

Shadow Memory for Tag Management

Shadow memory shows the taint status of a specific memory area or register that are used during program execution. Mapping granularity between shadow memory and application can be different based on usage. In SeeC tool they have used byte precision to improve performance and reduce space requirements. Each byte of memory

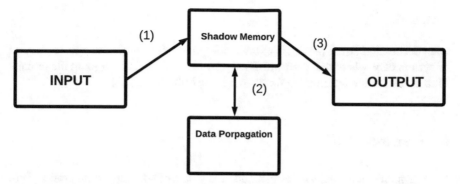

Fig. 5 Dynamic taint analysis (1) initial marking, (2) propagation, and (3) assertion [22]

application is unit in shadow memory. Change in any bit result in change in taint the tag for whole byte. 4 byte enables a tag to gain more information and it combines the data [22].

Tag Propagation

To capture the copy and transformational dependencies, propagation and clearance policies used to propagates taint markings.

- If at least one source operand is tainted, then the destination operand(s) should be also tainted.
- If the all inputs to an operation are clear, the destination operand(s) should be also cleared.

A sequence of assignments causes transitive propagation thereby realizing flow tracking. When two or more input operands are involved, all the bit-fields of the two source operands are preserved at the destination tags for 4 byte tags. Some instructions that perform transformations may involve mixed flow of information is an important features [22].

Taint Sources

Taint sources refers to starting points for newly introduced data with a proper color value where tool marks a tag. Each color value represent the data source. It may describe from where data is arrive. In this part malicious file can be given. System calls are main source points where read operation introduced a data from outside the process like network and other system. User can define filename or IP address for those file. Memory area can be tainted for application that can also be specified by user [22].

Taint Sinks

Data destination points are refers to taint sink points. It is similar to write operation of system calls where tool perform validity checking for every outgoing file or data. Sink check the tag information for the outgoing buffer data [22].

6 Research Work

From dynamic taint analysis, we can obtain key data of software configuration files. This process is divided into two part. One is taint source determination and taint propagation tracking. We have used window XP system, scada file system and configuration system software.

6.1 Taint Source Determination

Taint source configuration file is monitored in dynamic taint analysis. Taint data consists of multiple bytes which collectively used for taint sources. Collection of bytes represents offsets. Then we monitor the functions of file operation class and mark it as a taint source.

6.2 Taint Propagation Tracking

After tainting source data, taint sources are propagated. We need to track this propagation. It will check that all the data are tainted or not and then it stores the taint results. We have used sample configuration file for dynamic taint analysis to obtain results. We got the results as shown in Table 1 result set of taint data and those key

Table 1 Results of dynamic taint analysis

Program	Taint file	Maximum frequency settings	Offset
StartUp.exe	PassWord.psd	4	[1, 2, 3, 4]
	CommPtList.ini	4	[4, 80, 120, 208, 257]
	Comset.ini	4	[4, 73, 156, 180, 223, 266, 420]
	Rdb file	4	[1, 2, 3, 4, 5, 6, 106]
CodeSys.exe	Pro project file	50	[1, 2, 3, 4, 5, 6, 7, 8, …, 23,009]

Table 2 Results comparison on files

	Taint analysis on start-up file	Taint analysis on Codesys file
Number of test cases	400	1000
Test time	40 m	2 h
Number of abnormality found	50	120

bytes are then mutated. In small amount time it will generate faster response. In Table 2 we show that how many samples are examined for dynamic taint analysis and their running time and number of abnormalities found in taint analysis is shown.

7 Conclusion

The main motive of this research work is to provide various ways to defend ICS from malignant attacks. Because nowadays, attackers are more aware to find ways to attack on systems, once they found loopholes, they easily attack on various system and penetrate into the systems. ICS consists of large Power plant, Oil plant, Gas plant, Nuclear plant of any nation. To deal with attacks on this infrastructure, we need to analyze various malwares, through dynamic taint analysis. It takes small amount of time to analyze the large number of files accurately. As well as we have done taint analysis on various files of configuration software. In which tainting resources are mutated, so it becomes hard to target for attack. Traditional ICS cannot easily detect malicious attack. So, more powerful way to restrict those kind of attacks are required. To secure ICS we have done dynamic malware analysis of various configuration files and achieved security against malware attack.

References

1. Stouffer, K., Falco, J., Scarfone, K.: Guide to industrial control systems (ICS) security. NIST Spec. Publ. **800**(82), 16 (2011)
2. Bennett, S.: A brief history of automatic control. IEEE Control Syst. Mag. **16**(3), 17–25 (1996)
3. Gicsp, E.H., Assante, M., Conway, T.: An abbreviated history of automation & industrial controls systems and cybersecurity (2014)
4. Stouffer, K., Falco, J., Proctor, F.: The NIST Process Control Security Requirements Forum (PCSRF) and the future of industrial control system security. In: TAPPI Paper Summit, Atlanta, GA, 2004
5. Daniels, A., Salter, W.: What is SCADA? In: International Conference on Accelerator and Large Experimental Physics Control Systems, pp. 339–343 (1999)
6. Coates, G.M., et al.: A trust system architecture for SCADA network security. IEEE Trans. Power Del. **25**(1), 158–169 (2010)
7. Gligor, A., Turc, T.: Development of a service-oriented SCADA system. In: Emerging Markets Queries in Finance and Business, vol. 3, pp. 256–261 (2012)

8. Rrushi, J., Bellettini, C., Damiani, E.: Composite Intrusion Detection in Process Control Networks. Università degli Studi di Milano (2009)
9. Tan, K., Lee, T., Soh, C.Y.: Internet-based monitoring of distributed control systems—an undergraduate experiment. IEEE Trans. Educ. **45**(2) (2002)
10. Modbus.org. [Online]. Available: http://www.modbus.org/docs/ModbusNews_Dec2009.pdf. Accessed 15 Mar (2019)
11. Berge, J.: Fieldbuses for Process Control: Engineering, Operation, and Maintenance. ISA (2002)
12. DNP Users Group: Distributed Network Protocol Specification (2007)
13. Rockwell Automation. DeviceNet Adaptation of CIP. ODVA Website (2017)
14. Massioni, P., Verhaegen, M.: Distributed control for identical dynamically coupled systems: a decomposition approach. IEEE Trans. Autom. Control **54**(1), 124–135 (2009)
15. Quinton, B.R., Wilton, S.J.E.: Post-silicon debug using programmable logic cores. In: Proceedings of Conference on Field-Programmable Technology (FPT), pp. 241–248 (2005)
16. Irfan, M., Saad, N., Ibrahim, R., Asirvadam, V.S.: Development of an intelligent condition monitoring system for AC induction motors using PLC. In: IEEE Business Engineering and Industrial Applications Colloquium (BEIAC), pp. 789–794, 7–9 Apr 2013
17. Chumachenko, K.: Machine Learning Methods for Malware Detection and Classification. XAMK (2017)
18. Pirscoveanu, R., Hansen, S., Larsen, T., Stevanovic, M., Pedersen, J., Czech, A.: Analysis of malware behavior: type classification using machine learning. In: International Conference on Cyber Situational Awareness Data Analytics and Assessment (CyberSA), London, pp. 1–7 (2015)
19. ICS Malware—NJCCIC. NJCCIC (2019). [Online]. Available: https://www.cyber.nj.gov/threat-profiles/ics-malware-variants/. Accessed 19 Mar 2019
20. Available: https://www.fireeye.com/blog/threatresearch/2016/06/irongate_ics_malware.html. Accessed 19 Mar 2019
21. Industroyer: ICS were developed decades ago with no security in mind. WeLiveSecurity. [Online]. Available: https://www.welivesecurity.com/2017/06/19/industroyer-interview-ics-developed-decades-ago-no-security-mind/ (2019). Accessed 19 Mar 2019
22. Kim, H.C., Keromytis, A.D., Covington, M., Sahita, R.: Capturing information flow with concatenated dynamic taint analysis. In: Proceedings: International Conference on Availability Reliability and Security, pp. 355–362 (2009)

Multi-stage Cyber-Attacks Detection in the Industrial Control Systems

Tomáš Bajtoš, Pavol Sokol and Terézia Mézešová

Abstract Industrial Control Systems are a prestigious target for attackers and the attacks are becoming more sophisticated. Intrusion detection systems can uncover suspicious activity and point towards steps of attacks. Detection systems raise an overwhelming number of alerts, so their aggregation and correlation are necessary. It is important for the security analysts to correlate the alerts raised by detection systems and project the next steps of the attack to better protect critical resources. In this chapter, we search for attack patterns in the correlated alerts from industrial control systems network. Our correlation approach is similarity-based according to IP addresses and ports. We construct a directed graph that describes all possible attack paths between multiple attack stages. Several interesting patterns are discussed.

Keywords Alert · Aggregation · Correlation · Multi-stage · Attack patterns

1 Introduction

Our society is becoming more dependent on computers and networks. Industrial control systems are no exceptions as they govern the most critical systems, without which people cannot continue with their daily lives. Supervisory control and data acquisition (SCADA) systems are used in industrial control systems (ICS) to run automated processes to perform various industrial tasks. SCADA manages and controls a variety of systems, including cooling, ventilation, and power distribution and generation in addition to sensitive processes such as nuclear fusion. However, the landscape of the ICS creates plenty of security challenges that must be addressed. When the devices in ICS are connected to the Internet, they must be properly protected. One example

T. Bajtoš · P. Sokol (✉) · T. Mézešová
Faculty of Science, Institute of Computer Science, Pavol Jozef Šafárik University in Košice, Košice, Slovakia
e-mail: pavol.sokol@upjs.sk

T. Bajtoš
e-mail: tomas.bajtos@student.upjs.sk

T. Mézešová
e-mail: terezia.mezesova@student.upjs.sk

© Springer Nature Switzerland AG 2020
E. Pricop et al. (eds.), *Recent Developments on Industrial Control Systems Resilience*,
Studies in Systems, Decision and Control 255,
https://doi.org/10.1007/978-3-030-31328-9_8

of a sophisticated attack towards ICS is Stuxnet [1]. The statistics by Kaspersky Lab published in [2] clearly shows that there is an increase in the number of connected ICS components and the vulnerabilities in those components are diverse and exploitable by low-skilled attackers.

An **intrusion** is defined as any set of actions that compromise the integrity, confidentiality, or availability of a resource [3]. The main line of defence in critical computer networks are intrusion detection systems (IDS). They mostly work by recognizing known attacks from signatures or detecting anomalies in the network traffic. As network traffic increases, the alerts produced by IDS are also increasing exponentially. Sophisticated attacks, however, evade IDS systems by splitting the attack into several consequential phases and carrying out each phase independently. Moreover, current cyberattacks show a tendency to become more precise, distributive, and large-scale. The consequences of such attacks being undetected are severe. Establishing a description and projection of the attack and documenting the attacker's behaviour is useful for immediate use to protect critical resources [4] or for later use, such as planning of patches.

To detect the individual stages of the attack, and assign them to a broader context, we need to look at relationships between individual security events sent from the IDS. This chapter contributes to the detection of multi-stage attacks by presenting a method for finding attack patterns in a dataset of raw individual security events captured from network industrial detection systems.

To formalize the scope of the chapter, we state the following research objectives:

- to survey related works about searching for attack paths, applicable to ICS;
- to establish an appropriate method for searching for attack paths in industrial control networks.

This chapter is organized into 5 sections. Section 2 presents the main terminology and introduces the functionality of the components needed for multi-stage attack detection. In Sect. 3 we discuss the related works in the area of finding attack patterns. Section 4 focuses on the specifics of 4SICS dataset with events from industrial control systems. Section 5 presents our approach to finding relationships between security events. Finally, in Sect. 6, the results of our approach run on the dataset 4SICS are presented and discussed.

2 Theoretical Background

In this section, we present the basic theoretical background of multi-stage attack detection. An **event** is a low-level entity that is analysed by an IDS and comes from various sensors in the network or network hosts themselves. IDS generate **alerts** to notify parties of interesting events. A single event might be the cause of many alerts, especially in a networked IDS environment. Events in the log correspond to actions executed during some process, e.g. a user registration system might record all its actions in a log. These recorded actions (create a user, update user, etc.) represent

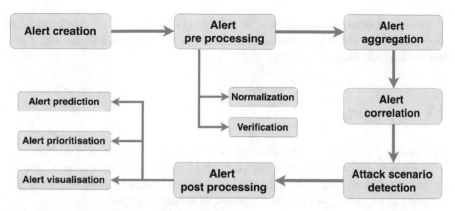

Fig. 1 Alert processing overview

activities in the process [5]. A single alert can describe a set or a sequence of events [6]. Events processed during the user registration process would constitute an alert. The phases of alert analysis are outlined in Fig. 1.

2.1 Alert Pre-processing

This stage includes normalization and verification procedures to clean and transform raw alerts into the formats expected by the aggregation and correlation modules. The alerts can be checked for false positives, filtered according to their severity or enhanced with any contextual information, e.g. spatial data. Davis and Clark [7] presented a review of the state of art on data pre-processing techniques used by anomaly-based IDS. It can be divided into normalization and verification phases.

2.2 Normalization

Normalization converts heterogeneous alerts from multiple sources into a standard format acceptable by correlation modules [8]. A common format for alerts is an XML-based format named The Intrusion Detection Message Exchange Format (**IDMEF**). It is a tree structure allowing for deep inner structure within the tree nodes. It is intended to be a standard data format for reporting of suspicious alerts between automated intrusion detection systems [9].

Another deep tree-structured format is Incident Object Description Exchange Format (**IODEF**) [10], intended for sharing information between various Computer Security Incident Response Teams. It is a robust scheme capable of including detailed incident data.

Intrusion Detection Extensible Alert (**IDEA**) [11] is a format for sharing security alerts created by the organization CESNET. It is a JSON-based scheme with a well human-readable object hierarchy.

2.3 Aggregation

The sensors placed in computer networks, e.g. in the industrial control systems generate big amounts of log data that might or might not be of interest for security-related purposes. Clustering and thus reducing the alerts are necessary steps in minimizing fatigue and improving situational awareness of security analysts.

Aggregation is done on the commonality of values in attributes that are present in every single event. Most prominent attributes are the source and destination IP address, source and destination port, type of the intrusion or attack class, and timestamp of the event. Alerts are considered similar if they match in some or all these attributes. Most common aggregation strategies are [5, 12, 13]:

- *one-to-many*: alerts are aggregated according to the source IP address to form groups with single attacker attempts to compromise several targets;
- *many-to-one*: alerts are aggregated according to the destination IP address to form groups of multiple attackers attempting to compromise the same single target.

The aggregated alerts can on top of the common attributes contain also additional statistical information, such as a count of events, their size, or a list of the individual events themselves.

2.4 Correlation

Correlation extracts useful and high-level information from the security events. Its aim is to provide understanding the logical relationships between them [14]. Correlation algorithms are, in practice, applied to aggregated alerts to construct all possible multi-stage attack paths.

There are several approached to alert correlation. Three major categories can be identified from literature [6, 8], shown in Fig. 2. Results from the aggregation

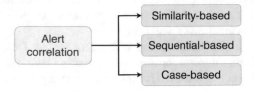

Fig. 2 Types of alert correlation

and correlation procedures are often called **hyper-alerts**. They can be fed back to pre-processing modules to improve their accuracy.

Similarity-based correlation methods aim to reduce the total number of alerts by clustering and aggregating them by their similarities. Each generated alert has several associated parameters, such as source and destination IP addresses, source and destination ports, timestamps, service that generated the alert, alert message, etc. Alert correlation algorithm finds similarity between alerts that match closely in these parameters [15]. The occurrence of the alerts within time windows can also be considered as a similarity—i.e. alerts caused by the same fault are likely to be observed within a short period of time after that fault occurrence [8].

Sequential-based correlation methods find causal relationships between the alerts, such as pre- and post-conditions, graphs, Markov models, Bayesian networks, etc. In these approaches, the key parts are the identification of requirements that trigger an alarm (pre-conditions) and the state of the system after the alarm was raised (post-conditions) [8].

Knowledge- or **case-based** correlation matches alerts with prior knowledge based on fixed patterns of alerts [6]. The biggest disadvantage is that the expert knowledge must be periodically updated, otherwise, the detection system is unable to correlate alerts of previously unknown attacks.

2.5 Attack Scenario Identification

Reconstruction of attack scenarios is a process complicated by the presence of false-negatives and false-positives, that stem from limitations of IDS. False-positives lead to incorrect attack scenarios, while false negatives (i.e. attacks missed by IDS) either make the reconstruction of the attack impossible or lead to an incomplete scenario.

There are three main aspects to a typical attack scenario reconstruction process: identifying related alerts, mapping an adequate subset into a relevant scenario, and meaningfully ordering the sequence of alerts. Identification of relationships between alerts is a challenging aspect [16]. The latest topic of interest in this area is discovering multi-stage attack scenarios from a sequence of alerts for which the template is unknown [17].

2.6 Alert Post-processing

Alert post-processing are procedures focused on aiding security analysts to determine how to behave during a security incident identified as a multi-step attack, predict the next steps of the attacker and what the probability of each possible step according to the current contextual information. The aim of post-processing is to provide administrators with situational awareness of the ongoing incidents in the monitored network [17].

Alert prioritization utilizes contextual information about the network and organization's setup and environment to assign a criticality level. The purpose is to take
appropriate actions with each criticality level. This component should consider various domain information, such as network topology, installed software, vulnerability
assessment, security policies.

Alert prediction methods should be capable of determining what types of alerts
and wherein the network they will be raised based on the history of previous attacks.
It is important, but a complex aspect of situational awareness. Countermeasures for
intrusion prevention systems can be optimized according to the prioritization and
prediction outcomes.

After the alerts are processed and appropriately annotated by the prioritization
procedures, they can be **visualized** depending on the needs of the security analysts
in the operational centres. Most used types of visualization in this field are various
graph visualizations, plots, charts, or heatmaps to quickly show the statistical changes
in the alerts over time.

In this chapter, we do not present any post-processing procedures after the attack
pattern is computed.

3   Related Works

In this section, we discuss similar works in the area of detecting multi-stage attacks
with a focus on the correlation and pattern searching modules.

Shittu et al. [18] present a method that after correlation prioritizes the meta-alerts
based on anomaly detection. After meta-alerts are correlated, a set of representative
features is extracted with frequent pattern mining, which adds contextual information
to true-positives. The parameters needed for correlation were set experimentally.
Their prioritization component reduced false-positives significantly in both tested
scenarios. Both an online and offline correlation is possible with their approach.

Ramaki et al. [19] provide a real-time algorithm for alert correlation. It uses causal
correlation to extract critical episodes from alert sequences. The processing of alerts
is done in batches and starts when the time window is completed. Offline mode acts
as a learning model for when the alerts from sensors are processed in real-time. This
aids in also predicting the next steps of active attacks.

Paper [20] presents a hybrid model for alert correlation. The first module consists
of correlating alerts for known attacks into an attack graph. If this is unsuccessful,
a similarity-based method is used to correlate the unknown attacks and update the
attack graph. The attack graph is generated with a depth-first search algorithm with
assumptions that raised alerts to correspond to exploits. The similarity is calculated
from IP addresses. They evaluated their approach with the DARPA2000 dataset.

In [21] authors correlated from rules reconstructed from an ontology about known
attacks. Then, they use the frequent itemset mining algorithm with attribute-oriented
induction algorithm to find attack patterns of unknown scenarios. The rules are written

in a query-based language and thus can be easily updated. As in [20], this was also evaluated with the DARPA2000 dataset.

DARPA2000 dataset is also used by [22]. Authors extract attack strategies by using equality constraints and records them into a correlation matrix. No predefined knowledge nor training data are needed for this method.

Detection of security scenarios is implemented in [23]. Each scenario model is previously defined with corresponding indicators. After a security warning is captured, a matching algorithm is applied to calculate possible scenarios which are then reported to the analysts.

Another approach utilizing reconstruction of scenarios within DARPA2000 dataset is [24]. It works as follows: each alert is represented as a member of the formal intrusion ontology and mapped into the instances. Two instances a, b are considered similar if in the graph the instance a is referenced by c and instance b is referenced by instance d, and if the instances c, d are similar. Then, attack causality identifies the sequence of the attacks and thus the scenarios are reconstructed without prior expert knowledge.

A Bayes-based correlation is presented by Kavousi and Akbari in [25]. History of observations is used to automatically extract the transition patterns in an attack with a Bayesian network learning algorithm. Only known attack strategies are detectable with this approach.

Bahareth and Bamasak [26] discovers multi-stage attacks using a sequential pattern mining algorithm. The algorithm processes the alerts in an online mode. Each time an incoming alert is being matched to a scenario sequence, a severity level of the sequence is updated. The matching algorithm takes this level into account when processing the next incoming alerts.

In [27] focus is on the detecting patterns in the security alerts and extracts descriptive statistics for them. These can later be used for arbitrary detection methods. Security alerts are grouped into time series and statistics are calculated separately for each.

The pre-processing stage of [28] includes an initial generation of attack sequences based on attack classification and statistics. A destination IP address is the base for the correlation. Lastly, they apply pattern mining to find and define rules for the most prominent attack sequences.

de Alvarenga et al. [5] builds on the traditional models of extracting attacks from security events by hierarchically clustering them into simpler attack models and design a method for evaluating a number of clusters required for attack models. In their evaluation scenario, the clustered models correctly indicated that the attack was pre-planned.

A self-adaptive plugin for SIEM systems is proposed by [29]. Self-adaptation is achieved by maintaining a context-based knowledge with computed attacks used to continuously generate correlation rules.

An ICS specific dataset is used only in [23]. Most approaches choose the DARPA dataset for testing their correlation approaches [19, 20, 24–27]. There is a trend to move towards online or even real-time algorithms. They require an offline component that is computed first and the online component updates the learnt model according to

Table 1 Overview of related works

Attribute	Works
Time-window aggregation	[18, 27, 28]
Offline only correlation .	[5, 20–25, 27]
Online/real-time correlation	[18, 19, 26, 28, 29]
Statistical correlation	[18, 19, 25–28]
Ontology-based correlation	[20–23]
Post-processing included	[5, 18, 24, 26, 28]

the metadata from the new incoming events raised by the intrusion detection systems. Overview of related works is shown in Table 1.

4 Dataset

In this chapter, we use the public dataset called 4ICS Geek Lounge [30]. The 4SICS dataset consists of network data, which were collected from the ICS Lab's environment in 2015. This environment consists of industrial equipment including Hirchmann EAGLE 20 Tofino, Allen/Bradley Stratix 6000, Moxa EDS-508A, automation direct DirectLogic 205 PLC, and Siemens S7-1200. Dataset consists of 3 PCAP files (3,773,984 packets, 247.2 MB), each for 1 day of packet capture:

- 4SICS-GeekLounge-151020.pcap (246,137 packets, 17.2 MB)
- 4SICS-GeekLounge-151021.pcap (1,253,100 packets, 96.5 MB)
- 4SICS-GeekLounge-151022.pcap (2,274,747 packets, 133.5 MB).

There is the largest number of ICS-related protocols present in this dataset (e.g. S7Comm, Modus/TCP, EtherNet/IP, and DNP 3.0). It "can be considered a priority for the research of ICS protocol" [31]. It is one of the datasets used in [32] for detecting timing-based anomalies of the communication patterns. Hansch et al. [33] use it to present methods for non-interfering recording, compression, and transmission of industrial network packet captures. Muller et al. [34] used it to show to what extent a training attack (and its more sophisticated variants) impacts machine learning-based detection schemes, and how it can be detected.

For alert processing, we used the intrusion detection system SNORT. Lenny Hansson has created SNORT rules [35] based on this dataset. These rules can be used for asset detection and some provide security alerts. There are:

- general rules for NMAP identification
- general rules for possible brute force
- rules for remote shell
- rules for bad login and login failed in the telnet protocol, etc.

5 Creating Multistep ICS Attack Scenarios

This section describes our proposed method for creating multistep attack scenarios in industrial control systems. We provide an explanation of the used formulae and provide details on algorithms for relevant steps. Our proposed method consists of 4 steps:

1. Aggregation
2. Frequency calculation
3. Correlation
4. Pattern searching.

5.1 Aggregation

The first step is to aggregate the event logs from the dataset into aggregated alerts. Algorithm 1 shows that security event logs are split into evenly t-sized windows and all events in each window are added to one aggregated alert a_i. Such aggregated alert consists of the following parameters:

- a message string msg, representing the alert type
- a set of source IP addresses $srcIPs$
- a set of destination IP addresses $dstIPs$
- a set of source ports $srcPorts$
- a set of destination ports $dstPorts$
- the timestamp when the first event log in the window appears $start\_time$
- the timestamp when the last event log in the window appears end_time
- total number of event logs in this alert cnt
- an array of the original event logs in the window.

The aggregation algorithm (Algorithm 1) takes as input the events directly from an intrusion detection system, such as SNORT, and a parameter t defining the size of the aggregation window. It outputs a sequence of aggregated alerts as described above. The incoming events are split into distinct time windows. All events with the same msg parameter in the given time window are aggregated together into one alert and this alert is added to the returning sequence.

Algorithm 1

Aggregation into alerts

Require:
    E: sequence of event logs from IDS with n alerts
    t: size of the time window for aggregation
Returns:
    A: sequence of aggregated alerts a_i
1:   **function** AggregateAlerts(E, t)
2:      Initialize A as an empty list
3:      Let W be a list of event logs from E split according to the size of time window t
4:      **for all** alert windows $w \in W$ **do**
5:         $j \leftarrow$ empty list
6:         **for all** event logs $e \in W$ **do**
7:            add event log e to aggregated alert $j[e.msg]$
8:         **end for**
9:         $A \leftarrow A \cup j$
10:     **end for**
11:     **return** (A)
12: **end function**

5.2 Correlation

As an intermediary step before correlation, we provide definitions on frequency matrix and similarity formula. An element $fr[msg_i][msg_i]$ of a **frequency matrix** fr represents how many times the alert type msg_i is followed by alert type msg_j in a sliding time window. In our dataset, we had 51 distinct alert types, so the frequency matrix is 51 rows by 51 columns. The time window in which the frequency is computed is configurable. We set it to 30 s.

Similarity represents the degree of two aggregated alerts a_1, a_2 being similar to each other in their parameters. We compute the similarity using source and destination IP addresses and source and destination network ports. Following variables occur in the formula for similarity, given below:

- $a_i[parameter]$ returns the given parameter from the aggregated alert
- w_i is an arbitrarily set weight

$$similarity(a_1, a_2) = w_1 \frac{|a_1[srcIPs] \cap a_2[srcIPs]|}{|a_1[srcIPs] \cup a_2[srcIPs]|}$$
$$+ w_2 \frac{|a_1[dstIPs] \cap a_2[dstIPs]|}{|a_1[dstIPs] \cup a_2[dstIPs]|}$$
$$+ w_3 \frac{|a_1[srcPorts] \cap a_2[srcPorts]|}{|a_1[srcPorts] \cup a_2[srcPorts]|}$$
$$+ w_4 \frac{|a_1[dstPorts] \cap a_2[dstPorts]|}{|a_1[dstPorts] \cup a_2[dstPorts]|}$$

where

$$0 \leq w_1, w_2, w_3, w_4 \leq 1$$
$$w_1 + w_2 + w_3 + w_4 = 1$$

Algorithm 2 computes a correlation matrix C for all available alert types in line 7 according to the following correlation formula:

$$C[m_1][m_2] = \alpha \cdot C[m_1][m_2] + \beta \cdot similarity(a_1, a_2)$$
$$\cdot \frac{fr[m_1][m_2]}{\sum_{i \in T} fr[m_1][i]}$$

where

$$0 \leq \alpha, \beta \leq 1$$
$$\alpha + \beta = 1$$
$$m_1, m_2 = a_1[msg], a_2[msg]$$
$$T = \text{set of all message types from alerts}$$

The alert correlation algorithm (Algorithm 2) takes as input the sequence of aggregated alerts (returned by Algorithm 1), the length of the sequence and the configurable parameters: the size of the time window, as well as arbitrary weights. At first, the correlation matrix is initialized with zero values.

The algorithm computes the new correlation value according to the formula above between the first alert in the time window and all following alerts in the same time window. In the next step, the value in the matrix respective to the two alert types is updated rationally according to the weights α, β.

Algorithm 2

Alert correlation

Require:
    A: sequence of aggregated alerts
    k: length of A
    t: size of the time window for correlation
    α, β: configurable weights
Returns:
    C: Correlation matrix
1:   **function** CorrelateAlerts(A, t, α, β)
2:      Initialize C with zero values
3:      **for** $i \leftarrow 1$ until k **do**
5:         $j \leftarrow i + 1$
6:         **while** $\left(a_j[start\_time] - a_i[end\_time]\right) < t$ **or** a_j is not the last element of A **do**
7:            update element $C\big[a_i[msg]\big]\big[a_j[msg]\big]$ of correlation matrix C
8:            $j \leftarrow j + 1$
9:         **end while**
10:     **end for**
11:     **return** (C)
12: **end function**

5.3 Pattern Searching

The correlation matrix is used for generating a directed graph. Each distinct message type becomes a single graph node. A directed edge (m_1, m_2) is added between two nodes if the corresponding element in the correlation matrix $C[m_1][m_2]$ is higher than the threshold θ (an arbitrarily set value). If $m_1 = m_2$, an edge is not created, to avoid loops on a single node. In the generated graph, we search for weakly connected components and find simple paths between each pair of nodes in each component. A **simple path** is a path in a graph where each node occurs exactly once. These simple paths are the found patterns and are usable for further prioritization or analysis. For example, they can be used for tracking steps of an attack in real-time.

6 Results and Discussion

6.1 Dataset Pre-processing

Firstly, the dataset 4SICS was processed with an intrusion detection system. No security events were returned for the first day. The next 2 days were put into 1 common file. IDS returned 803,686 event logs. The processed parameters were a timestamp, source, and destination IP address, source and destination port, and an event message.

6.2 Results of the Correlation Algorithm

In this section, we present the resulting graph (Fig. 3) from the pattern searching phase on the prepared dataset. These alert types are not present in the final graph, because their correlation value with all other nodes is insignificant:

- '1'—SCAN NMAP SIP Version Detect OPTIONS Scan
- '6'—SCADA System detection—MB-Gateway from Automation direct
- '9'—SCADA Modbus—Function Code Scan
- '16'—Remote Shell port connected
- '25'—SCADA System detection—Siemens SIMATIC S7-1200
- '33'—SCADA S7—Connection established
- '44'—FTP login—Client sending username to server.

We describe below 3 weakly connected components from the resulting graph shown in Fig. 3. The graph was generated with Algorithm 2 using the following parameters:

- all $t = 30s$
- $\theta = 0.03$

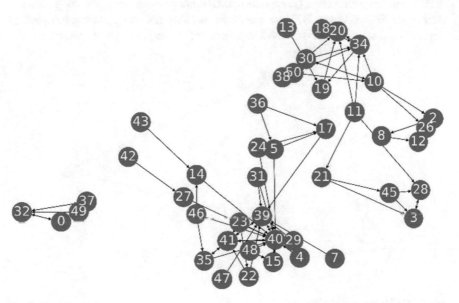

Fig. 3 Correlated alerts

- $w_1 = w_2 = 0.4$
- $w_3 = w_4 = 0.1$
- $\alpha = 0.4, \beta = 0.6$.

All the time window values were empirically set to 30 s. A too short time window cannot contain all the event logs, on the other hand too long window may contain too many of them.

The threshold value was set empirically after a series of experiments with threshold values $0.01 \leq \theta \leq 0.1$. When the threshold was set to 0, only 2 alert types from the dataset did not appear in the graph. The threshold of 0.03 maintains strong relationships between nodes while discards alerts which have no relationship between each other.

When aggregating the security events into alerts, more weight is given on the likeness in source and destination IP addresses, reflected by weights w_1, w_2, rather than the likeness in ports (weights w_3, w_4). A higher value for correlation coefficient β reflects the importance of the newly computed value of the correlation over the stored value in the matrix.

6.3 Weakly Connected Component 1

The first component of the graph, that we describe comprises of these alert types '0', '32', '37', '49', as shown in Fig. 4. The corresponding elements from the correlation matrix are presented in Table 2. From the node '49' there is a path to all other nodes, but it has no incoming edge, thus a return back is not possible. From the loop between alert types '0', '32', and '37', we can conclude they are interchangeable as attack steps and the attacker does not have to be careful about the order of execution.

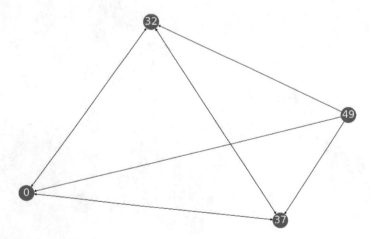

Fig. 4 Weakly connected component 1

Table 2 Correlation matrix for component 1

	'0'	'32'	'37'	'49'
'0'	0.027	0.086	0.041	0
'32'	0.034	0.025	0.05	0.001
'37'	0.11	0.11	0.11	0.01
'49'	0.08	0.08	0.08	0

Table 3 A simple path in component 1

Node	Alert	Security events
'49'	SCADA S7—Error Code—Function unavailable	• alert tcp $HOME_NET 102 -> any any • content: "|d4 01 0a 00|"
'0'	Known OS—Web server Running NET + ARM 1.00	• alert tcp $HOME_NET 80 -> $EXTERNAL_NET any • flow:to_client,established; • content: "NET+ARM Web Server/1.00"; nocase; http_header
'32'	SCADA System detection—Phoenix FL IL 24 BK-PAC	• alert tcp $HOME_NET 80 -> $EXTERNAL_NET any • content: "class=|22|devicetyp|22 3e|FL|20|IL|20|24|20|BK-PAC|3c|"
'37'	SCADA System detection—Phoenix System	• alert $HOME_NET 80 -> $EXTERNAL_NET any • lowbits:isset,NF-Phoenix; • flow:from_server,established; • content: "NET+ARM Web Server/1.00"; nocase; http_header;

One simple path from this weakly connected component is sequence <'49', '0', '32', '37'> and describes reconnaissance scanning and detection of concrete devices by an attacker. The alerts and security event logs aggregated within are presented in Table 3.

This simple path in component 1 shows, that attacker from the industrial network attempts to execute an unavailable function. In the next stage, the attacker attempts to identify devices with NET + ARM Web Server/1.00. In the last stage, the attacker tries to identify SCADA Phoenix Systems.

6.4 Weakly Connected Component 2

The second component, shown in Fig. 5, contains nodes '2', '3', '8', '10', '11', '12', '13', '15', '18', '19', '20', '21', '26', '28', '30', '34', '38', '50' and there are several interesting paths. The starting nodes for the paths are '11', '13', '18', '38'. There

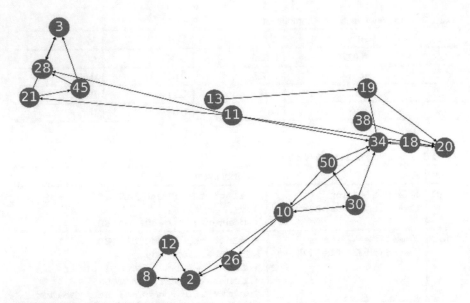

Fig. 5 Weakly connected component 2

are strong relationships between '2', '12', '8', '26'. All paths starting in '18' or '38' proceed to node '20' (their correlation is higher than 0.1).

We present 3 simple paths from this component and the correlation matrices for the respective nodes, path <'30', '10', '26', '2', '8', '12'> is shown in Tables 4 and 5, path <'50', '10', '30', '34', '19', '20'> in Tables 6 and 7 and path <'11', '21', '3', '28'> in Tables 8 and 9.

The path 1 in component 2 shows, that attacker executes firstly several types of scanning targeting the industrial network. The attacker attempts to identify the operating system and executes XMAS scanning (-sX). In the next stage, the attacker attempts to identify devices used Modbus protocol (502/TCP). Subsequently, the attacker sends a payload to the device, which is not defined in the Modbus communication protocol. In the last stages, the attacker executes write and read operations in devices via Modbus communication protocol.

Table 4 Correlation matrix for path 1 in component 2

	'2'	'8'	'10'	'12'	'26'	'30'
'2'	0.105	0.044	0.014	0.06	0.04	0.01
'8'	0.137	0.005	0.015	0.157	0	0.014
'10'	0.034	0	0.025	0.01	0.034	0.055
'12'	0.12	0.004	0.013	0.019	0.023	0.013
'26'	0.167	0	0	0	0.04	0
'30'	0.023	0	0.057	0.009	0.023	0.022

Table 5 Path 1 in component 2

Node	Alert	Security events
'30'	SCAN nmap XMAS	• `alert tcp any any -> $HOME_NET any` • `flags:FPU,12`
'10'	SCAN nmap fingerprint attempt	• `alert tcp any any -> $HOME_NET any` • `flags:SFPU`
'26'	SCADA Modbus—Read Slave Device Identification	• `alert tcp $EXTERNAL_NET any -> $HOME_NET 502` • `content: "|00 00|"`
'2'	SCADA Modbus—Non-Modbus Communication on TCP Port 502	• `alert tcp $EXTERNAL_NET any -> $HOME_NET 502` • `pcre: "/(\S\s]{2}(?!\x00\x00)/i"`
'8'	SCADA Modbus—Unauthorized Write Request to a PLC	• `alert tcp ![192.168.2.166, 192.168.2.42, 192.168.2.44] any -> $HOME_NET 502` • `pcre: "/[\S\s]{3}(\x05|\x06|\x0F|\x10|\x15|\x16)/iR"`
'12'	SCADA Modbus—Unauthorized Read Request to a PLC	• `alert tcp ![192.168.2.156, 192.168.2.42, 192.168.2.44] any -> $HOME_NET 502` • `pcre: "/[\S\s]{3}(\x01|\x02|\x03|\x04|\x07|\x0B|\x0C|\x11|\x14|\x17|\x18|\x2B)/iR"`

Table 6 Correlation matrix for path 2 in component 2

	'10'	'19'	'20'	'30'	'34'	'50'
'10'	0.025	0	0.021	0.055	0.031	0.026
'19'	0.012	0	0.065	0.012	0	0.011
'20'	0.006	0.019	0.251	0.005	0.016	0.006
'30'	0.057	0	0.024	0.022	0.033	0.03
'34'	0.003	0.047	0.083	0.003	0.014	0.003
'50'	0.073	0	0.02	0.078	0.031	0.019

Table 7 Path 2 in component 2

Node	Alert	Security events
'50'	SCAN NMAP OS Detection Probe	• `alert udp any 10000: ->` `$HOME_NET 10000:` • `content:` `"CCCCCCCCCCCCCCCCCCCC";`
'10'	SCAN nmap fingerprint attempt	• `alert tcp any any ->` `$HOME_NET any` • `flags:SFPU`
'30'	SCAN nmap XMAS	• `alert tcp any any ->` `$HOME_NET any` • `flow:stateless;` `flags:FPU,12`
'34'	SCAN PING NMAP	• `alert icmp any any ->` `$HOME_NET any` • `dsize:0; itype:8`
'19'	SCAN Potential VNC Scan 5800–5820	• `alert tcp any any ->` `$HOME_NET 5800:5820` • `flags:S,12; threshold:` `type both`
'20'	SCAN NMAP—sS 1024 Window	• `alert tcp any any ->` `$HOME_NET any` • `fragbits:!D; dsize:0;` `flags:S,12; ack:0;` `window:1024;` `detection_filter:track` `by_dst, count 1`

Table 8 Correlation matrix for path 1 in component 2

	'3'	'11'	'21'	'28'
'3'	0.066	0	0.019	0.034
'11'	0	0	0.034	0.034
'21'	0.054	0	0.11	0.026
'28'	0.054	0	0.02	0.03

Table 9 Path 3 in component 2

Node	Alert	Security events
'11'	TELNET—SCAN Behavioral unusually fast inbound Telnet Connections, Potential Scan or Brute Force	• alert tcp $EXTERNAL_NET any -> $HOME_NET 23 • flags: S,12; threshold: type both, track by_src, count 50, seconds 60;
'21'	SSLv2 Client Hello attempt	• alert tcp $EXTERNAL_NET any -> $HOME_NET [443,465,536,636,989,992,993,994,995,25] • flow:to_server,established; ssl_state:client_hello; ssl_version: sslv2; content: "|01 00 02|"
'3'	SSL—Possible SSL Bruteforce attempt	• alert tcp $EXTERNAL_NET any -> $HOME_NET 443 • flow:to_server,established; ssl_state:client_hello; detection_ filter:track by_src, count 50, seconds 60
'28'	SCAN Nmap Scripting Engine User-Agent Detected (Nmap Scripting Engine)	• alert tcp any any -> $HOME_NET any • flow:to_server,established; content: "User-Agent|3a| Mozilla/5.0 (compatible|3b| Nmap Scripting Engine"; fast_pattern:38,20; http_header; nocase;

The second path in component 2 shows, that attacker also executes several types of scanning. At first, the attacker attempts to identify the operating system. In the next stages, the attacker executes XMAS scanning (-sX) and generates ICMP traffic, which has an empty data payload and has the ICMP type field set to 8.

In the next stage, the attacker executes scanning of the industrial networks and searches for devices communicating on the TCP ports 5800–5820, to identify potential VNC service. In the last stage, the attacker executes TCP SYN (Stealth) scanning.

The third simple path in component 2 shows, that attacker executes unusually fast inbound connections via telnet protocol (potential scan or brute force) from the external network to industrial network. Subsequently, the attacker executes an SSLv2 Client Hello attempt and an SSL brute force attempt (e.g. for using vulnerability CVE-2015-3197). In the last stage, the attacker uses nmap scripting engine (e.g. use the shellcode for vulnerability CVE-2015-3197).

6.5 Weakly Connected Component 3

The most complex component contains almost half of all alert types. As Fig. 6 shows, the starting nodes for the paths are '4', '7', '23', '29', '31', '36', '42', '43', '46', '47', '48'. Nodes '39' and '40' are the middle nodes for many of the paths. This

170   T. Bajtoš et al.

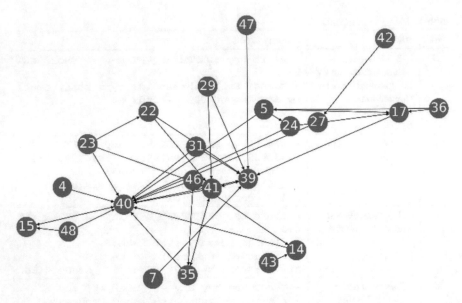

Fig. 6 Weakly connected component 3

suggests that even though a different starting point is used, the attackers still reached
the same place and attempted the same further steps.

Correlation matrices and alerts for the path <'36', '5', '24', '40', '39', '41'> is
shown in Tables 10 and 11.

The simple path for component 3 shows that attacker executes NetBIOS query
from industrial network to any network in order to get NetBIOS statistics. The next
stage is the identification of the webserver (WinCE, WinCE with InduSoft, GoAhead-
Webs). Subsequently, the attacker executes a DoS attack from HTTP ports to the
external network. The last stage is the detection of device usage SSH (RUGGEDCOM
RS910).

Table 10 Correlation matrix for a path in component 3

	'5'	'24'	'36'	'39'	'40'	'41'
'5'	0.018	0.064	0.013	0.027	0.038	0.024
'24'	0.027	0.004	0.017	0.024	0.037	0.024
'36'	0.039	0.019	0	0.025	0.022	0.025
'39'	0.021	0.008	0.005	0.057	0.063	0.061
'40'	0.018	0.012	0.004	0.05	0.052	0.02
'41'	0.024	0.013	0.004	0.06	0.065	0.03

Table 11 A simple path for component 3

Node	Alert	Security events
'36'	SCAN NBTStat Query Response to External Destination, Possible Windows Network Enumeration	• `alert udp $HOME_NET 137 ->` `any any` • `content: "|20 43 4b 41 41` `41 41 41 41 41 41 41 41 41` `41 41 41 41 41 41 41 41 41` `41 41 41 41 41 41 41 41 41` `41 00 00 21|"`
'5'	Known OS—Web server Running Microsoft-WinCE	• `alert tcp $HOME_NET 80 ->` `$EXTERNAL_NET any`
'24'	SCADA System detection—Server using WinCE with InduSoft Web Studio 6.1—Possible Beckhoff CX1010	• `alert tcp $HOME_NET 80 ->` `$EXTERNAL_NET any` • `content: "InduSoft Web` `Studio 6.1"`
'40'	Known OS—Web server Running GoAhead-Webs	• `alert tcp $HOME_NET 80 ->` `$EXTERNAL_NET any` • `content: "GoAhead-Webs";`
'39'	DOS—Large amount of TCP Zero Window from web server—Possible Dos attack—SCADA equipment too busy	• `alert tcp $HOME_NET` `$HTTP_PORTS -> $EXTERNAL_NET` `1023:` • `flags:A; window:0`
'41'	SCADA System detection—RUGGEDCOM RS910 device using SSH	• `alert tcp $HOME_NET 22 ->` `$EXTERNAL_NET any` • `content: "SSH-2.0-Mocana` `SSH"`

7 Conclusion

The increasing number of industrial control system (ICS) components being connected to the Internet leaves them open to attacks if they are not properly secured. There is a high number of exploits available for publicly known vulnerabilities and the difficulty of patching those vulnerabilities means that ICS must be prepared to detect attacks. Intrusion detection systems (IDS) are a common tool for monitoring computer networks. IDS raise security events corresponding to actions executed by attackers. ICS are especially vulnerable to attacks consisting of several steps and so the security events must be correlated into attack paths to be able to detect such multi-stage attacks.

In this chapter, we presented a method for aggregating security events into more complex alerts and correlating the alerts. Correlation is based on the similarity of the alerts in the source and destination IP address and ports parameters. Furthermore, alerts are only correlated if they occurred shortly after each other in the specified time window. We presented the results of the algorithm on data from an ICS dataset 4SICS and described in detail several found paths. Searching for attack patterns showed that often attackers do not execute their steps consecutively, but that some steps can be done in parallel and interchangeably. The next step in reacting to the raised alerts would to further prioritize the patterns and evaluate them under constraints of the particular industrial control system.

Acknowledgements We would like to thank our colleagues from the Czech chapter of The Honeynet Project for their comments and valuable input. This paper is funded by the Slovak APVV project under contract No. APVV-14-0598 and the Slovak APVV project under contract No. APVV-APVV-17-0561.

References

1. Langner, R.: Stuxnet: dissecting a cyberwarfare weapon. IEEE Secur. Priv. **9**, 49–51 (2011). https://doi.org/10.1109/MSP.2011.67
2. Andreeva, O., Gordeychik, S., Gritsai, G., Kochetova, O., Potseluevskaya, E., Sidorov, S.I., Timorin, A.A.: Industrial Control Systems Vulnerabilities Statistics (2016)
3. Debar, H., Dacier, M., Wespi, A.: A revised taxonomy for intrusion-detection systems. Ann. Télécommun. **55**, 361–378 (2000). https://doi.org/10.1007/bf02994844
4. Husak, M., Komarkova, J., Bou-Harb, E., Celeda, P.: Survey of attack projection, prediction, and forecasting in cyber security. IEEE Commun. Surv. Tutor. 1–21 (2018). https://doi.org/10.1109/comst.2018.2871866
5. de Alvarenga, S.C., Barbon, S., Miani, R.S., Cukier, M., Zarpelão, B.B.: Process mining and hierarchical clustering to help intrusion alert visualization. Comput. Secur. **73**, 474–491 (2018). https://doi.org/10.1016/j.cose.2017.11.021
6. Al-Mamory, S.O., Zhang, H.L.: A Survey on IDS Alerts Processing Techniques (2007)
7. Davis, J.J., Clark, A.J.: Data preprocessing for anomaly based network intrusion detection: a review. Comput. Secur. **30**, 353–375 (2011). https://doi.org/10.1016/J.COSE.2011.05.008
8. Salah, S., Maciá-Fernández, G., Díaz-Verdejo, J.E.: A model-based survey of alert correlation techniques. Comput. Netw. **57**, 1289–1317 (2013). https://doi.org/10.1016/J.COMNET.2012.10.022
9. Debar, H., Curry, D., Feinstein, B.: The Intrusion Detection Message Exchange Format (IDMEF). https://www.ietf.org/rfc/rfc4765.txt (2007)
10. Arvidsson, J., Cormack, A., Demchenko, Y., Meijer, J.: Incident Object Description and Exchange Format Requirements. https://www.ietf.org/rfc/rfc3067.txt (2001)
11. Intrusion Detection Extensible Alert. https://idea.cesnet.cz/en/index (2018)
12. Cipriano, C., Zand, A., Houmansadr, A., Kruegel, C., Vigna, G.: Nexat: a history-based approach to predict attacker actions. In: ACSAC'11 Proceedings of the 27th Annual Computer Security Applications Conference, p. 383. ACM Press, New York, USA (2011)
13. Lee, S., Chung, B., Kim, H., Lee, Y., Park, C., Yoon, H.: Real-time analysis of intrusion detection alerts via correlation. Comput. Secur. **25**, 169–183 (2006). https://doi.org/10.1016/J.COSE.2005.09.004
14. Soleimani, M., Ghorbani, A.A.: Multi-layer episode filtering for the multi-step attack detection. Comput. Commun. **35**, 1368–1379 (2012). https://doi.org/10.1016/j.comcom.2012.04.001
15. Heigl, M., Doerr, L., Almaini, A., Fiala, D., Schram, M.: Incident reaction based on intrusion detections' alert analysis. In: 2018 International Conference on Applied Electronics (AE), pp. 1–6. IEEE (2018)
16. Saad, S., Traore, I.: Semantic aware attack scenarios reconstruction. J. Inf. Secur. Appl. **18**, 53–67 (2013). https://doi.org/10.1016/j.jisa.2013.08.002
17. Ramaki, A.A., Rasoolzadegan, A., Bafghi, A.G.: A systematic mapping study on intrusion alert analysis in intrusion detection systems. ACM Comput. Surv. **51**, 1–41 (2018). https://doi.org/10.1145/3184898

18. Shittu, R., Healing, A., Ghanea-Hercock, R., Bloomfield, R., Rajarajan, M.: Intrusion alert prioritisation and attack detection using post-correlation analysis. Comput. Secur. **50**, 1–15 (2015). https://doi.org/10.1016/J.COSE.2014.12.003
19. Ramaki, A.A., Amini, M., Ebrahimi Atani, R.: RTECA: real time episode correlation algorithm for multi-step attack scenarios detection. Comput. Secur. (2015). https://doi.org/10.1016/j.cose.2014.10.006
20. Ahmadinejad, S.H., Jalili, S., Abadi, M.: A hybrid model for correlating alerts of known and unknown attack scenarios and updating attack graphs. Comput. Netw. **55**, 2221–2240 (2011). https://doi.org/10.1016/j.comnet.2011.03.005
21. Wang, Q., Jiang, J., Shi, Z., Wang, W., Lv, B., Qi, B., Yin, Q.: A novel multi-source fusion model for known and unknown attack scenarios. In: Proceedings of 2018 17th IEEE International Conference on Trust, Security and Privacy in Computing and Communications/12th IEEE International Conference on Big Data Science and Engineering (TrustCom/BigDataSE), pp. 727–736 (2018). https://doi.org/10.1109/trustcom/bigdatase.2018.00106
22. Wang, C.-H., Chiou, Y.-C.: Alert correlation system with automatic extraction of attack strategies by using dynamic feature weights. Int. J. Comput. Commun. Eng. **5**, 1–10 (2015). https://doi.org/10.17706/ijcce.2016.5.1.1-10
23. Liang, L.: Abnormal detection of electric security data based on scenario modeling. Procedia Comput. Sci. **139**, 578–582 (2018). https://doi.org/10.1016/j.procs.2018.10.207
24. Barzegar, M., Shajari, M.: Attack scenario reconstruction using intrusion semantics. Expert Syst. Appl. **108**, 119–133 (2018). https://doi.org/10.1016/j.eswa.2018.04.030
25. Kavousi, F., Akbari, B.: Automatic learning of attack behavior patterns using Bayesian networks. In: 2012 6th International Symposium on Telecommunications (IST), pp. 999–1004 (2012). https://doi.org/10.1109/istel.2012.6483132
26. Bahareth, F.A., Bamasak, O.O.: Constructing attack scenario using sequential pattern mining with correlated candidate sequences. **II** (2013)
27. Pierazzi, F., Casolari, S., Colajanni, M., Marchetti, M.: Exploratory security analytics for anomaly detection. Comput. Secur. **56**, 28–49 (2016). https://doi.org/10.1016/j.cose.2015.10.003
28. Lu, X., Han, J., Ren, Q., Dai, H., Li, J., Ou, J.: Network threat detection based on correlation analysis of multi-platform multi-source alert data. Multimed. Tools Appl. (2018). https://doi.org/10.1007/s11042-018-6689-7
29. Suarez-Tangil, G., Palomar, E., Ribagorda, A., Sanz, I.: Providing SIEM systems with self-adaptation. Inf. Fusion **21**, 145–158 (2015). https://doi.org/10.1016/j.inffus.2013.04.009
30. SICS Geek Lounge: SCADA/ICS PCAP Files From 4SICS. https://www.netresec.com/?page=PCAP4SICS (2019)
31. Choi, S., Yun, J.-H., Kim, S.-K.: A Comparison of ICS Datasets for Security Research Based on Attack Paths, Sept 2019
32. Lin, C.-Y., Nadjm-Tehrani, S., Asplund, M.: Timing-Based Anomaly Detection in SCADA Networks, Oct 2018
33. Hansch, G., Schneider, P., Plaga, S.: Packet-wise compression and forwarding of industrial network captures. In: 2017 9th IEEE International Conference on Intelligent Data Acquisition and Advanced Computing Systems: Technology and Applications (IDAACS), pp. 66–70. IEEE (2017)
34. Muller, S., Lancrenon, J., Harpes, C., Le Traon, Y., Gombault, S., Bonnin, J.-M.: A training-resistant anomaly detection system. Comput. Secur. **76**, 1–11 (2018). https://doi.org/10.1016/J.COSE.2018.02.015
35. Hansson, L.: Scada SNORT Rules. https://networkforensic.dk/SNORT/ (2019)

Using Honeypots for ICS Threats Evaluation

Nitul Dutta, Nilesh Jadav, Nirali Dutiya and Dhara Joshi

Abstract Industrial Control System (ICS) is an integration of hardware and software with a sophisticated network connection that supports instrumentation in industry. These systems are weak and prone to be exploited easily by an attacker due to its simple architecture which uses low processing power and memory. In recent years, the cyber-attack on ICS goes very vigorously and lures high amount of damage in terms of cost and time, it is difficult to prevent ICS from different malicious activity as the components of ICS will not be able to take any updates or patches due to its simple architecture. Certainly, we can prevent those attacks by detecting any defamatory activity by the intruder using some defense techniques such as Intrusion Detection System (IDS). Honeypots are useful in such scenarios, they are the subtle traps that are configured to detect any unauthorized access to a legitimate system, with an intention to know and learn the behavior of a hacker or its activity to mitigate the risk of any loss. Traditionally, we have network bases defense detection techniques such as IDS, Intrusion Prevention System (IPS), firewall and some encryption techniques, however, these systems are not that intelligent as honeypots are Honeypot poses the power of capturing the data, aptness to log, create an alert and detect everything the intruder is doing in the system. Researchers are finding new ways to trap those attackers using honeypots in order to secure ICS, not only defended ICS but, also it disturbs the attacker, using their "Camouflage Net", which is a reconfigurable honeypot. There is a need for a preventive measure which provides early detection and alert mechanism for ICS, provides a multi-stage attack detection using honeypot which generates signatures to unveil any invader in the ICS. Uses the improved configurable honeypot based on SNAP7 and IMUNES, these honeypots

N. Dutta (✉) · N. Jadav · N. Dutiya · D. Joshi
Department of Computer Engineering, Faculty of PG Studies, MEF Group of Institutions (MEFGI), Rajkot 360003, India
e-mail: nituldutta@gmail.com

N. Jadav
e-mail: nileshjadav991@gmail.com

N. Dutiya
e-mail: nirali.dutiya@marwadieducation.edu.in

D. Joshi
e-mail: dhara.joshi@marwadieducation.edu.in

© Springer Nature Switzerland AG 2020   175
E. Pricop et al. (eds.), *Recent Developments on Industrial Control Systems Resilience*,
Studies in Systems, Decision and Control 255,
https://doi.org/10.1007/978-3-030-31328-9_9

are configured and deployed rapidly in the ICS system. Supervisory Control and Data Acquisition systems (SCADA) is another type of ICS system, SCADA honeypots such as conpots not only detect the outside attack but it also detects any malign tampering within its network. With this intention to secure ICS, this chapter focuses on threat detection using reliable and confined honeypots to evaluate and analyze the dilemma of ICS security. A comparison among different preventive measures of low interaction and high interaction honeypots and certain tools and methodologies which helps in intercepting any tampering activity will be the foremost focus of this chapter.

Keywords ICS · Honeypots · SCADA · Conpot · PLC

1 Honeypot Introduction

Honeypot is a security mechanism which is set to detect attempts at unauthorized utilization of information, with the goal of learning from the previous attacks utilize the information for future improvement of security. Honeypots are used to identify attacker's attack and defend against future attacks. If honeypot is effectively applicable, the attacker will have no clue that he is being monitored. Most honeypots are introduced inside firewalls so they can more readily be controlled, however it is conceivable to introduce them outside of firewalls. It works in totally opposite manner as compared to normal firewall; rather than confining what comes into system from the Internet, the honeypot firewall enables all traffic to roll in from the Internet and limits what the framework sends pull out. Deploying a physical honeypot is frequently time serious and costly as various OS require specific hardware and each honeypot requires its own physical system. Honeypot is actually a system on your network that goes about as a temptation and draws attacker like bears get attracted to Honey. It contains false information instead of live information. A well configured honeypot ought to have a significant number of similar features. It may consist graphical user interface, warning message, login display etc. An attacker shouldn't almost certainly distinguish that he is on a honeypot and that his activities are being checked. Multiple honeypots are set together and create Honeynet. Honeypot can secure network through deception. It can detect earlier and protect our system from different kind of attack. It is also useful to better understand the motive or goal of attacker, it can be helpful for system administrator to get some additional valuable information. So, Honeypots can distract attacker from important systems on network, alert about new attack and allow thorough examination of attacker. Honeypot don't fix anything, they just provides valuable information.

1.1 Types of Honeypots

Honeypots are wide stream, can be characterized dependent on their deployment and dependent on their dimension of association.

(A) Based on Dimension of association (between System and Intruder), Honeypot can be categorized as:

The extent to which an attacker can interact with honeypot and underlying operating system. We can say that, the more the attacker can interact with your honeypot, the more information we can collect from that incident.

(1) Low Interaction Honeypot

As name suggest, these kind of honeypots have limited extend of interaction with external system. It simulate the activities often requested by attacker. Low interaction honeypots require less number of resources, so it can be possible that many Virtual Machines (VMs) can be hosted on one system. It can be benefitted like getting response from the same system will result into short response time, reduce the level of complexity and code is also less. It is very easy to install. It only provides some emulated (fake) services. Low interaction Honeypot minimize the risk as attacker cannot operate on real operating system. It is used to analyze spammers (Fig. 1).

(2) Medium Interaction Honeypot

Medium Interaction Honeypot may fully implement the HTTP protocol to emulate a well-known vendor's implementation, such as Apache. As compared to Low interaction Honeypot, it provides more interaction. In this Honeypot, services are still emulated, Scripts provides more information. To develop this kind of honeypot requires high knowledge as it is more complex and time consuming. It is also called 'Mixed Interactive Honeypot'. It provides attacker with better illusion of the operating system so, complex attacker can be identified and analyzed.

(3) High Interaction Honeypot

High Interaction Honeypot is more advanced honeypot. This kind of honeypot [1] have higher level of interaction with systems. It provides more realistic experience to attacker and collect higher amount of data about planned attacks. It also includes higher risk of capturing entire honeypot. They are very complex and time consuming

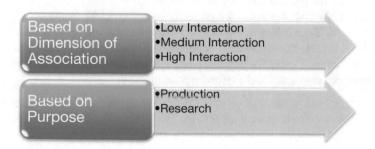

Fig. 1 Honeypot types

to design and maintain. This honeypots are best in case of "0-Day attacks", they are complex as they involves real OS. In this honeypot, everything is real, nothing is emulated. It gives precise information related to how attack is going to be perform or how malware executes in real time.

(B) Based on Purpose, Honeypot can be categorized as:

(1) **Production Honeypot**

Honeypot can be set up to imitate your own production environment. It can allow you to identify vulnerability in your environment. The main aim of production honeypot is to protect the network. They are very easy to build and deploy as they need less functionality. Honeypot can detect attacks and giving alert message to system administrator. Production honeypot can capture limited amount of information and basically it is used within the organizations or companies. Production honeypots are set up inside the production network with other production servers to enhance overall security. As compared to Research Honeypots, Production honeypots can provide less information about attack or attackers. Working of Production Honeypot is: It mirror the production network of company, thus invites attackers and expose them to organization vulnerabilities.

(2) **Research Honeypot**

Research Honeypot is more focused on researching the motive of attacker. They collect information regarding intention and strategies of attacker targeting different systems. Its primary objective is to pick up information about the manner by which the attackers intend to perform attacks. They are utilize to inquire about the threats organization face and to figure out how to more likely ensure against those threats. They are complex to deploy and maintain and are utilized by Research, Military or government association. It provides platform to learn about cyber-attacks. Research Honeypot do study of attacker, Pattern of attacks and motive of attacker. Role of Honeypots in Security:

- Prevention: Honeypot add little valuable information, it also introduce the risk (i.e. it can detect and prevent the planned attack).
- Deterrence Method: Advertise the presence of honeypot to attackers.
- Deception Method: It wastes the time of attacker and occupies that attacker.
- Detection: It collects information and displays in some analytical format which provides summary form, for purpose like statistical analysis (data aggregation) so that honeypot can detect and perform accordingly.

However, if your system is vulnerable, no Honeypot can prevent your system against attacks carried out by attacker (Fig. 2).

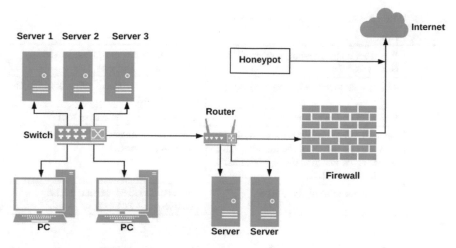

Fig. 2 Deployment of Honeypot

1.2 Difference Between IDS and Honeypot

IDS—Intrusion Detection System: Any device or application that actively monitors your network for malicious activities and alert you when it detect an attack. The estimation of a honeypot is determined by the information that can be acquired from it. Observing the information that enters and leaves a honeypot gives us a chance to assemble data that isn't accessible to IDS [2]. To recognize malicious situation, IDS [3] need signature of well-known attacks and mostly neglect to identify compromises that were unknown at the time it was sent. On the other side, honeypots can recognize vulnerabilities that are not yet previously known. IDS depends on signature marking to identify attacks. This implies unknown or unique attack cannot be identified. Interestingly, honeypots are intended to catch all known and unknown attacks. Types of IDS: Network Intrusion Detection System (NIDS), Host based Intrusion Detection System (HIDS), Perimeter Intrusion Detection System (PIDS), Virtual Machine based Intrusion Detection System (VMIDS), on the other hand Types of Honeypot: Low Interaction, High Interaction, Medium interaction, Research honeypot, Production Honeypot (Table 1).

1.3 Advantages and Disadvantages of Honeypot

- Advantages of Honeypot:

  a. Low assets Requirement: Honeypot requires low assets, it only captures unusual activity.

Table 1 Difference between IDS and Honeypot

Sr. No.	Honeypot	IDS
1	If you set up a honeypot, you want it to be attacked	If you set up IDS, you want to know when you are attacked
2	All the data received is unauthorized	Can't cope up with the network traffic on large network
3	It can be used at offensive tool	It can be used as defensive tool
4	Honeypots are used to attract intruders to visit the system. Once the intruders have accessed the machine (honeypot), their details can be stored	In IDS the intruders are also detected but they can be blocked
5	There is No false positive alarm	Untuned IDS alert too many false positive

b. New Tools and Strategies: Honeypot intended to capture anything even though it is never previously seen by honeypot.

c. Small Information index: Honeypot gathers small amount of information. Honeypot can utilize 1 MB in logged book instead of using 1000 MB. It provides small amount of data but it is valuable data.

d. Encryption: Honeypot work in IPv6 environment. Whatever attacker throws at Honeypot, it will identify and capture it.

e. Simplicity: Honeypot is very simple. It doesn't involve complex algorithms creation, state table maintenance or signature updating.

f. Occupies Hacker: Honeypot occupies attacker, it wastes time of attacker. Instead of real system, attacker tries to attack on emulated one.

Thus, Honeypot has different kind of benefits like: It observer attacker and learn about their behaviour, creates profile of attacker who are trying to get access of our system, it wastes attacker's time and resources.

- Disadvantages of Honeypot:

a. Attack Identification: Honeypot can only identify the activities which is directly interacts with it. If attacker tries on different system and honeypot is untouched attack cannot be detected.

b. Narrow Vision: Honeypot alarm only when it is attacked.

If Honeypot is deployed outside the firewall, it may not be able to trap internal attacks. If Honeypot is deployed in internal network alongside with server and workstation, it may be possible attack can able to reach at other internal system from the Honeypot. Thus, Honeypots never used as a replacement, but play important role in providing security. Industrial Control systems are a set of systems that govern/controls automation of the industrial process. Special software's are integrated along with hardware for the process control and automation. They are developed to reduce human effort and improve the overall efficiency of the industrial process. Control systems can be integrated with any industry such as power grids, pharmacy industry, water and sewage management, food processing etc. The industrial control system can be implemented using any of these, programmable logic controllers (PLCs), Distributed control system (DCS), Supervisory Control and Data Acquisition (SCADA) [4], programmable automation controllers (PACs), remote terminal units (RTUs), etc. Using various such technologies industrial control system can be scaled from small applications to very large-scale distributed applications.

2 Programmable Logic Controllers (PLCs)

A programmable logic controller is a controller used for automation of a specific industrial process, some machine function or sometimes for the whole production line. PLC controllers are adaptable to any application and are flexible and robust. Sensor or input devices connected with PLCs provide input data, based on the input received and pre-defined/programmed parameters output is generated [5]. PLCs can be used for starting or stopping any process, for example, it can be used to start and stop motors of conveyor belt. Other applications include recording and monitoring various parameters such as temperature, machine productivity, pressure etc. It can also be used for triggering various alerts and alarms during emergency situations or malfunctioning of machines.

2.1 Working of PLCs

The programmable logic controller is very much similar to computers [6]. PLCs are mainly used for controlling tasks and industrial environment, while former is used for calculation and displaying results. PLCs uses a programmable memory for storing instructions and functions that can implement logic and perform mathematical calculation including counting, sequencing, etc. PLCs are microprocessor-based controller whose primary use is to control machines and processes' can be pre-programmed using a very simple ladder and functional block programs for implementing basic logic such as AND, OR, XOR, NAND and NOR, and latching. Various switching operations such if A occurs or B occurs switch to C, id A occurs and B occurs switch to D, are programmed using such basic logic. To modify functionality in PLC one has to only change set of instructions installed in it. With is, ability usage of PLCs is very cost-effective and flexible (Fig. 3).

2.2 Architecture

PLC contains a processor unit, memory, input/output interface, power supply, the communication interface, and a programming device. CPU contains a microprocessor

Fig. 3 Programmable logic
controller

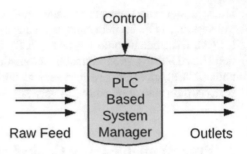

which is responsible for interpreting the signals and executing various operations as
stored in memory. The communication interface is used for transfer of data between
remote PLCs. It deals with verification of data, synchronization between various
applications. Memory unit contains program and data used for execution. Various
memory elements such as Read Only Memory (ROM), Random Access Memory
(RAM) Erasable and Programmable (EPROM) are present in PLC. Where ROM
is used for store data permanently, While RAM will store the data temporarily.
The data which get stored in RAM comprises the status of various devices, counter
and timer values. EPROM is an extra module which contains memory that can be
erased and reused. Input interface includes input devices from where the processor
will receive input data for processing. Input devices can be temperature sensors,
flow sensors, switches, etc. Output devices include motor coils, valves, etc. Using
programming device program is developed as well as entered into the memory of
PLC. Program devices can be of several types such as hand-held, desktop consoles,
personal computers. Hand-held devices are portable, and has enough memory for
retaining the programs. Desktop consoles are the devices having a display unit and
keyboard connected with it. The personal computer is greatly used as it has a hard
disk to store program and also has optical drives for making copies of those programs.
Although as per the requirement of application some special communication cards
needs to be integrated along with software running on personal computers. Power
Supply component is responsible for converting a.c voltage to d.c voltage (5 V)
(Fig. 4).

3 Supervisory Control and Data Acquisition (SCADA)

The Supervisory Control and Data Acquisition is a system containing software and
various hardware components. SCADA is used for control and monitoring various
industrial process locally as well as at remote locations. It also collects the real-time
data and records the readings in the log file. It collects data from various devices such
as valves, sensors, pumps, motors, etc. A software component installed in SCADA
act as a bridge between SCADA and its connected devices. SCADA systems can
be used in any industrial domain such as water and waste management, recycling,
transportation, manufacturing, power management, etc. (Fig. 5).

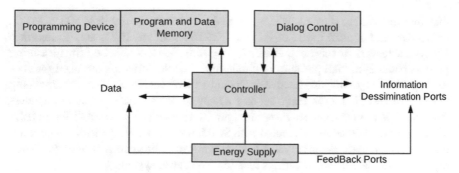

Fig. 4 Architecture of PLC

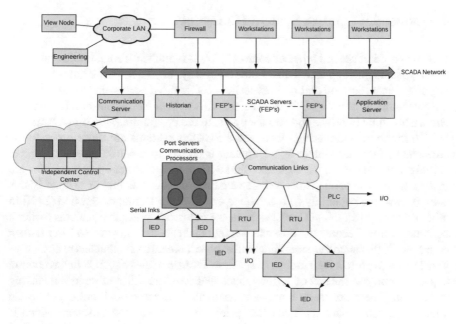

Fig. 5 Architecture of SCADA

3.1 Architecture

SCADA system contains various connected devices such as Remote Terminal Devices (RTUs), Programmable Logic Controller (PLCs), Motor Control Centres (MCCs), Master Terminal Units (MTUs), Intelligent Electronic Devices (IEDs), Front End Processors (FEPs), etc. SCADA system contains Human Machine Interface (HMI) that performs analysis of data and as well as storage of data [7]. RTUs are used for one or more IEDs which in turn unites data collection and control traffic. It acts as a gateway between MTU or SCADA and IED.IEDs are small devices that are connected to sensors. MTUs are used for connecting multiple RTUs. MTUs unites data and control traffic of RTU to SCADA systems. PLCs acts as interface between actuators or sensors and SCADA system. Sensors are used for measuring process parameters like pressure or temperature, while actuators are used for controlling actions like opening a valve or closing a valve. MCCs are used for governing working of electric motor. FEPs act as gateway for sensor network and computers that runs SCADA systems. As shown in figure, various RTUs connect different IEDs at one end and at other are connected with SCADA system having historian, communication server, application server, etc. Communication links consist of tools needed to transfer the data between various locations to central system.

4 Industrial Control System Honeypots

ICS composed of several types of control systems regulated from a micro controller to a large scale interconnected distributed control system. Large scale industries such as power generation, oil and fuel refineries, chemical industries, electrical grid, and telecommunication uses SCADA. The SCADA system reads the inputs from the Human Machine interface and sets the points to the Programmable Logic Controller (PLCs) in the manufacturing plant. These SCADA systems are more prone to the attackers as they are simple to manipulate and easy to create a chaos. A typical paradigm to understand the attack would be a nation cyber war—two countries are in disputes from several years on some valuable asset such as fuel or water crisis. A country have the capability to manipulate and create a Denial of Service (DoS) to other countries' [8]. SCADA system then the industrial system which was earlier a commodity now become the possible weapon that can start a cyber-war in between countries. To find such vulnerabilities and expose it is a rare skill that has to be involve in industries to protect their industrial systems. From last few years ICS honeypots are capable to mitigate the risk of various attack that are more susceptible in vandalizing the system. There are many honeypots available that can reveal and cover up the attackers ideology such as it can foot print the entire SCADA software which is running in the control room building. It can discover weak access points and control points, also we can identify the vulnerabilities, exploit and various misconfiguration systems.

SCADA Honeynet—With growing security awareness about various vulnerabilities and motives of the attacker, we are still lacking behind to protect the industrial systems with newly released attacks and Zero day vulnerabilities in ICS. Although there are many research ongoing in this specific area, but there are no particular tool or framework available that can certainly entertain different attacks of an attacker, however there are many IPS and IDS available that can analyses the attack but not effectively as honeypot. The SCADA project provides several countermeasures from an attack such as—Gather data of an attack, it allow scriptable environment where we can write our own scripts to test a real live protocol, other features such as— device hardening, obfuscation, network segregation [9]. SCADA Digital Bond, and

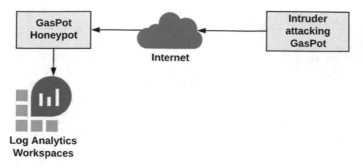

Fig. 6 Gaspot

ICS based Research Company, had created the Honeynet that consist of two virtual machines available publically. One machine consist of PLC honeypot and the other one is a honeywall that analyses the inbound and outbound network traffic. Honeywall has a Snort IDS and other specific services related to the PLC that includes Telnet, HTTP, SNMP, Modbus TCP and FTP.

GridPot—It consist of conpot honeypot and power distribution system simulator GridLABD, both can be embedded and come along with conpot honeypot. It is an open source electric grid simulator that provides data acquisition with simulation environment. Moreover it imitate SCADA/HMI and ICS protocols that can detect anonymous and real time attacks.

GasPot—it greatly deals with Guardian AST gas-Tank monitoring systems which are common to all oil and gas industry which suffered with many attacks and affects inventory control, data collection and unavailability of gas in various stations (Fig. 6).

This honeypot detects those threats by creating virtualized environment of Guardian AST tank-monitoring system that function exactly the original system, attackers easily gets lured with the virtual system and gets trapped in the Pot. The pot can capture the data, logged it in a file, can alert and detect the activity by the intruder that can expose the intruder (Table 2).

4.1 Conpot Honeypot

As the computing world is getting more advance day by day, hackers are also motivated in indulging to attack the system in more advance way, there are no isolated system which are completely protected, there always going to be one single loophole in the system which can be the entry point of the attacker. To understand the topography of such attacks it is necessary to conduct some data collection and research and therefore we plot effective honeypots. Conpot is an open source ICS server side low interaction honeypot, simple to deploy, extend and modify. The project was established under The Honeynet Project released in May 2013. It can simulates a Siemens SIMATICS S7-200 PLC, Modbus TCP, SNMP and HTTP protocols same

Table 2 Types of ICS Honeypots

Honeypots	Descriptions
SCADA Honeynet	It is a Programmable Logic Controller that supports management points and interfaces consist of default parameters, which were decided at the time of installation
Conpot	It is an ICS honeypot that is capable to collect intelligent information related to the attacker's motives and methods
GridPot	It is an open source electric grid honeypot
GasPot	In Oil, Fuel refineries and gas stations have tank gauges that are common to every gas stations. This honeypot is designed to simulate Veeder Root Guardian AST which distinguish that no two instances look exactly the same
Nozomi TriStation Basic Honeypot	It is a tristation dissector that includes pcap and honeypot simulator
HoneyPoint Security Server	It is a commercial distributed honeypot that consist of IT and SCADA emulators

as SACADA Honeynet. Many research is ongoing with conpot and its templates, Arthur F Jicha has done some incredible work with conpot, and they host multiple conpot honeypots using Amazon Web Services and ran for almost a month to gather as much data from the honeypot for analysis purpose. They can access those hosted honeypots using SSH (Secure Shell), they then perform various scans such as nmap and SHODAN against the honeypots. It has templates that can simulates SIEMENS S7-200, guardian AST and Kamstrup 382 PLC. These templates are configurable and contrive as per our requirements and also it provides a way to create our own templates simulates other PLC's. As SIEMENS S7-200 PLC is more vulnerable to be exploited by intruder, researchers are more focused in that direction. Conpot records all network activity and stores it in log file, which can be used for further analyses to understand the attack. The main drawback of conpot is it never notifies the user that a honeypot is been attacked by the attacker, however the problem can be resolved using Open Source Host-based Intrusion Detection System (OSSEC-HIDS) (Fig. 7).

Siemens S7-200 PLC
It is the cheapest PLC from Siemens and easy to be programmed it, S7-200 is the simplest and smallest PLC model in Siemens PLC series. In addition to that it is easy to connect, customize, programmed, compiled and debug your program. We can use Siemens STEP 7 Micro/WIN to program the PLC system. There are many modules we can get with Siemens S7-200 such as CPU module, input-output module and sensor specific module. CPU module has PLC program [5], stored and execute it. Siemens S7-200 supports CPU module such as CPU- 221, 222, 224, 224XP, 224XPSi, and 226.

Guardian AST
It is a monitoring system that is designed for compliance and inventory control for storage tank. The system can handle to monitor one or two tank and up to 6 sensors, has 3 output relay and RS-232 interface and also console interface for monitoring

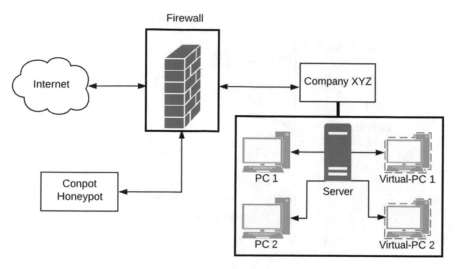

Fig. 7 Conpot Honeypot

input and output ability. Guardian AST console panel consist of—24 character LCD, 13-key keyboard used for programming, operating and reporting, internal warning and external alarm indicator, Report printer to print specific information such as— inventory, alarm, sensor or any warning sign.

Kamstrup 382 PLC
This PLC is an electric meter, to observe the precise measurement of energy utilization, irrespective of mounting direction. It also consider the shunt measuring convention, that ensures the meter and its metering value doesn't influenced by magnetism, also it is configured to quantify imported and exported energy, it comes with three independent and galvanically separated measuring system that takes definite mensuration, whether it measures 1, 2 or 3 system. Moreover it comes with inbuilt data logger which sheltered the data for a long time.

4.2 Conpot Protocols

Conpot comes up with 7 default template and each has its own protocol to operate, by analyzing each protocol we can have a large analyses of the attack that is performed on the ICS [10].

(1) HTTP

Hypertext Transfer protocol (HTTP) is an application layer protocol, through which a client can send HTTP request to a HTTP server and server responds to a client as HTTP response. Server uses default port 80 or 8080 for incoming connections.

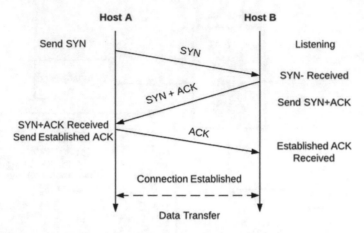

Fig. 8 TCP 3way handshake

Additionally, client has to first establish a connection with server using TCP via 3-way handshake before making an actual connection (Fig. 8).

HTTP request message consist of GET, POST, HEAD, PUT and DELETE method, GET method extracts the response of a request from various resources, POST method forward or sends the extracted output in terms of webpage, HEAD method appeal for HTTP headers that have necessary information regarding that package, PUT method upload the pages and DELETE method simply used to get rid of that particular page. HTTP response status code indicated a successful operation or not, such as 200 range indicates success, 300 range indicates redirection, 400 specifies error or resource not found.

(2) MODBUS over TCP/IP

It is also an application layer open source protocol, developed by Modicon in 1979 which uses serial communication. Furthermore it is used for transmitting information between electronic device using serial lines. Modbus consist of master and slave entity, where a device requesting a resource known as Modbus Master and the device that provides that resource is known as Modbus Slave. Modbus with Conpot can communicate over TCP/IP using port 502, it handles four types of messages—request, response, indication and configuration between connected devices. Modbus request and response are encapsulated with Modbus TCP/IP header known as Modbus Application Data Unit (Fig. 9).

(3) S7Comm

S7Communication protocol are used for Siemens programmable logic controllers, which again uses the same model as Modbus of master-slave or client-server. It communicate through port 102. S7Comm used for PLC programming, exchange data and resources among other PLCs, retrieve PLC data from SCADA environment. S7 Packet Data Unit (PDU) header consist of 10 bytes that is composed of Protocol ID,

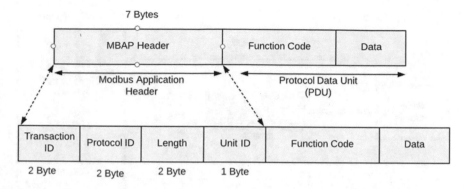

Fig. 9 Modbus application data unit

Message Type, Reserved, Parameter length, Error Class and Error Code shown in Fig. 10. Parameter defines the structure of the message and the type of PDU. Data field consist of data that it carries, it could be code, memory block, firmware or any other value. S7Comm uses three components to communicate—client, server and partner. Client can request for the resource using S7 request and based on that server will respond using S7 answer. Server doesn't need any configuration as it automatically handles by the firmware of the communication processor. Partner could by any peer who sits in between client and server and it can communicate both sides.

(4) **SNMP**

Simple Network Management Protocol (SNMP) is a network management protocol used to collect and organize information about devices on IP network. OSI layer has many protocols stayed on every layer and every layer has network devices such as—hub, bridges, routers, switches and many others, to mitigate any issues and constraints among these devices and the network, it has to be monitored and analyzed by network administrator. SNMP protocol encompassed in application layer of TCP/IP as stipulate by the Internet Engineering Task Force (IETF). It consist of SNMP manager, SNMP agent, Management Information Base (MIB) that involve information of resources that has to be managed (Fig. 11).

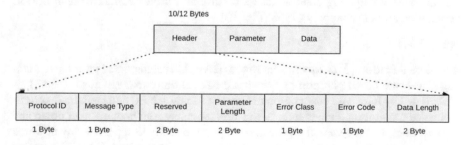

Fig. 10 S7Comm packet data unit

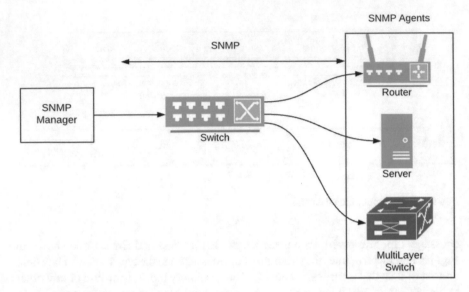

Fig. 11 SNMP architecture

SNMP manager is accountable to communicate with the SNMP agent consist of various network devices, functions included such as—it inquires the agent, get respond from agent, it can set the variables and parameters in agents, and it monitors the asynchronous events. SNMP agent is a software program within the network that uses to collect the management information from devices and passes it to the SNMP manager whenever it is requested by the manager. Every SNMP agents retain one database that describes device variables, manager uses this database to queries the agents for particular information from the devices. The database that is shared between an agent and a manager is known as MIB.

(5) **BACNet**

Building Automation and Networking protocol (BACNet) is again a communication protocol by ASHRAE used to connect different devices for different services. It is mainly used for Heating, Ventilating, and air conditioning (HVAC) in ICS, moreover it also used for lighting control, access control and fire detection. Default port to communication in conpot is 47808 (Fig. 12).

(6) **IPMI**

Intelligent Platform Management Interface (IPMI) is a standardized protocol to manage hardware based platform by which we can centrally manage our servers. It is developed by Intel with cooperation of Hewlett Packard, NEC and Dell. Its functions includes monitoring temperature, power utilization, system boot up, shut down and log event of the hardware. It can be used to provide a way to supervise any powered off computer which is not responsive due to network connections. In addition to that it also uses when we need to install custom operating system remotely, without any

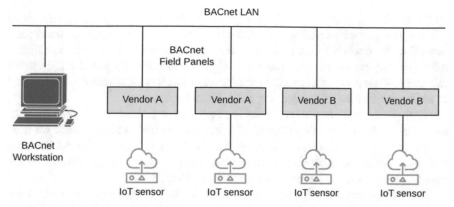

Fig. 12 BACnet architecture

physical entity that uses CD, USB or any other installer to install the OS. Using IPMI, we can mount and ISO image, simulate the installer and can perform the installation remotely.

(7) **EtherNet/IP**

It is thrived by Control Net International and Open Device Net Vendor Association (ODVA). It comprises explicit and implicit messages—wherein explicit message will communicate between client and server using port 44818 and implicit message will communicate between devices using port 2222.

4.3 Research Work

This section specifies a miniscule work with conpot and then the results will be evaluated and analysed to understand the attacker's intention. Possibly there are many solution in terms of IDS and IPS and other honeypots that can mitigate the risk of such attacks in ICS [11], however with conpot we can easily integrate it over the system and can examine it in a great detail. ICS's system has common and many open ports that can be manipulated to a big intervention shown in Table 3. Ports are an essential entity of the communication network through which client can access the service that is offered by the server such as to access the webpage using 8080, or

Table 3 Conpot templates and its respective open ports

Honeypot templates	Respective open ports
IPMI	623
Kampstrup smart meter	1025, 50100
Guardian AST	10001
Siemens S700	80, 102, 161, 502, 623, 47808

FTP using 21. Ports are of two types—Open and Closed—Open ports are vulnerable as it can accept packets without any obstacle, whereas the closed port can accept and can reject the packets based on some rule assigned in the firewall. The key idea is not to just deploy conpot on Ubuntu machine (act as our ICS), but also to hide the existence of the honeypot from the intruder and to analyse it we are comparing port scanning with various fingerprinting tools, IP and port scanner.

Open ports are the major concern of any operable organization and when it comes to ICS it leads to major devastations. To prevent and analysed the attack we used conpot honeypot, which have certain open ports shown in Table 3. Conpot is placed inside an ABC organization's system and three other systems are used to analyse the installed honeypot. The architecture is shown in Fig. 13.

The above architecture has an internal network consist of three machine—Ubuntu, Kali Linux and Windows 10, where conpot is installed and running on Ubuntu machine. Conpot has many templates discussed in above section, from that we used default template—*conpot –f –template default*. Once it's up and running, it will log all minute details that you or an attacker performing on this specific IP address. To emphasize on open ports, nmap—a port scanner is used from Kali machine and on regular basis we scanned Ubuntu machine that have an IP address of 192.168.202.135. As the scan get started conpot will start notify the scan and gets logged in the log file. We had almost used 10 nmap scripts shown in Table 3 to analyse the conpot machine, and founded our honeypot open ports. This could be an advantage to the attacker, as scan shows various open ports associated with its vulnerable services, and now it's easy for an attacker to find the exploit and access the system. Meanwhile, Zenmap—another port scanner with GUI functionality—scanned the same

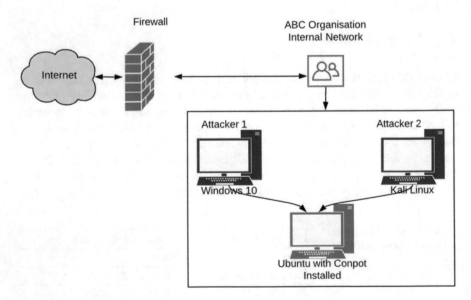

Fig. 13 Conpot installation

IP address and founded the same open ports. Zenmap gets the same scan results, as Nmap scan shown us, furthermore *nmap -p -A -v -Pn 192.168.202.139* scripts shown us maximum conpot ports, that can be exploited. To enhance the work, we also used Nessus Vulnerability Scanner, SolarWinds Port Scanner, and Advanced Port Scanner to scan the same ports, however, we couldn't get the higher efficiency in finding the same open ports as nmap does for us (Tables 4 and 5).

These results show that not all ports are accessible with these automated port scanner software, however using nmap it fetches all open ports. The services related to the open ports can be exploited such as ccproxy-ftp has an exploit—*ccp_telnet_ping*, EtherNetIP has an exploit—*multi_cip_command*. Metasploit is a framework where we can test these services, based on their exploits can find out various vulnerabilities associated with that system. Cumulatively, if someone finds out our honeypot open ports than they certainly can enter into our legitimate system, however with a risk of exposing themselves, as conpot logged everything that an attacker is doing. As conpot starts, we open a webpage with a port 8800 that is of sunwebadmin in another

Table 4 Nmap Scripts to Scan Conpot Open Ports

Nmap scripts	Open ports founded	Services
nmap 192.168.202.139	21, 80, 2121, 8800	ftp, http, ccproxy-ftp, sunwebadmin
nmap -T4 -A -v 192.168.202.139	21, 80, 2121, 8800	ftp, http, ccproxy-ftp, sunwebadmin
nmap -sS -sU -T4 -A -v 192.168.202.139	21, 80, 2121, 8800, 161	ftp, http, ccproxy-ftp, sunwebadmin, snmp
nmap -p 1-65535 -T4 -A -v 192.168.202.139	21, 80, 2121, 8800, 5020, 10201, 44818	ftp, http, ccproxy-ftp, sunwebadmin, zenginkyo, rsms, EtherNetIP-2
nmap -T4 -A -v -Pn 192.168.202.139	21, 80, 2121, 8800	ftp, http, ccproxy-ftp, sunwebadmin
nmap -sn 192.168.202.139	No Ports Founded	No Services Founded
nmap -T4 -F 192.168.202.139	21, 80, 2121	ftp, http, ccproxy-ftp
nmap -sV -T4 -O -F –version-light 192.168.202.139	21, 80, 2121	ftp, http, ccproxy-ftp
nmap -p -A -v -Pn 192.168.202.139	21, 80, 2121, 8800, 5020, 10201, 44818	ftp, http, ccproxy-ftp, sunwebadmin, zenginkyo, tcpwrapped, EtherNetIP-2

Table 5 Port scanner with Conpot open ports

Port scanner	Open ports
Nessus Vulnerability Scanner	631, 5353, 6969, 47808
SolarWinds Port Scanner	44818
Advanced Port Scanner	21, 80

```
2019-03-27 21:31:12,325 Starting TFTP server at ('0.0.0.0', 6969)
^@2019-03-27 22:07:33,194 New http session from 192.168.202.1 (cb174955-2738-4d6
d-96ed-1bfa715e289c)
2019-03-27 22:07:33,796 HTTP/1.1 GET request from ('192.168.202.1', 14144): ('/'
, [('Host', '192.168.202.135:8800'), ('User-Agent', 'Mozilla/5.0 (Windows NT 6.3
; WOW64; rv:62.0) Gecko/20100101 Firefox/62.0'), ('Accept', 'text/html,applicati
on/xhtml+xml,application/xml;q=0.9,*/*;q=0.8'), ('Accept-Language', 'en-US,en;q=
0.5'), ('Accept-Encoding', 'gzip, deflate'), ('DNT', '1'), ('Connection', 'keep-
alive'), ('Upgrade-Insecure-Requests', '1')], None). cb174955-2738-4d6d-96ed-1bf
a715e289c
2019-03-27 22:07:33,799 HTTP/1.1 response to ('192.168.202.1', 14144): 302. cb17
4955-2738-4d6d-96ed-1bfa715e289c
2019-03-27 22:07:35,136 HTTP/1.1 GET request from ('192.168.202.1', 14144): ('/i
ndex.html', [('Host', '192.168.202.135:8800'), ('User-Agent', 'Mozilla/5.0 (Wind
```

Fig. 14 Conpot logger

machine, once the webpage opens, the IP address of the other machine gets logged in conpot terminal and we can see everything that an attacker is doing with that webpage. The same thing we found with FTP with port 2121, we started vsftpd, manipulated with several things such as wrong user credentials, importing and exporting different extensions of a file, simultaneously conpot logging the activity an intruder is performing on the other machine. Figure 14 shows us the logging of conpot when an attacker of IP address 192.168.202.1 had tried to open a webpage using port 8800, the activity is jotted inside the terminal as—New HTTP session from 192.168.202.1, also it specifies the browser fingerprinting and other necessary points that an invader is trying to do such as website fingerprinting or SQL injection. It may happen that SCADA system is hosted on a website, in that case this conpot logging could be useful in finding out the real culprit.

Host Based Intrusion Detection System (HIDS) is an intrusion system that observe a system to find and detect any intrusion or any misuse through logging and notifying the network administrator. Conpot and HIDS are very similar in nature however there is no functionality to notify the network administrator about any intrusion. Open-Source Host Based Intrusion Detection System (OSSEC-HIDS) [12] is an open source HIDS that will be useful in log-analysis, integrity inspection, rootkit detection, alerting and windows registry monitoring. We installed OSSEC on the same machine where conpot is installed therefore we got one additional feature of notification which can enlighten the administrator about the intruder. Above snippet shows us all the open ports that are linked up with conpot, same scan as of Nmap.

ossec: output: 'netstat -tan |grep LISTEN |egrep -v '(127.0.0.1| \\1)' | sort':

tcp 0 0 0.0.0.0:10201 0.0.0.0:* LISTEN
tcp 0 0 0.0.0.0:2121 0.0.0.0:* LISTEN
tcp 0 0 0.0.0.0:44818 0.0.0.0:* LISTEN
tcp 0 0 0.0.0.0:5020 0.0.0.0:* LISTEN
tcp 0 0 0.0.0.0:8800 0.0.0.0:* LISTEN
tcp 0 0 127.0.1.1:53 0.0.0.0:* LISTEN
tcp6 0 0 ::: 21 :::* LISTEN
tcp6 0 0 ::: 80 :::* LISTEN

Previous output:

ossec: output: 'netstat -tan |grep LISTEN |egrep -v '(127.0.0.1| \\1)' | sort':
tcp 0 0 127.0.1.1:53 0.0.0.0:* LISTEN
tcp6 0 0 ::: 21 :::* LISTEN
tcp6 0 0 ::: 80 :::* LISTEN

5 Conclusion

The intention behind this paper is to convolute various ways to defend our ICS system as from last few years the attacks are more severe and untamed towards our ICS. The attackers are more prone to find loopholes inside the system and once it founds, they can easily create exploits, furthermore penetrate the system and can do further devastation. ICS is not only handling small-medium sized systems it handles nuclear power stations and share markets, which can be comprised results into havoc which cannot be overruled easily. To deal with such progressive attack, we installed conpot over the ICS system, to analyze attackers attack, furthermore, we also channelize our work towards hiding our conpot from various vulnerability scanners and port scanner. In addition to that, we summarize conpot, protocols and its template to a greater extent. We also observed that port scanner scans the open ports, however, the ports are safe and currently there are no exploits which can be opted in compromising the system. Furthermore, we also compared the scan with OSSEC-HIDS with conpot and other port scanner to get the same open ports. The analyses could be helpful in protecting our ICS system and can leverage the attacker activity that can further profitable in research and development.

References

1. Sadasivam, G., Hota, C.: Scalable honeypot architecture for identifying malicious network activities. In: International Conference on Emerging Information Technology and Engineering Solutions, Pune, India, pp. 27–31, 20–21 Feb (2015)
2. Dongxia, L., Yongbo, Z.: An intrusion detection system based on honeypot technology. In: International Conference on Computer Science and Electronics Engineering, Hangzhou, China, pp. 451–454, 23–25 March (2012)

3. Mahajan, V., Peddoju, S.: Integration of network intrusion detection systems and honeypot networks for cloud security. In: International Conference on Computing, Communication and Automation (ICCCA), Greater Noida, India, pp. 829–834, 5–6 May (2017)
4. Smith Sidney, C.A.: Survey of research in supervisory control and data acquisition (SCADA). No. ARL - TR - 7093. In: Army Research Lab Aberdeen Proving Ground Md Computational and Information Sciences Directorate (2014)
5. Hackworth, J., Hackworth, F.: Programmable Logic Controllers: Programming Methods and Applications. Pearson, New Jersey (2004)
6. Bolton, W.: Programmable Logic Controllers. Elsevier, Amsterdam (2003)
7. Ahmed, M., Soo, W.: Supervisory control and data acquisition system (SCADA) based customized Remote Terminal Unit (RTU) for distribution automation system. In: IEEE 2nd International Power and Energy Conference, Johor Bahru, Malaysia, pp. 1655–1660, 1–3 Dec (2008)
8. Rosa, L., Cruz, T., Simoes, P., Monteiro, E., Lev, L.: Attacking SCADA systems: a practical perspective. In: Proceedings IFIP/IEEE International Symposium on Integrated Networks Manage, pp. 741–746 (2017)
9. Zhang, Y., Di, C., Han, Z., Li, Y., Li, S.: An adaptive honeypot deployment algorithm based on learning automata. In: IEEE Second International Conference on Data Science in Cyberspace (DSC), Shenzhen, China, pp. 521–527, 26–29 June (2017)
10. Patrick, D., Fardo, S.: Industrial Process Control Systems. Delmar Publication, Albany, N.Y (1997)
11. Vlad, A., Obermeier, S., Yu, D.: ICS threat analysis using a large-scale honeynet. In: 3rd International Symposium for ICS & SCADA Cyber Security Research 2015, Ingolstadt, Germany, pp. 20–30, 17–18 Sept (2015)
12. Kuman, S., Gros, S., Mikuc, M.: An experiment in using IMUNES and Conpot to emulate honeypot control networks. In: 2017 40th International Convention on Information and Communication Technology, Electronics and Microelectronics (MIPRO), Opatija, Croatia, pp. 1262–1268, 22–26 May (2017)

Intrusion Detection on ICS and SCADA Networks

Marian Gaiceanu, Marilena Stanculescu, Paul Cristian Andrei, Vasile Solcanu, Theodora Gaiceanu and Horia Andrei

Abstract Recent attacks on Industrial Control Systems (ICS) show the vulnerabilities of the existing ICSs. One emergency solution is to detect the anomalies and to defend the ICS/SCADA systems. Currently, on-line and off-line intrusion detection solutions are delivered in the specified technical literature. In this chapter, the authors provide Defence-In-Depth architecture with demilitarized zone based on the security standards. The use of the machine learning on intrusion detection into ICS and SCADA networks are emphasized and implemented in this chapter. At the same time, the existing security tools are envisaged and comparative analysis is provided. In order to extend the availability of the missing non-anomalies data the forecast of the energy consumption model is built, the obtained results are introduced in the chapter. The existing Intrusion detection algorithms are studied, some of them are implemented through the specific software and the obtained results are provided. At the end of the chapter different case studies of machine learning approach for Intrusion detection are introduced, the obtained numerical results being available in this chapter.

M. Gaiceanu (✉)
Department of Control Systems and Electrical Engineering, Dunarea de Jos University of Galati, Galati, Romania
e-mail: marian.gaiceanu@ugal.ro

M. Stanculescu · P. C. Andrei
Department of Electrical Engineering, University Politehnica Bucharest, Bucharest, Romania
e-mail: marilena.stanculescu@upb.ro

P. C. Andrei
e-mail: paul.andrei@upb.ro

V. Solcanu
Dunarea de Jos University of Galati, Galati, Romania
e-mail: vasilesolcanu@dedeman.ro

T. Gaiceanu
Gheorghe Asachi Technical University of Iasi, Iasi, Romania
e-mail: gaiceanu.theodora@ac.tuiasi.ro

H. Andrei
SM-IEEE, Bucharest, Romania
e-mail: hr_andrei@yahoo.com

© Springer Nature Switzerland AG 2020   197
E. Pricop et al. (eds.), *Recent Developments on Industrial Control Systems Resilience*,
Studies in Systems, Decision and Control 255,
https://doi.org/10.1007/978-3-030-31328-9_10

Keywords Intrusion detection · Industrial control systems · SCADA · Machine learning · Algorithms · Optimization

1 Introduction

The main ideas of this chapter arise from the vulnerability of data provided by industrial control systems (ICS) and Supervisory Control and Data Acquisition (SCADA) to intrusion and deliberate attacks. Both systems, ICS and SCADA, can be integrated in the complex set of cyber-physical systems (CPS). Starting from the technological current state of ICS, SCADA equipment and data transferred through them, the authors are going to highlight the specific security features, machine learning and intrusion detection algorithms. Robust defend methods against deliberate or involuntary perturbation can be developed, and therefore the ICS resiliency can be improved.

Cyber security (CS) is concerned with identifying the risks involved in using computer networks, both public and private, and offering solutions for removing them. Cyber security has in view the normal operation assurance through the proactive and reactive strategic measures. The triad of the *information security* consists of confidentiality, integrity, and availability. The CS ensures the information security in cyberspace, authenticity (unaltered data exchange) and non-repudiation of digital data used in public and private resources and services. Proactive and reactive measures can include security policies, concepts, standards and guides, risk management, training and awareness-raising activities, implementation of cyber-protection technical solutions, identity management, and consequences management. The domain of cyber security has begun to gain new dimensions with the increase of ICS' technological level, and while the expansion and diversification of threats/cyber-attacks. A greater importance on CPS appears in the last decade. The main reason is the destructive attacks of the modern processes, equipment into power plants [1–4].

The second Section of this Chapter includes the modern ICS architectures. The general architecture of the ICS contains three layers (enterprise management, supervisory, and field). The security features of the ICS (open system) are different from the conventional information systems because the system is closed and the security refers to safety. At the same time, the comparisons between the conventional and modern industrial control systems are highlighted. The real-time restriction, the limited computing resources in ICS, the industrial protocol vulnerabilities are some of the requirements on the ICS.

The guidelines for designing the CPS are emphasized in the third Section entitled "Standards". The cyber-security for ICS becomes a priority for normal operation assurance of the industrial processes. The architectural framework of SCADA systems is depicted in [5]. The guidelines for the operators could be found in the specific standards [6, 7]. The standard IEC 62443-2-1 shows the necessary elements to be developed into ICS. Other considered standards are based on the IEC 27000 and NIST (National Institute of Standard and Technology) SP 800-82 in order to secure the ICS with SCADA into distributed control systems. The SCADA guideline for normal operation is highlighted in IEEE (Institute of Electrical and Electronics Engineers, Inc.) Std C37.1-1994 [8], and the SCADA security guide could be found on API 1164 standard. Recommendation of the National Institute of Standards and Technology on Intrusion Detection and Prevention Systems can be found in [9].

Next Section is dedicated to Machine learning on ICS and SCADA networks. Artificial Intelligence (AI) aims at studying and designing intelligent agents, systems that perceive the environment and maximize the chances of their own success through behaviour. A common feature of AI is that the system is capable of learning, with or without external aids, with the goal of constantly improving. One application of the AI is the Network Intrusion Detection and Prevention. Using AI it can detect whether a software is malware or normal software. In order to develop an IA application that performs "Malware Detection", it is first necessary to establish some distinctive features. In addition to harmless and malware features, the system will be trained. Machine Learning is used in many applications. In most of these, the goal of detection is defined. Conversely, in cyber security, what we want to detect is not defined by default. At the same time, Cyber Security is asking for the latest information and getting it is one of the challenges [10].

The characteristics and applied algorithms for Intrusion Detection Systems are analysed and emphasized in the fifth Section of Chapter. One important tool in cyber-security is the Intrusion Detection Module (IDM). The IDM can be found in the third phase of the Risk Management Process (RMP), named detection [11, 12]. The identification and protection are the first two phases, response and recovering are the last phases of the RMP. The main purpose of IDM is to detect the abnormalities. There are different techniques of developing IDM, where most of them use only one reference model for comparison with the real time system. In this Section different techniques of intrusion detection on the ICP are implemented and specific results are delivered. One of the challenges for the IDM is the respond way to alarms. Another challenge is to find the best intrusion detection technique in real time systems in order to maintain the detection accuracy.

In Section six the authors provide solutions to detect the intrusions based on Machine Learning algorithms through the specific case studies. For IDM in ICS different learning techniques can be used from the IT computer systems. The IDM use the state variables of the physical system (PS). The intrusion or a signature can be detected by comparing the measured state variables with the reference ones. The different learning (supervised or unsupervised) methods applied to PS are used in IDMs. The reference model used in IDS should take into consideration the changes in the variables of the PS, and the sequence of the exchanged messages between IDM and network, the trace of the control software execution. The IDM is used to automatically detect the cyber-attacks. It is the most important part for maintaining the security in CPS. The conclusion has drawn in Sect. 7. At the end of the Chapter the list of references is presented.

2 ICS Architecture. The Security Features. SCADA Networks

Industrial control systems (ICS) comprise certain control systems, like SCADA systems, Programmable Logic Controllers (PLC) and distributed control systems (DCS) mostly used in industry branches. Some examples of ICS implementation sectors are: wastewater, natural oil and gas, chemical, and electrical plants, food and manufacturing such as automotive, aerospace so on [13]. The basic diagram of an ICS system is presented in Fig. 1.

An ICS network operating plan is mainly to ensure reliability, to avoid risks and threats and to offer safe operation.

For security, when speaking about network architecture of the ICS, it is advisable to separate it from the company network, because network data trafficking on the two networks should not interfere. For example, if the traffic on the ICS network is done on the company network, it can become subject to cyber-attacks. Most often, because of certain practical reasons, one needs a connection between the ICS and company network, and this connection may pose a potentially high security risk which requires a careful approach to its implementation. Thus, if there must be a link between the two networks, one recommends only minimal connections to be granted through firewalls and a perimeter network (a separate network part which is connected to the firewall). This means that ICS information which must be accessed by the company network is stored on this network part [14].

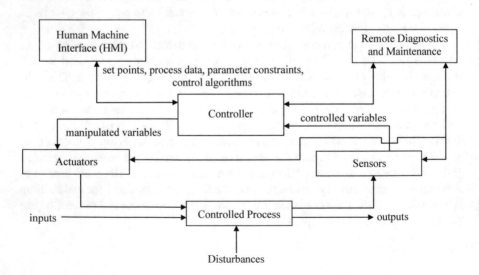

Fig. 1 ICS basic diagram

For external connections, the firewall must allow minimum access. Just the ports needed for specific communication should be opened. Logical separation between the ICS and the company network must be done on physically separated network devices.

The presence of enterprise connectivity requires the following [15]:

- minimal and documented points of access between the networks of the company and of the ICS
- only explicitly authorized traffic should be allowed, and the rest must be blocked through firewall placed between the two networks
- Filtering for: source and destination, ports and code should be allowed by firewall rules.

For keeping the data integrity and confidentiality on ICS systems, one implements security controls divided into operational, technical and management classes which combine security policies for securing the ICS.

For example, an efficient cyber security technique is called defence-in-depth and consists in layer security mechanisms which minimize the impact produced by a failure of a system part.

One should supervise security controls through activities such as configuration management and controlling system components, analysing security impact when changes occur in the system, continuous security controls evaluation and status reporting.

A good strategy is to conceive a plan for securing an ICS system to continuously enhance security. The system security must be thought on the entire ICS system life-cycle, starting with architecture-procurement-installation-maintenance-dismantling.

It is advisable for companies to define techniques to counteract security incidents which may occur on the network, minimizing the possible resulting consequences by reducing the risks which may produce these incidents.

An important aspect regarding risk is to establish whether the information which flows from the ICS to the company network is valuable. In this respect, a high security level can be achieved, but this may lead to the decrease of functionality.

Thus, the security must be well designed, putting into balance risks and function-ality. A company's general security plan should consider the cyber components and ICS network data physical protection and also to guard personnel from dangerous situations while exerting their jobs.

If one wants to access physically a control system or room, this means also to logically access the process control. This permits an attacker to achieve control on the physical access. Security measures would imply the locking or removal of the computers media drives, the disabling of power buttons, the placing of servers and ICS network devices like switches, routers and controllers in secured areas.

Many threats can be avoided by ensuring a proper physical security for the con-trol centre, which often has dashboards permanently linked to the server, where data transmission, response speed and the general network view processes is very important.

For identifying a potential attacker trying to hack the system, one can implement an intrusion detection system (IDS), which monitors network traffic, log entries and file access. An efficient IDS presumes both host-based and network-based IDS. The former are mostly implemented on machines that use general-purpose operating systems or applications like SCADA servers while the latter are mostly implemented between the company and ICS control networks, corroborated with a firewall. IDS can considerably improve security when properly configured and also they can improve control network performance by detecting traffic which is not important.

Current IDS can detect and prevent most common Internet attacks. However, just recently they started to take into consideration ICS protocol attacks. Thus, IDS developers began incorporating attack types for protocols used by ICS such as Modbus, Distributed Network Protocol (DNP), and Inter-Control Centre Communications Protocol (ICCP) [16].

Although beneficial from a reduced risk point of view, the security programs implemented for ICS can alter the access of programs and applications by human operators. Thus, companies should organize training sessions for employees to explain them how these methods of control and access diminish risks and what could be the possible impact on the company if these features are not introduced, highlighting their utility.

The most common authentication mode of systems connected to ICS networks consists on classic passwords, which are mostly default, given by the system supplier and also changed very rarely, which can inflict security risks.

Moreover, because of certain difficult moments, like the occurrence of a major crisis, a human operator may fail to correctly remember the password when his intervention is critical for controlling the process.

Biometric authentication may also exhibit flaws, so companies have to take into account the security needs for using such authentication modes on these systems. If the companies' executives do not take actions for implementing authentication procedures in ICS for certain reasons, such as negative impacts upon safety, reliability or performance, then strict physical controls must be used to ensure a high equivalent protection level for the ICS.

On some ICS systems is hard to set secure passwords, due to their small length and because for level access the systems grant group instead of individual passwords. Some industrial protocols use plaintext, such as ASCII to send passwords, thus increasing the chance of intercepting them. So, it is advisable for operators to use different and distinct passwords combined with encrypted/unencrypted protocols.

Further are presented some recommendations in what concerns the use of passwords:

- The responsible person with maintaining the passwords must be available during emergencies and very trustworthy. It should store all password copies in a limited access secured location.
- Only very trusted persons can be authorized to change passwords besides the responsible person.

- A balance should exist between security and system ease of access, given by the password attributes such as complexity, strength and length.
- Passwords must have the proper complexity for their required security level and should not contain anticipated characters
- It is advisable for a device which has a password to have a secure connection with the network. Passwords can be used to log on local devices. Playback attacks can be prevented by not allowing password transmission over the network.
- If passwords are inserted on interface devices like control terminals on critical processes, this could produce security risks when operators are not allowed access in case of critical events.
- If the ICS network presents an increased risk of cyber-attack, the companies should consider other authentication types, such as biometric, challenge/response or tokens.
- More secure methods are advisable for network service authentication, like public key authentication or challenge/response.

The security of data provided by ICS and by SCADA system can be analysed by taking into consideration its vulnerabilities. To analyse the security of the systems with application in industrial security control (ISC), the main aspects should be taken into consideration and understood as general functioning principles. These general elements are: SCADA systems, distributed control systems (DCS), SCADA server or Master Terminal Unit (MTU), Programmable Logic Controller (PLC), Intelligent Electronic Device (IED).

SCADA systems are distributed systems used to control pieces of equipment spread on a certain geographical area and usually spread at distances that can cover hundreds of kilometers, where data control and monitoring represent critical actions for the system functioning.

SCADA control centres are used to monitor and control, in a centralized manner, the so-called field sites (FS) or remote stations, through long distance communication networks including the actions such as monitoring alarms and data status processing.

Based on the information received from the FS, automatic commands or commands supervised by an operator can be sent to the field devices (FD). FD controls the local operations such as: the control of switches, data collection from different sensors and environment monitoring to assess the alarm conditions.

DCS are used to control major industrial processes with a high impact on our every-day life: energy production, water treatment plants, refineries, food industry, automobile industry, etc. DCS are systems which are integrated within the control architecture which contains a control level responsible for controlling the details of a local process. The control of the process is usually realized using control loops (feed forward and feed backward), while the states corresponding to key processes are automatically maintained around pre-set values. In order to ensure a pre-set tolerance for the addressed process in the respective field, one uses specific PLCs. The corresponding specific PLCs settings (which can be proportional, integral or derivate) are modified such that to obtain a pre-set tolerance and the self-correction rate during the reset process. DCS are widely used on large scale in the process-based industry.

PLCs are small computer-like devices, which possess a microprocessor and they are used for industrial processes automation, such as controlling equipment from the assembly line. PLCs can be, on one hand, components of the control system used in all SCADA and DCS systems and, on the other hand, they can be primary components in smaller control configurations used to control local processes (assembly lines, soot blower). PLCs are widely used in almost all industrial processes.

A first classification of industry divides it into two main parts: process based (manufacturing) industry and discrete industry.

Manufacturing industry involves two major processes [17]:

1. Processing that requires continuous processing. These processes function continuously, often with transitions towards different product grades. Examples of such processes can include the steam flux or the fuel flux from a power plant, the petroleum from a refinery or the distillation process from a chemical factory.
2. Processing that implies different processing steps or stages of a given material quantity—also called batch processing. Examples of such processes are often found in food industry. There is a clear distinction between initial and final step of such a process, with the possibility of obtaining steady states, limited in time, during the intermediary steps.

Discrete industry implies a series of steps that should be performed by a single piece of equipment or device with the purpose to obtain the final product. Examples include mechanical and electrical part assembly or piece's processing industry.

All types of industry (process-based and discrete) use the same types of control systems, sensors and networks. Some industry types can by hybrid, which means they are a combination of the two. While the control systems look similar for different industry types, there are some aspects which are different. A primary difference is the sub-systems controlled either by DCS or PLC, are usually places in a central part of the plant or the factory, while SCADA FS are spread over a geographical area.

DCS and PLCs communications uses LAN technologies which more reliable and faster than the long-distance communications provided by SCADA systems. In fact, SCADA systems are specifically designed to deal with the challenges raised by the long-distance communication, such as delays or data losses, due to the various media involved in the communication.

SCADA systems are used to control "goods" which are spread over an area. These are used in various systems, such as dirty water collection, systems which implies pipelines (gas, petroleum), energy transmission and distribution systems, transport systems, etc. SCADA systems integrate the data acquisition and systems with the software called Human Machine Interface (HMI) [18]. The purpose of this integration is to offer a central supervision and a control system for numerous inputs and outputs of a process. SCADA systems are designed to collect the information form the site and to ensure their transfer to a central computer, followed by its graphical representation (text, image), such that an operator to be able to monitor and control the whole system form a central location in a useful time. Function of the used configuration which corresponds to each individual system, its control and its functioning can be done in automatic manner or through the commands received from and operator.

SCADA is a modern computer-based system with the function of controlling and monitoring the technological processes. It implies the use of special software to be installed on the computer with the role of sending commands and monitor the technological process through some local pieces of equipment (e.g. PCL). The advances in this domain allowed SCADA systems to be used in many domains such as: large consuming goods, metallurgical, chemical and energy industry up to the nuclear domain.

SCADA systems consists of both hardware and software components. Usually, the hardware component includes an MTU placed at the control centre, communication equipment (telephone line, satellite, radio, etc.) and one or more FS which can be remote terminal units (RTU) or PLCs—with the main task to control the and/or to monitor the data. MTU stores and process the information collected form the RTU inputs and outputs while locally, a process is controlled by either RTU or a PLC. The communication hardware allows data transfer forward and backward between MTU and RTU or PLC. The software is designed such that to tell the system the following: what and when to monitor, which is the acceptable ranges for the parameters and what is the action/answer to be initiated in case the parameters does not fit within the pre-set domain and exceeds the acceptable values. An intelligent electronic device (IED), such as a protection relay, can be used to ensure the communication with SCADA server. IED can offer a direct interface for the control and monitoring of the devices and the sensors. IEDs can be directly controlled by SCADA server, without the need of any further instructions from the SCADA control centre. SCADA systems are in general designed in such a manner that they are error tolerant and having a significant redundancy encapsulated within the system's architecture. The general structure in term of the layout for a SCADA system is given in Fig. 2.

A SCADA system consists of two major hardware components, listed below [19]:

1. SCADA Server (one or many)—generically addressed by the term SCADA-MTU.

This is connected to the FS through various data acquisition systems. Data acquisition systems are realized based on microcontroller and they have the task to acquire data from the corresponding process and to supervise and control the process functioning. Data acquisition is realized by the intelligent sensors which can connect directly to a computer or by some intermediate devices called communication units or communication masters which concentrated data from many intelligent sensors. Data acquisition and process control devices are also called PLC. The server is responsible for all data collected from the FS, with specific functions such as: data base management, ensuring the communication between PLCs and the process, etc.

2. The client or the viewer. This is connected to the server through the network and ensures the communication with the operator. In small systems its presence is optional because the server can provide also the viewer function.

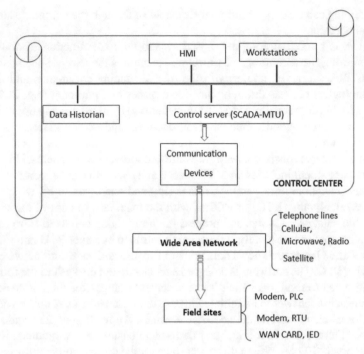

Fig. 2 SCADA system—general scheme

The servers are connected to PLCs using various communication drivers. Practically, on the market, there are hundreds of available drivers for all PLCs provided by known companies. One server can communicate with many protocols. New drivers and communication protocols can be further developed. The servers and the viewers are connected using the Ethernet protocol. The web adopted technology allows the visualization of a process using the internet access. SCADA systems are implemented in various technological systems and they are designed by considering the advantages given by this kind of access (internet). Among the principal functions provided by SCADA systems, there are the following:

– Control and supervision
– Alarm
– Event logs and production reports
– Post-damage analysis.

The communication architecture MTU-RTU can vary, function of the intended features of the corresponding implementation. Therefore, in practice there are several architectures in use, such as: point-to-point, series, series-star and multi-drop, as it is presented in Fig. 3.

Point-to-point architecture, from the functional point-of-view, is the simplest architecture. Because for each connection one needs a communication channel, this

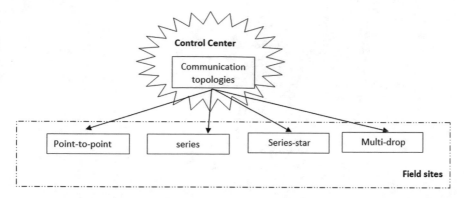

Fig. 3 Control centre architecture

architecture is an expensive one. In the series configuration, the number of the necessary communication channels is reduced because, by sharing a channel, the impact on efficiency or on the operation's complexity increases. In a similar manner, series-star or multi-drop architecture use a single communication channel, and therefore a decreased efficiency and an increased system complexity. The four architecture types can be further enhanced by using communication devices dedicated to managing these communications as well as managing the other functions such as message buffering and switching.

Large SCADA systems (hundreds of RTU) imply the use of sub-MTU which takes over some of the tasks form the primary MTU. Such an example is given in Fig. 4.

A schematic view of SCADA system implementation example is presented in Fig. 5.

3 Standards

Techniques used for design and analysis of Supervisory Control and Data Acquisition systems (SCADA)/Distributed Control Systems (DCS)/ICS with taking into account the safety are significantly different from techniques used for design and analysis of general network communication equipment due mainly to the following:

(1) Low computational capacity. For example, the RTUs (Remote Terminal Unit) equipment has limited computational capacity.
(2) Reduced rate data transmission due mainly to the low bandwidth of internal network communication.
(3) Real-time processing.
(4) High degree of accuracy- any errors in computation or delay in the transmission of data may be very serious consequences.

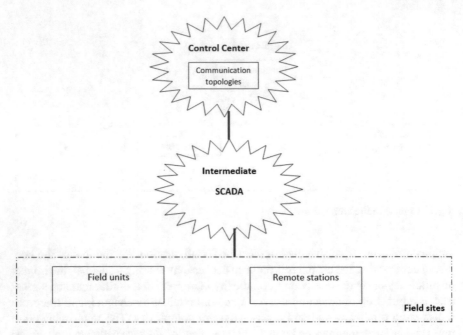

Fig. 4 Communication topology—large SCADA

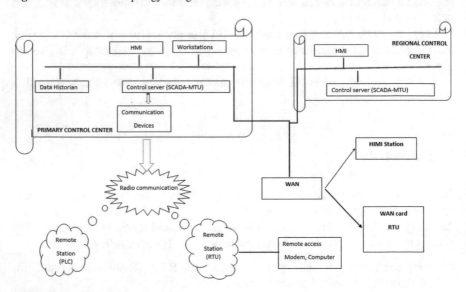

Fig. 5 SCADA system implementation—schematic view

(5) Requirements for total reliability and availability, the vast majority of these systems manage the continuous processes that cannot be turned off.
(6) Work in the industrial environment.
(7) Major financial, health, safety, and environmental implications in case of malfunction even for short periods of time.

In designing phase of the system, it is interesting to determine the interactions between different systems. As a result, it is not necessary to know in detail the internal characteristics of the equipment. Thus, in the analysis of security or cyber-security of systems, the individual elements can be considered as black boxes that have defined input and output characteristics. In the equipment analysis, the characteristics of the components and circuits must be taken into account in detail which contains the equipment.

Regardless of the type of system and field of work, 3 distinct life cycles can be identified [20, 21]

1. Definition Phase of the System
2. Design and Development of the System
3. Operation of the System.

1. *Definition Phase of the System*

The life cycle of a system starts with the Definition Phase. During this phase, the system evolves from mission (operational requirement of potential users) or idea (originated from the research environment) to the basic form of it. This can be either an idea that originated at a resarch laboratory or the. Since this phase it is essential to pay attention to the security of the system because it leads to the definition and specification of major features such as the type of architecture (open or closed, centralized or distributed), the type and standard of communications and data transmission used between system elements or between the system and the outside, data transfer rate, bandwidth, transmitter power, receiver sensitivity, antenna gain, the rejection factor when using wireless communication systems, etc.

During the definition phase, the system designer must provide and analyse the security issues that may arise. Thus, modern SCADA systems are designed using the features of open architectures to enable connectivity with other systems and products from different vendors.

In order to ensure the character of open architecture, design must be made in accordance with international standards in the domain such as POSIX or IEC 60870-6 for master station or IEC61850, IEC60870-5 or DNP 3 for substation communications.

At this level, the designer must predict that one of the ways in which a cyber-attack can be performed on the system is precisely using the access point of the communications system:

– Within or between system elements;
– Between the elements of the system and the elements of another system;
– Between the elements of the system and the electromagnetic environment in which the system will operate by using other vulnerable input paths such as electromagnetic interference with data blocks or data lines or secondary antenna lobes.

Choosing the use of wireless transmissions in ICS/SCADA systems is a risky decision from a security point of view and this should be taken jointly by both system planners and beneficiaries. As a general rule, wireless networks can only be used when financial resources, health, safety, and environmental implications are low.

In the case of adopting wireless communications solutions, security measures need to be considered at the design stage, including system definition phase. In this regard, NIST SP 800-48 and SP 800-97 can be consulted to provide valuable information on the security of these networks. [22–26].

An important standard that can be adopted to make secure communications is IEEE 802.11i. This standard corrects the defects in the IEEE 802.11 predecessor security specifications based on Wired Equivalent Privacy (WEP) technology, that uses RC4 (Rivest Cipher 4) cipher (in cryptography). For example, one of the huge problems for WEP, was that it was impractical to manage key distribution once you had more than a few tens of users. IEEE 802.11 RSN (*robust security network*) is more complex, involving multiple devices, protocols, and standards. It provides a solution that is both secure and scalable for use in large networks. IEEE 802.11i standard introduces a new type of wireless network named RSN, known as Wireless Protected Access 2- WPA2 (successor of WPA with two security types: authentication by using passphrase and encryption). 802.11i task group makes use of the Advanced Encryption Standard (AES) *block cipher* (the data is processed in blocks), whereas both WPA and WEP use the RC4 *stream cipher* (the data is serial stream processed, bit by bit). During on a system upgrade, IEEE 802.11i makes *transitional security network* (TSN), in which both RSN and WEP devices operates in derivation [27–29].

It should be borne in mind that the main difference between wired and wireless communications networks in terms of security is the relative ease of interception and intrusion of the latter. Therefore, even if the wired communication solution (copper or optical fibre) is chosen, all measures must be taken to protect the system against possible cyber-attacks. The designer should take care about the set of IEC 61784 standards (which include 36 standards) that use as protocol framework the series of IEC 61158 for the Communication Profile Families (CPF), very useful in the design of the communications devices from industry manufacturing, and industrial control. Additionally, based on ISO/IEC 8802-3 fieldbus profiles for real-time networks are defined [30].

Another important decision to be taken in these phase is the choice between the type of centralized architecture and the type of distributed architecture.

The centralized architecture model is commonly encountered in old/traditional systems and simple applications such as control and monitoring of an electrical distribution network. This model is also used for cost-related implementation and maintenance as:

– All applications run on a single hardware platform including software for the operator display and the interfaces between different software applications become relatively simple,
– Maintenance must be carried out at a reduced number of stations/computers.

Due to the above-mentioned reasons, systems using the centralized architecture type are particularly vulnerable to a cyber-attack. To prevent total deployment of these systems in the event of an attack, backup solutions must be provided from the design stage to the creation of a redundant hardware platform.

Distributed architecture is widely used in applications requiring increased computing power and data processing because data processing is shared between multiple server systems. Using multiple servers to process data, there are several benefits:

- Less processing power for each server,
- The cost of each server/platform decreases,
- It's a lot easier and cheaper to make upgrades,
- Taking out of service a server does not make the whole system inoperative.

In fact, the last enumeration makes distributed networks safer and more robust in the face of a cyber-attack than centralized architecture networks. Systems with distributed architecture are also used for other reasons. For example, different application-based operating systems can be used: real-time operating systems for data acquisition and processing such as UNIX or VMS, and operating systems that provide the end-user graphical user interface such as Microsoft Windows or Linux. Distributed architecture systems also share processing and data acquisition tasks across multiple servers/platforms that control all communications functions with substations and other external systems, creating additional access points in the system. This increases the vulnerability of these systems to cyber-attacks, which requires more safety precautions.

At the design stage of the system architecture it is recommended to think of a scheme that would allow the corporate network (CN) separation from the ICS network. In fact, the nature of traffic in the two networks is fundamentally different. On the one hand, in the corporate network, we have access to the Internet, FTP, remote access email, and on the other hand the ICS network must be governed by strict rules and procedures. As a matter of fact, the vast majority of ICS/SCADA network security publications recommend that the ICS/SCADA network be not connected to the Internet. However, if connection is required, it is recommended that this is done by as few connections/links as possible, if possible through a single connection. In this case, connections must be made through a firewall and a DMZ [22, 31].

DMZ (demilitarized zones) is a physical and logical network inserted as a neutral zone that interposes between the internal and the external network [32]. Thus, the ICS data servers that need to access the corporate network are installed in this network segment called DMZ that is directly linked to the firewall. It is also very important that any connection to an external network has to go through a firewall.

Separation of the CN from the ICS network can be done in several ways, including:

(a) Packet Filter between CN and ICS/SCADA control network (CtN) [33].
(b) Gateway between CN and ICS/SCADA CtN.

The major disadvantage of the Gateway Separation Method is that it can only be used when using standard HTTP and SMTP protocols and cannot be used with other industry-standard communications protocols such as EtherNet/IP or Modbus/TCP [33].

(c) Firewall between corporate network (CN) and ICS/SCADA control network (CtN)
(d) Firewall and Router between CN and ICS/SCADA CtN
(e) Firewall with DMZ between CN and ICS/SCADA CtN
(f) Paired Firewalls with DMZ between CN and ICS/SCADA CtN [22].

In [22] it looks like the most secure, manageable, and scalable control network and corporate network segregation architectures are typically based on a system with at least three zones, incorporating one or more DMZs, like in Fig. 6.

In Fig. 6, paired Firewalls with DMZ between corporate network and ICS/SCADA control network is shown [22].

The other method to avoid a cyber-attack is based on encrypting communication data through a *Virtual Private Network* (VPN), The VPN overlay on a public infrastructure, being a private network so that the private network can function parallel to public network. By following this idea in [31] is proposed an "ideal" architecture against cyber-attacks (Fig. 7).

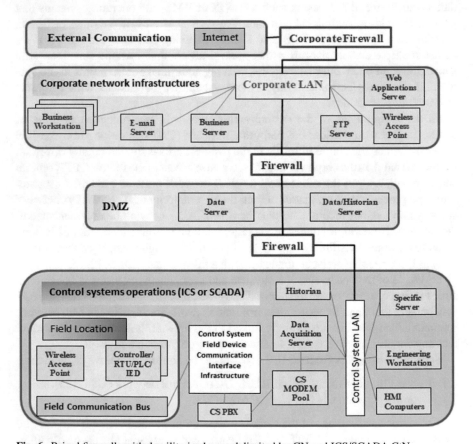

Fig. 6 Paired firewalls with demilitarized zone delimited by CN and ICS/SCADA CtN

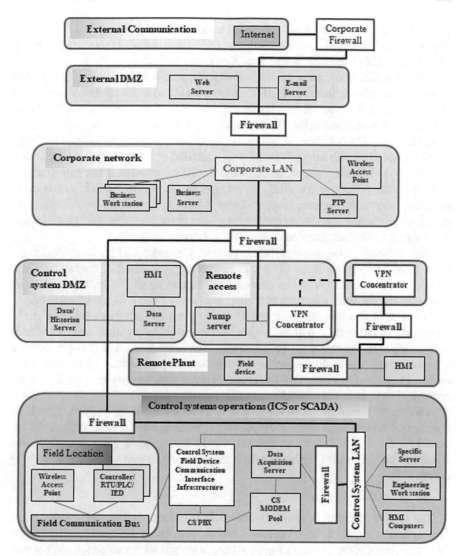

Fig. 7 Ideal ICS network configuration

Therefore, for greater efficiency, it is recommended to use a multiple layers strategy involving two (or more) different overlapping security mechanisms. This technique is known as defence-in-depth architecture [34]. To minimize the impact of a failure in any one mechanism the defence-in-depth architecture strategy includes the use of firewalls, the creation of demilitarized zones, intrusion detection capabilities along with effective security policies, training programs, incident response mechanisms, and physical security [22].

In Fig. 8 an ICS defence-in-depth architecture (DIDA) is shown [34]. The defini-tion of ICS DIDA comes from the developed DHS Control Systems Security Program NCCIC/ICS-CERT Recommended Practices Committee13 [34].

There are different kind solutions for network monitoring against an unusual or unauthorized activity. Intrusion Detection Systems (IDS) is one of the solutions. The main role of an IDS into a network is to watch and assess the traffic or network activity without impacting that traffic. Therefore, the IDS has a passive role. By using the collected data, IDS compares it against a pre-defined rule set, as well as against a set of known attack 'signatures'. In the next step, IDS has duty of both port numbers investigation and data payload. In this way, the IDS determine if any unfavourable activity (attack pattern) is occurring. By recognizing an unfavourable activity (attack pattern, or any deviation from what has been defined as normal/allowable traffic) the systems will dispose by a set of instructions, one of them can be alerting a systems administrator. Another function of the IDS is extensive logging.

Also in this stage, depending on the type of configuration and architecture chosen, the way of limiting unwanted effects in case of a cyber-attack or terrorist act will be decided. In order to ensure a continuous functioning of the system, redundant hardware and software solutions as well as backup systems must be considered.

This can be the solution of a master backup server that can take over major operations in the system without major interruptions. In practice, such a solution means doubling the primary control centre functionality even if it is provided at a reduced capacity. In this regard, a communication path should be considered to ensure a permanent synchronization between the emergency backup system and the main field devices of the control centre [35].

Another solution that can be adopted is to think of the backup system in the form a dedicated second or possibly multiple server platforms for dual or up to quad hot standby redundancy. This mean that one of the servers is in the active mode and it handles all the system functions while the other(s) are running, but not working online [33].

2. *Design and Development of the System*

The second phase of a system's life cycle is system design and development. During this stage, the system evolves from the previously set specifications to the final hardware details. At this stage, the design should be done taking into account the main security requirements:

- *Access control*: Access control to selected devices, information, or both in order to protect the system against unauthorized interrogations,
- *Use control*: Use control of selected devices, information or both to protect against unauthorized operations on devices or use of information,
- *Data integrity*: Ensure data integrity on selected communication channels to pro-tect against unauthorized changes,
- *Data confidentiality*: Ensuring data confidentiality on selected communication channels to protect against data leakage,
- *Restrict data flow*: Restricting data flow to communication channels to protect against publishing information to unauthorized sources,

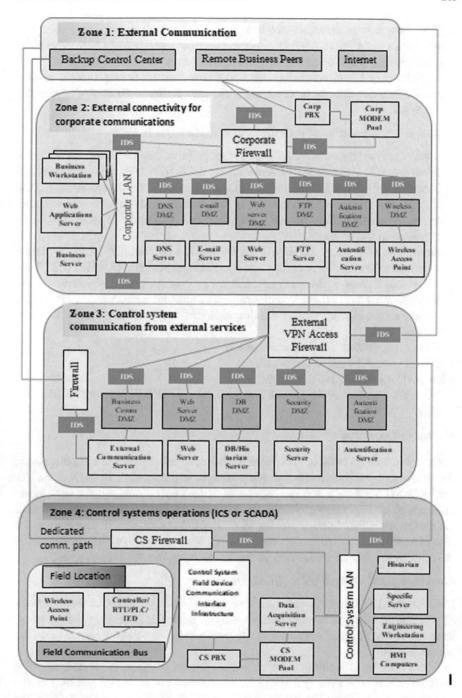

Fig. 8 CSSP recommended defence-in-depth architecture

- *Timely response to event*: Responds to security violations by notifying your own authority, presenting the evidence of violation and taking timely corrective measures in security situations or safety critical missions
- *Availability of the Network*: The networks structures must be protected against denial of service attacks. In this manner the availability of all network resources is assured.

At this stage, selection of the following items must be done:

- *Intelligent Electronic Device (IED)*. For this purpose, can be consult Standard for Substation IED Cybersecurity Capabilities—IEEE 1686-2007. The functions and features to be provided in substation IEDs)to accommodate critical infrastructure protection programs are defined in this standard. Security regarding the access, operation, configuration, firmware revision, and data retrieval from an IED is addressed in this standard.
- Human Machine Interface (HMI). ISA's first human-machine interface (HMI) standard has been approved as an American National Standard (ANSI). ANSI/ISA-101.01-2015, Human Machine Interfaces for Process Automation Systems, covers the philosophy, design, implementation, operation, and maintenance of HMIs for process automation systems, including multiple work processes throughout the HMI life cycle.
- Software, firmware, and hardware issues,
- External communication interfaces,
- Internal communication interfaces,
- Protocols. The Standards that can be consulted are IEEE P1711, and IEEE 1815-2012.

3. *System Operation*

System operation is the final phase in the life cycle of a system. During this phase, the system that was designed and developed is put into operation. In the present, it is necessary to consider the security aspects to various operational aspects such as the influence of space placement, frequency management, encryption key management, the maximum radiated power and the directional characteristics of antennas in the case of wireless communications, the shielding of the rooms in which the important elements of the system are located, including securing the earth lines, so on.

Electromagnetic Compatibility (ECM) issues

If a DCI/SCADA system is very well designed against cyber-attacks the execution of such an attack requires considerable efforts. Basically, by taking the proposed measures above we do not only prevent cyber-attacks but also discourage the initiation of such attacks.

There may be situations in which such systems of type ICS/SCDA must be spied/monitored and even out of use by economic opponents and not only at the lowest cost and without the highly qualified and specialised personnel. This can be easily achieved if it is exploited/speculated by the attackers by discounting the principles of electromagnetic compatibility EMC/EMC in the three phases of the design of ICS/SCADA systems. Thus, we can offer great opportunities to supervise data traffic not only in the case of wireless communications, but also in the case of using their copper feeder. Even if it opts for fibre optic communications it is almost impossible not to exist portions of data transmission through copper wires or, until the conversion of data for fibre optic traffic, it will first be in the form of an electrical size. They can be intercepted, for example, by speculating imperfections in the design and/or realization of equipment carcases or connectors. In these cases all the aforementioned measures (firewall, DMZ, VPN, so on) are useless. Therefore, unauthorized supervision is done without waking up the slightest suspicion of the ICS/SCADA network administrators. Basically, in this case it should be recalled the principle that if we can measure something, it means that we can control it. In other words, we can also offer opportunities for untimely intervention in the system at the right time.

The Data can be collected through the ground (Earthing) network if it is not at zero volts. A Network of Earthlings that does not have 0 V offers even intrusion opportunities in the system through common mode currents with the purpose of removing the system operation outside the projected parameters, which can have particularly serious consequences.

Also, great importance must be given to buildings/premises where these networks are located because:

(a) Can be permanently removed from use by practicing a NEMP (nuclear electromagnetic pulse) type of impulse in their vicinity. Devices that can produce such impulses (which are right to modest powers) can be made with components in trade. By using the NEMP type impulses several (near vicinity) buildings can be neutralised in terms of computing and control systems. The handiest measure is the build the rooms like Faraday cages. This measure cannot provide results in the case of physical intrusion of the impulse-triggering apparatus inside the building.

(b) Can be permanently removed from use by practicing a high power impulse reverse discharge through the electricity supply network. In this case, classical eclatory protections can be totally ineffective if the ascending front of the pulse is very small (Nano-picoseconds). The most effective protections are obtained by means of pre-ionized gas.

The standards that can be consulted are: IEEE Standard 1100-1999 Recommended Practice for Powering and Grounding Sensitive Electronic Equipment, IEEE Standard 142-1982 Grounding of Industrial and Commercial Power Systems, IEC 61000-5-2 (EMC—Earthing and Cabling). Voltage surges can exceed the test limits specified in IEEE Std 1613.

Standards that can be used in the safe design and operation of SCADA/ICS systems in terms of cyber security.

By going forward, the standards that can be used in the safe design and operation of SCADA/ICS systems in terms of cyber security will be presented. The Data is obtained by the official pages of the organizations that issued them:

ISA/IEC 62443 standard specifies security capabilities for control system components

The ISA/IEC 62443 set of standards, developed by the ISA99 committee and adopted by the *International Electrotechnical Commission* (IEC), provides a flexible framework to address and mitigate current and future security vulnerabilities in *industrial automation and control systems* (IACSs). The committee draws on the input and knowledge of IACS security experts from across the globe to develop consensus standards that are applicable to all industry sectors and critical infrastructure.

A new standard in the series, ISA-62443-4-2, Security for Industrial Automation and Control Systems: Technical Security Requirements for IACS Components, provides the cybersecurity technical requirements for components that make up an IACS, specifically the embedded devices, network components, host components, and software applications. The standard, which is based on the IACS system security requirements of ISA/IEC 62443 3-3, System Security Requirements and Security Levels, specifies security capabilities that enable a component to mitigate threats for a given security level without the assistance of compensating countermeasures.

The ISA-62443-4-2 standard follows the ISA/IEC 62443-4-1 standard, Product Security Development Life-Cycle Requirements, which specifies process requirements for the secure development of products used in an IACS and defines a secure development life cycle for developing and maintaining secure products. The life cycle includes security requirements definition, secure design, secure implementation (including coding guidelines), verification and validation, defect management, patch management, and product end of life.

Another key ISA/IEC 62443 standard expected to be completed in the coming months is ISA/IEC 62443-3-2, Security Risk Assessment, System Partitioning and Security Levels, which is based on the understanding that IACS security is a matter of risk management. That is, each IACS presents a different risk to an organization depending upon the threats it is exposed to, the likelihood of those threats arising, the inherent vulnerabilities in the system, and the consequences if the system were to be compromised. Further, each organization that owns and operates an IACS has its own tolerance for risk [36].

IEC (TC65,57) Standards

The IEC standards organization prepares and publishes standards for all kinds of electrical, electronic, and related technologies. These standards serve as a basis for creating national standards and as references for drafting international tenders and contracts [37].

Horizontal Standards

IEC TC 65 (TC 65) has created IEC 62443 for operational technology found in industrial and critical infrastructure, including but not restricted to power utilities, water management systems, healthcare and transport systems. These are horizontal standards, which are technology independent and can be applied across many technical areas [38].

Vertical Standards

Vertical Standards are designed to meet specific technical needs, for example in the energy sector, manufacturing, healthcare or shipping, among others. Several technical committees (TCs) and subcommittees (SCs) prepare International Standards that protect specific domains and keep industry and critical infrastructure assets safe (Table 1).

IEEE Standards

IEEE 1686-2007—Substation IED CS Capabilities. The standard contains the functions and features specific to substation IEDs in order to accomplish the critical infrastructure protection programs. The following security items are treated in this standard: access, operation, configuration, firmware revision, and data retrieval from an IED. Teleprotection and encryption for the secure transmission of data both within and external to the substation, including SCADA, are not discussed in this standard.

IEEE P1711—Standard treats cryptographic protocol to provide integrity, and optional confidentiality, for CS of serial links. This standard does not treats the hardware implementations, and it is decoupled from the underlying communications protocol.

IEEE 1815-2012—Standard for Electric Power System Communications-Distributed Network Protocol (DNP3) describes the DNP3 SCADA protocol, incorporating version five of the application-layer authentication procedure, named DNP3 Secure Authentication (DNP3-SAv5). DNP3-SAv5 uses a Hash Message Authentication Code (HMAC) process to verify that data and commands are received (without tampering) from authorized individual users or devices while limiting computational and communications overhead. SAv5 supports remote update (add/change/revoke) of user credentials using either symmetric or public key infrastructure (PKI) techniques. The confidentiality in SAv5does not exist (the encryption of the messages

Table 1 Series of IEC standards

TC 57	61850	Communication networks and systems for power utility automation	Electric power utilities
	60870	Telecontrol equipment and systems	
	62351	Power systems management and associated information exchange	
SC 62A	ISO/IEC 80001	Risk management for IT-networks incorporating medical devices	Healthcare
TC 65	62443	Specify security requirements for IACS	Industry
TC 80	61162	Maritime navigation and radio communication equipment and systems	Shipping

are not included), only authenticates. In order to take into account the confidentiality, SAv5 is used with encryption techniques such as transport layer security (TLS) or IEEE 1711 (standard for SCADA security) where confidentiality is required [39].

NERC Standards [40]

In the area of Cyber Security NERC they created the 1300 standards group that changed the/updated group 1200. The latest version of NERC 1300 is also in the form of CIP standards-002-3 up to CIP 009-3 (CIP—Critical Infrastructure Protection) [40]. Thus, the 1300 Standards Group covers the following areas [41]:

– Security Management Controls—1301
– Critical Cyber Assets—1302
– Personnel and Training—1303
– Electronic Security—1304
– Physical Security—1305
– Systems Security Management—1306
– Incident Response Planning—1307
– Recovery Plans—1308.

   The list of the CIP names is as follows:

– 002, CS—Critical Cyber Asset Identification.
– 003, CS—Security Management Controls.
– 004, CS—Personnel and Training.
– 005, CS—Electronic Security Perimeter(s).
– 006, CS—Physical Security of Critical Cyber Assets.
– 007, CS—Systems Security Management.
– 008, CS—Incident Reporting and Response Planning.
– 009, CS—Recovery Plans for Critical Cyber Assets.

NIST Standards

The NIST 800 series treats the information technology documents, reports and guidance in computer security. The subject of the NIST standards are related to cryptographic technology and applications, advanced authentication, public key infrastructure, internetworking security, criteria and assurance, and security management and support. There is the additional NIST SP 800-82 standard related to the ICS security community [42].

4 Machine Learning on ICS and SCADA Networks

This section is dedicated to machine learning on ICS and SCADA networks. Artificial Intelligence (AI) aims to study and design intelligent agents, systems that perceive the environment and maximize the chances of their own success through behaviour. A common feature of artificial intelligence is that the system is capable of learning, with or without external aid, with the aim of constantly improving. An AI application is to detect and prevent network intrusions. By using artificial intelligence, it can detect whether software is malware or normal software. To develop an AI application that performs "malware detection," it is necessary to first establish certain distinctive features. In addition to harmless and malware features, the system will be trained. Machine learning is used in many applications. In most of these, the detection target is defined. Instead, as far as cyber security is concerned, what we want to detect is not implicitly defined. At the same time, Cyber-security calls for the latest information and becomes one of the current challenges.

Cyber-physical systems (CPS) have the role of controlling and monitoring critical infrastructure. Currently, the CPS are mostly missing from the existing infrastructure. Due to the lack of awareness of the existing vulnerabilities by civilian owners of the systems, the risk to which they are subject, and the lack of knowledge of the impact on these systems when deployed, CPS are not widely encountered. There are now many tools that can detect the vulnerabilities of an existing physical system, like FAST-CPS environment [43].

The applications in which cyber-physical systems are used include industrial control systems (ICS), SCADA surveillance and acquisition control, machine building industry, transport, etc.

A CPS system is provided with physical inputs/outputs and contains processes that need to be controlled remotely. At the field site level, different sensors can be found, monitored and controlled locally by programmable logic controller (PLC), and remote by Remote Terminal Unit (RTU). In this way, in the centralized control the operators can monitor and control the process remotely. Into a CPS, the symbiosis between the computational tasks and the physical process can be found [44, 45].

Modern industrial systems are modular, interconnected and remotely accessed. The components from which such systems are built come from different vendors with different vulnerabilities. If, in traditional computer systems [46, 47], an attack cannot cause the injury or death of a person, in ICS these dangerous consequences can occur.

IT systems are based on three pillars: Confidentiality, Integrity, and Availability.

In these systems, the information flow is the most important. For data confidentiality, IT systems use encryption and cryptography [48]. Data contained in such systems must not interfere. Therefore, data integrity must be ensured. Validation of data integrity is a mandatory requirement. As Cryptographic tools can be used hash functions and HMACs [49].

The ultimate requirement of an IT system is introduced to ensure continuity. The continuous operation of the systems is carried out through reserve systems.

In the CPS schemes [50, 51] availability becomes more important than IT.

CPS deals with critical infrastructure such as traffic lights, power plants, water treatment plants.

In CPS Confidentiality is less important than in IT systems. The data flows for these systems contain the data from the sensors and the commands sent to the actuators. This is why communication protocols used in a CPS are not encrypted. In cases where data is confidential, it is recommended to use encryption.

People interact with physical systems; therefore, any attack on a CPS system can lead to injury to individuals. These systems have to take into account the safety, and from the point of view of control these systems are deterministic.

Modern, the autonomous vehicle has been introduced in transport domain. These vehicles are based on the information received from the GPS (for location and positioning), but also from the road markings, signs and traffic lights. This real-time information can be attacked. If a traffic light in train transport does not work, and the situation could become catastrophic. In this chapter the attack on the autonomous vehicle network will be shown.

Therefore, in CPS systems, the Confidentiality, Integrity, and Availability features become Safety, Reliability, and Availability [52].

Short history on industrial control attack: April 1999, Russia, Gazprom gas plant; January 2003, Ohio- Davis-Besse nuclear power plant; 2008 Australia, Maroochy Shire—water services [53]; January 2010, Iran- Stuxnet December 2015, Ukraine— the first electric power blackout causes by CrashOverride malware [54]; WannaCry ransomware and NotPetya threat attacked the worldwide control systems in shipping and manufacturing area [55]; TRISIS attack on petrochemical plant in Middle East (2017). In the latter three cases, vulnerabilities were exploited by the attackers, highlighting the importance of identifying known vulnerabilities in these systems and evaluating the possible consequences of an attacker exploiting them.

Industrial Risk Assessment Map [56] is a geospatial map tool which offers large details on the found worldwide industrial device vulnerabilities. In the developed tool, IRAM [56], the following industrial units and devices were taken into account: PDU- PowerDistribution Unit, BMS-Building Management System, SCADA, UPS-Uninterruptible Power Supply, PLCND—Programmable Logic Controller Network Device, PLC, ERP-Enterprise Resource Planning.

By using IRAM, the vulnerabilities on the above mentioned industrial units and devices can be found (Figs. 9, 10, 11, 12, 13, 14 and 15). Similar to IRAM, by using Python environment and adequate geospatial library the worldwide SCADA map extraction can be provided, as it is shown in Fig. 16.

The IPew is a Java-based visualization tool. For the source attack node the country of origin the tool uses the statistical model. In order to obtain the cartographic attack data, the Datamaps framework was used [57]. The application delivers the source of an initialized attack and the destination country or victim attribution of attack. The tool also identifies the names of the attacks (Fig. 17).

The short resulting output list of IPew tool

esp (145.103.54.40) attacks USA (198.223.187.215) (SYN FLOOD BA-BY)
ita (232.199.11.182) attacks USA (174.233.240.252) (SYN FLOOD BA-BY)
bra (216.173.12.216) attacks deu (22.48.107.88) (SNAILshock)
bra (36.19.40.247) attacks usa (127.65.119.114) (Spaghetti RAT).

I.1 FAST-CPS tool [43] automatically evaluates the security of a CPS using a CPS model. Practically, FAST-CPS detects system vulnerabilities. To make the FAST-CPS model, they use system models Systems Modelling Language, SysML [58] to

Fig. 9 PLCND

Fig. 10 PLC device

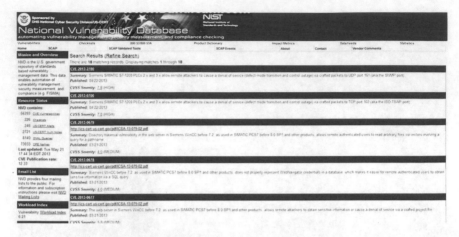

Fig. 11 National vulnerability database-.NIST

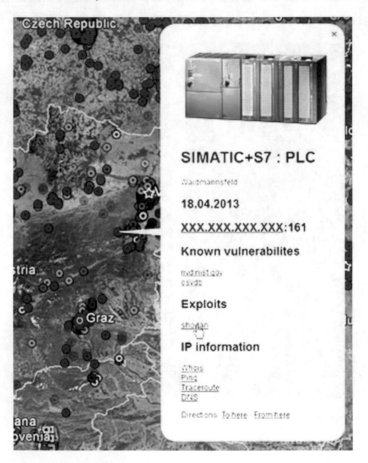

Fig. 12 PLC known vulnerabilities

Fig. 13 SHODAN exploits—database

Fig. 14 Access the vulnerabilities of the identified (Fig. 12) SIEMENS Simatic S7-300/400 CPU START STOP module through the information provided by SHODAN exploits (Fig. 13)

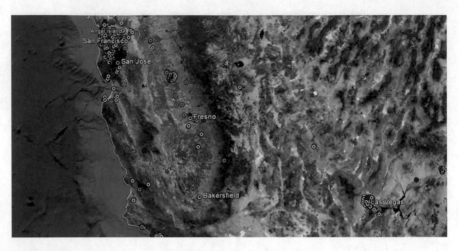

Fig. 15 IREM map of the known SCADA units

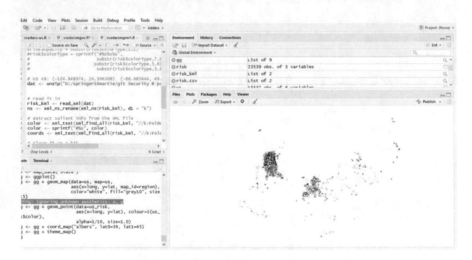

Fig. 16 Python based map extraction of the spreading devices into a specified area

create dynamic fault trees (DFT). In order to calculate reliability of a CPS the DFTs are analysed. Other tools are based on Unified Modelling Language, UML [59].

Other useful tools for building a CPS model are as follows:

I.2 Cyber Security Evaluation Tool (CSET) created by Homeland Security [59].

This instrument was assessed to questionnaires conformity of a system according to the security standard chosen.

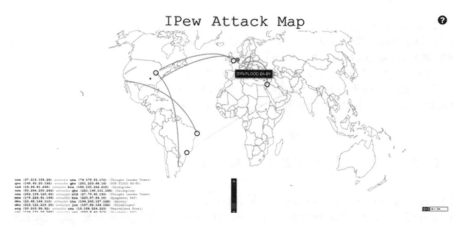

Fig. 17 The IPew attack map

I.3 ADVISE

ADVISE [60] determines the most likely way that an attacker attacks a UI system planning. Advise works at Möbius [61]. The modelled CPS in ADVISE is make from the attacker point of view, the CPS system architecture is not modelled, but only the attack path (defining a probability). For this, the attack and the opponent are modelled.

I.4 CyberSAGE

CyberSAGE [62] uses a threat agent workflow as input, this software tool providing the likelihood of attacking the attack, as well as the attack steps.

I.5 CySeMoL [63] provides the likelihood that an attacker achieves attack targets in a CPS system.

In the CySeMoL application, the evaluator cannot change the attack and the attacker, they are fixed. An attacker is believed to be a penetration tester, assessing probabilities of attack. CySeMoL works on probabilistic models (PRM), so the user has to build a relational probabilistic model. In order to include a CPS, we will use the Operating System block that relates to ApplicationServer or ApplicationClient tools that are required to interconnect PLCs, HMI interfaces, and other devices.

Most attacks on industrial systems are based on vulnerability detection and exploitation.

In order to build an SPC system based on existing software tools can be used: fast-SPC (environmental modelling extension Papyrus Eclipse) to determine their vulnerabilities system, useful information for ADVISE tools, and providing the most likely attack path, but not CPS system configuration solutions. To find a network security solution, you can use the CSET tool, providing solutions for introducing firewalls into the network architecture.

In order to evaluate their hardware and software vulnerabilities with the FAST-CPS tool, the CPS should be developed by the user in SysML. The database used to identify vulnerabilities is provided by ICS-CERT.

In the paper [58] there is a procedure to narrow the process vulnerabilities.

One major problem for machine learning is the availability of the dataset for training purpose. In the following subsection, one application in energy area is shown.

4.1 Forecast of Energy Consumption

The energy consumption forecasting is based on the deducted mathematical model of energy demand. The mathematical model must predict the energy requirement based on previous daily consumption data. An Excel data file is provided, providing the daily energy demand (expressed in GWh) over a period of 9 years, following an energy forecast consumed over the next 2 years.

ARIMA model

ARIMA represents the autoregressive integrated moving average. For the forecast, two types of ARIMA models are used: seasonal and non-seasonal.

The data in the Excel file is completed with two columns representing the day of the week and day of the year, then saved in the POSIX format ("2010–02–09" "2010–02–10"…) [64]. In order to create the test data, the original data (Fig. 18) is divided into the "Energy demand evolution" at a one-day frequency (Fig. 19). The result is shown in Fig. 20.

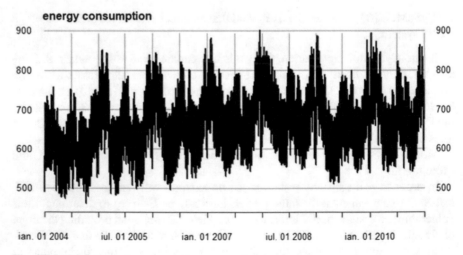

Fig. 18 Initial data: energy consumption per month

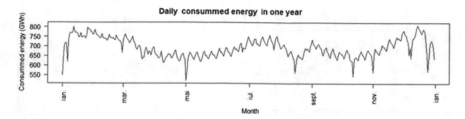

Fig. 19 The demand of energy is higher during winter (ian), comparing with summer (iul)

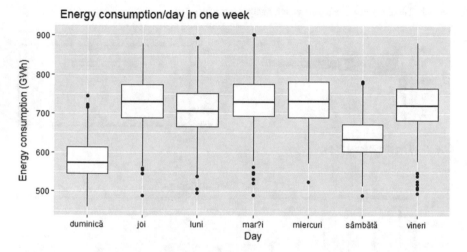

Fig. 20 Daily energy consumption in one week

Figure 20 shows obviously the real energy demand per day in one week (during weekend is low). The temperature is an important variable in energy demand. Therefore, a seasonal dependency of the demand for energy is very clear. For demand prediction the lagged electricity demand is taken into consideration, along with the time of day, and day of week variables.

In Fig. 21 the calculated daily average in one year consumed energy is shown.

The dynamical errors are shown as follows (Fig. 22). It could be noted that the biggest errors are negative.

Following the obtained autocorrelation (ACF) function, Fig. 23, the seasonal non-stationary process with high autocorrelated values is shown in Fig. 24. Therefore, the appropriate cycles are evidently (weekly, yearly).

From the ACF Fig. 23, the dependence of ARMA or AR processes could be finding. The correlation vanishes fast, having the lag. For the PACF (Fig. 25) evolution the AR models are better (Fig. 26). The seasonal components are evidently (yearly, weekly).

The initial time series is decomposed in seasonal time series with weekly cycles. Decomposition of additive time series (Fig. 27).

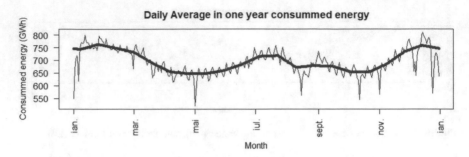

Fig. 21 The daily average in one year consumed energy

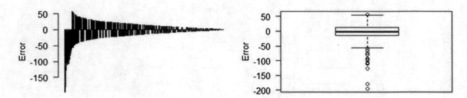

Fig. 22 The dynamical error

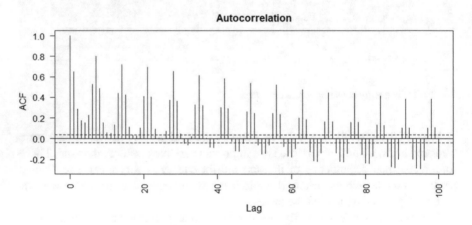

Fig. 23 ACF

1. Observed, 2. Trend, 3. Seasonal, 4. Random.

The following step is to calculate the difference between the demand and week seasonal. The result is a yearly seasonal time series (Fig. 28).

By subtracting the original time series from the weekly seasonal data series the yearly seasonal series is obtained (Fig. 28).

By subtracting the original time series from the seasonal weekly and the yearly series, a new data could be obtained (Fig. 29).

The average of the daily requested energy is plotted in Fig. 30.

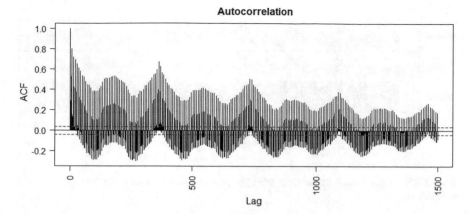

Fig. 24 Autocorrelation

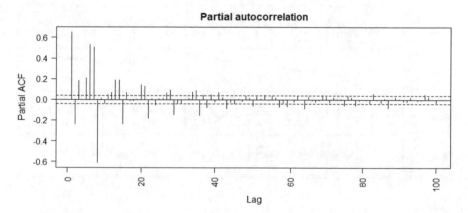

Fig. 25 PACF

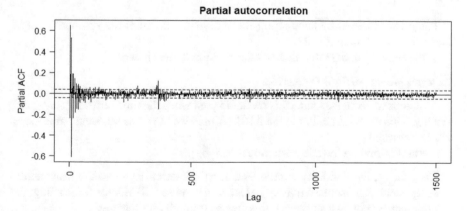

Fig. 26 Partial autocorrelation

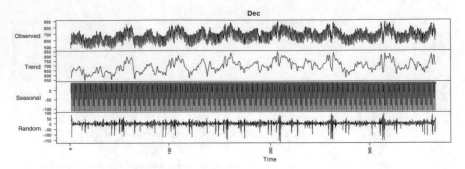

Fig. 27 Decomposition of additive time series (from top to down: 1. Observed, 2. Trend, 3. Seasonal, 4. Random)

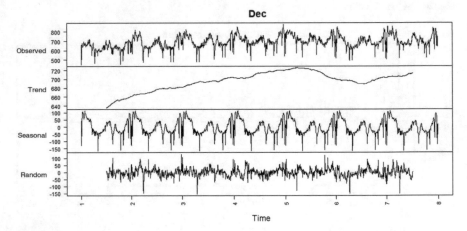

Fig. 28 The decomposition of the original time series into a yearly seasonal one

The error obtained by subtracting the daily demand from the seasonal data is lower (Fig. 31) than in Fig. 22.

In the Figs. 32 and 33 the fresher ACF and PACF are shown:

Seasonal Models ARIMA (SARIMA):

As it appears from the name, this model is used when the time series presents seasonality. This model is similar to the ARIMA models, but some seasonal parameters will be activated.

Method of finding the adequate model [65–69]:

1 Checking of the stationary component: If a time series has a trend or seasonality component, it must be stationary before you can use ARIMA for forecasting.
2 Stationarity: obtain a stationary time series through differencing.
3 Sampled series filtering validation: it will be used to validate how accurate the chosen model is.

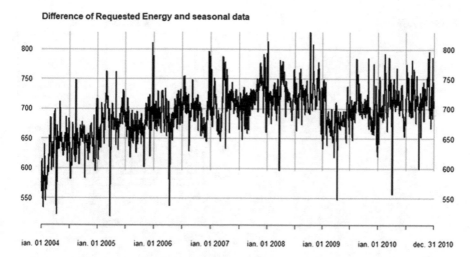

Fig. 29 The difference of requested energy and seasonal data during the observation interval (2004-2010)

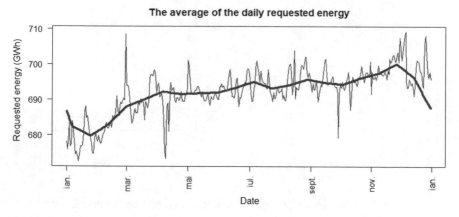

Fig. 30 The average of the daily consumed energy

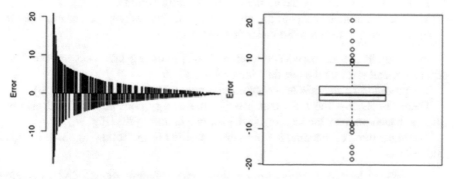

Fig. 31 The error obtained by subtracting the daily demand from the seasonal data

Fig. 32 ACF

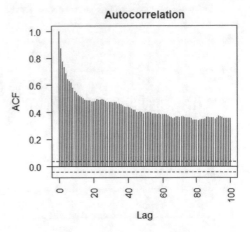

Fig. 33 PACF

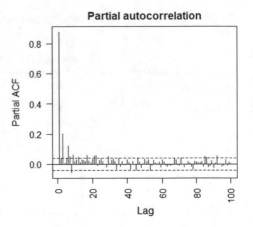

4. Selecting AR and MA terms: ACF and PACF will be used to decide whether to include AR, MA or individually.
5. Build the pattern and fix the number of forecasting periods.
6. Validate the model: compare the obtained errors (the values forecasting against the real ones) in the sampled validation series.

In the Fig. 34 the comparison between real and predicting data is shown [66]. The obtained mean error on the test data interval is **1.96%**.

The predicting process is tested based on the obtained model (Fig. 35) [66].

There are the on-line tools that provide the energy consumption, and energy forecast based on machine learning (ML) for one country [70] (Fig. 36).

The algorithms of machine learning are classified in the following three groups (Table 2):

1. Supervised learning: regression, decision tree. Random forest, KNN, logistic regression, so on.

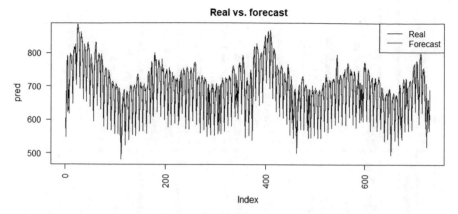

Fig. 34 The comparison between real data and forecast data

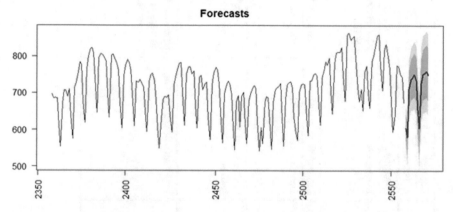

Fig. 35 Forecasting of the energy demand (GWh)

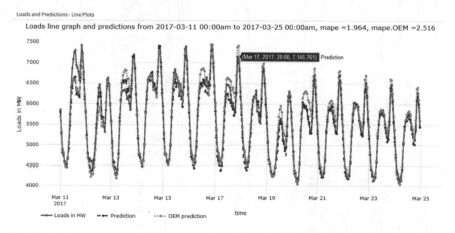

Fig. 36 Loads and predictions—line plots 11 Mar 2017–25 Mar 2017 [65]

Table 2 ML algorithms

Supervised learning		Unsupervised learning		Reinforcement learning
Classification	Regression	Clustering	Dimensionality reduction	
Random forest	Linear	K-means	Association Rules	Q-Learning
Support vector machines	Logistic	K-medians	*PCA*, Factor Analysis, ICA, t-SNE	Policy evaluation: – Temporal difference methods – Monte-carlo methods
Neural networks	Polynomial	Mean-shift clustering	Missing Value Ratio	Deep adversarial networks
Naïve Bayes classifier	Stepwise	Density-based spatial	Low variance filter	Policy gradients: – Actor-critic algorithm – Deep deterministic policy gradient
k-nearest neighbor	Support vector	Clustering of applications with noise	High correlation filter	Natural policy gradient – TRPO- trust region policy optimization – PPO- optimize natural policy gradient
Boosted trees: gradient boosting algorithm GBM; XGBoost; LightGBM.	Ordinary least squares regression	Expectation–maximization	Backward feature elimination Forward feature selection	Model-based reinforcement learning: – Model predictive control) – Policy search with backpropagation – Policy search with Gaussian Process (GP)

(continued)

Table 2 (continued)

Supervised learning			Unsupervised learning		Reinforcement learning
Classification	Regression		Clustering	Dimensionality reduction	
Discriminant analysis	Multivariate adaptive regression splines		Clustering using Gaussian mixture models	Principal Component Analysis	Guided policy search (GPS): – Deterministic GPS – Stochastic GPS
Decision Trees	Locally estimated scatterplot smoothing		Agglomerative hierarchical clustering	Methods based on projections: t-distributed stochastic neighbor embedding (t-SNE); UMAP	Imitating optimal control: – PLATO algorithm, Dyna-Q Cross-entropy method (CEM) Double gradient descent (DGD) ε-greedy policy Generalized advantage estimation (GAE)

Fig. 37 Cost function
evolution

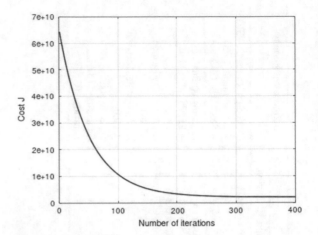

2. Unsupervised learning: a priori algorithm, K-means.
3. Reinforcement learning: Markov's decision-making process.

In the following, the specified characteristics of 10th most used machine learning algorithms are reminded.

1. Linear regression (LR)

Based on continuous variables LR is used to estimate real values. The relationship between the variables consists of the best line fitting. The LR is linear equation (*regression line*) described by $Y = b * X + c$, where the Y is dependent variable; X independent variable; coefficients c (constant), b (regression coefficient).

LR is divided in: *simple* LR and *multiple* LR. The *simple* LR is characterized by an independent variable. According to the name, *multiple* LR contains more than one independent variable. *Multiple* LR consists of finding the best match line. Other regression type can be *polynomial* or *curvilinear regression*.

LR with one variable

In Fig. 37 the learning rate evolution is shown.

The problem formulation consists in divide in consumers or not consumers of a population sample in rural area over 20.00 o'clock. Based on the existing data, the Machine Learning (ML) could be applied in order to predict the energy consumers in rural area after 20.00. By applying ML the surface plot delivers the minimum for the cost function vectorized. The obtained theta parameter value, the gradient descent algorithm will classify with high accuracy the output predicted data set (different from the trained set) (Figs. 38 and 39).

The contour plot (Fig. 40) delivers the optimal parameters values.

After calculating the parameter, the obtained model can be used to predict for different input data.

Fig. 38 The classification based on linear regression

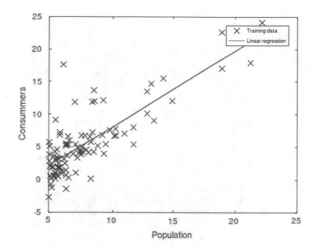

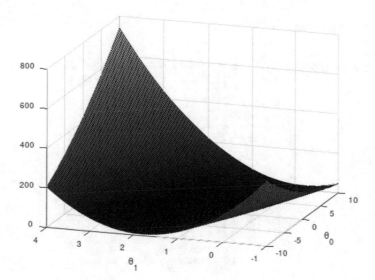

Fig. 39 The surface plot

Application: SCADA Water tank

The polynomial regression fit, with non-zero learning rate, predicts the water flowing of the dam by varying the water level parameter. The error between the initial (Fig. 41) and predicted (Fig. 42) data is zero (Fig. 43).

The training data (Fig. 44), the normalized training set (Fig. 44), and the learning rate (Fig. 44) are presented accordingly.

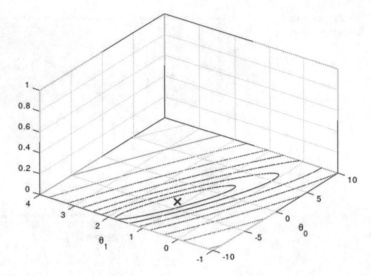

Fig. 40 The contour plot

Fig. 41 The real data

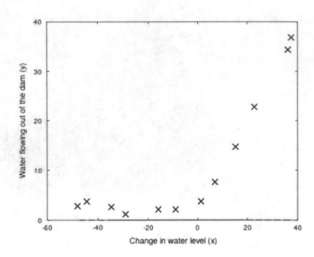

LR with multiple variables

In single and multiple cases the cost function is based on the calculation of the gradient descent algorithm.

2. Logistic regression

It is important to underline the fact that logistic regression is not a regression algorithm, as it is name, but a classification one. Logistic Regression estimates the binary outputs function (it is values lies between 0 and 1) on the input set of independent

Fig. 42 The polynomial regression fit

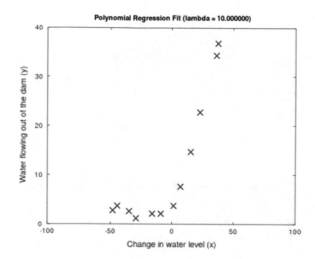

Fig. 43 The error of data predicting

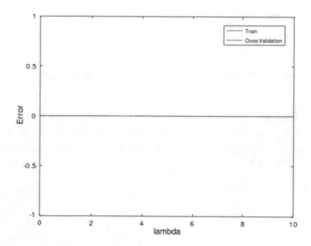

variables. In other words, logistic regression is used to predict the probability of event realization having logic function as the fitting output. In this way, the name of logistic regression is fully justified.

Logistic Regression is an extension of multiple LR, in which the dependent variable is binary, $Y \in \{0, 1\}$. Therefore, logistic regression is mainly used for binary classification.

Application: The conditional distributions of Y given X are modelled by using Bernoulli distributions. The output values of Y can be $\{0, 1\}$. The problem consists of a given set values input to be classified as the output event is performed or not. The event is the energy transfer.

Classification: Energy transfer: Yes/No;
$Y = 0$ or 1;
Logistic Regression Model- Sigmoid function;

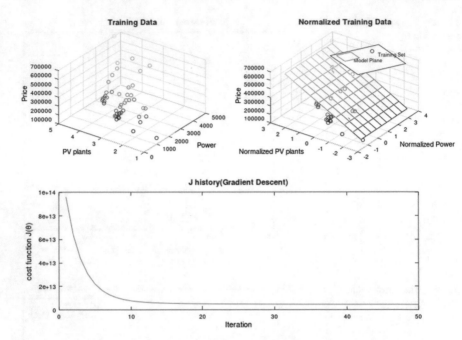

Fig. 44 The training data, the normalized training set, and the learning rate

Decision boundary: Non-linear decision boundaries;
Cost function: logistic regression Cost at initial theta (zeros): 0.693147.
By applying the logistic regression algorithm, the following Figs. (45 and 46) are
obtained:

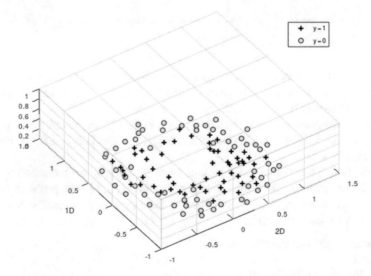

Fig. 45 Non-linear decision boundaries

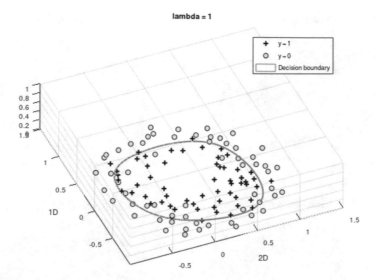

Fig. 46 Trained data with decision boundary

3. The decision tree (DT)

The DT is a nonlinear algorithm of the supervised learning technique, and it is mainly used for classification issues. The DT works for both types of dependent variables: categorically and continuously. As the application, the DT can divide the population into homogeneous data sets. To perform this, the DT uses the most significant independent variables (attributes) to perform different groups (Fig. 47).

4. SVM (Support Vector Machine)

This type of machine learning algorithm performs classification. By noting n—the numbers of the input data features, the SVM algorithm gives the value of each features (particular coordinate) into an n-dimensional space.

5. Naive Bayes (NB)

By using Bayes' Theorem (taking into account the predictor independence) the NB is a classification algorithm. The predictor independence supposes the independence of the features into a class (the presence of any feature into a class is independent to the presence of the other features. The NB algorithm is used to big data sets. The NB algorithm is simple and more efficient than the other complicated classification techniques.

6. k—Nearest Neighbors (KNN)

The KNN algorithm is useful for both categories of machine learning: classification and regression. The KNN is appropriate for industry classification issues. KNN or closest neighbors algorithm classifies the cases according to the k neighbors vote. By using a distance function (the mostly used is Euclidean distance, but for improving

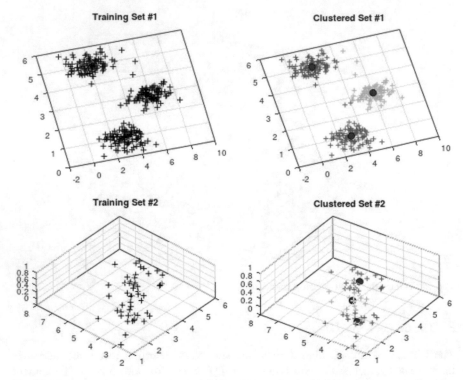

Fig. 47 The group classification of the training set. The results obtained by using k-means algorithm

purposes the Minkowski, Manhattan or Hamming distance can be used), the KNN algorithm classify the new case according to the nearest k neighbors to the specific class.

For categorical variable, the Hamming distance is used. Instead of the categorical variable, for the continuous variables the other three above mentioned distance functions are used. The closest neighbors' class is assigned for the case if the k value is unitary.

7. k-means

The unsupervised machine learning algorithms are used for clustering, and dimensionality reduction. The k-means algorithm is specific for clustering purpose. Having an input data, the k-means algorithm classifies the input data points according to k-data group (clusters). The data into a cluster are homogeneous and heterogeneous (Figs. 47).

8. Random Forest (RF)- an ensemble of decision trees
9. Dimensionality Reduction Algorithms

The algorithm of dimensionality reduction uses different learning methods, like DT, RF, Factor Analysis Missing Value Ratio, PCA, Identify Based on Correlation Matrix, so on.

10. Gradient Boosting Algorithms

Four versions of the gradient boosting algorithms are reminded as follows.

10.1. Gradient Boosting Machines (GBM)

The boosting algorithm GBM can be used for big data for classifying, regression and ranking (different objective functions). By applying GBM high prediction can be obtained. The robustness of the single estimator is highly increased through a set of learning algorithms that uses a prediction combination of several core predictors. The strategy of the GBM algorithm is to combine numerously low/medium predictors with a robust predictor.

10.2. eXtreme Gradient Boosting (XGBoost)

For ranking and classification purposes, the eXtreme Gradient Boosting algorithm is the best choice (it is used in Kaggle competitions) for event accuracy.

XGBoost increase the existing gradient algorithms being $10\times$ times faster. XGBoost is the best choice to discriminate the winning and losing competitors.

10.3. Light Gradient Boosting Machines (LightGBM)

One of the machine learning with tree-based algorithm is Light Gradient Boosting Machines. LightGBM is more powerful than XGBoost (speed up the training process for the same accuracy), it is a gradient improvement method based on tree learning algorithms.

10.4. Categorical + Boosting (Catboost)

Categorical + Boosting includes the implementation of the ML algorithms based on the gradient boosted decision tree (XGBoost, LightGBM). The CatBoost can handle different data formats and requires a small data training comparative with the other ML models.

Catboost manipulates categorical variables by helping to find the adequate model without displaying the errors.

5 Intrusion Detection Algorithms

The anomaly detection algorithm based on Machine Learning will be applied. The objective is to identify the failing servers into a SCADA network. The main features measure two variables: through-put (mb/s) and latency (ms) of response of each server. Because majority of the servers are supposed to be operating normally (non-anomalous data), some of them might acts as anomalously in the collected data. Therefore, a Gaussian model is used to find out the anomalous servers. The 2D dataset is used. A Gaussian distribution is fitting on the data; the resulted values with low probabilities will be considered anomalies. By going forward, the anomaly detection algorithm will be applied to a larger dataset with increased dimensions.

Anomaly detection based on Machine Learning

The Gaussian distribution contours fit to the existing data and the algorithm classify the anomalies, by detecting them (Figs. 48 and 49).

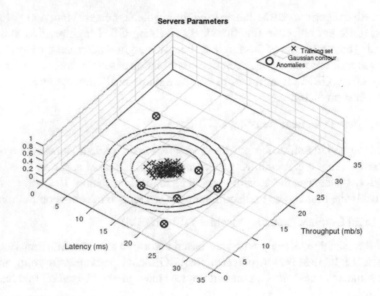

Fig. 48 The classifying of the anomalies

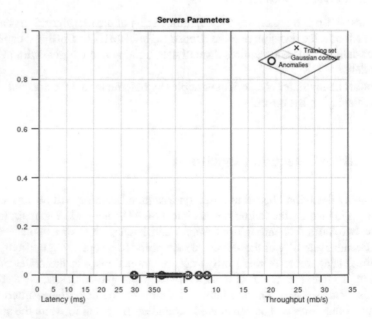

Fig. 49 Anomalies detection

Anomaly detection in classes

The blue + are normal points (placed near to origin), the red + is considered anomaly, being away from origin (Fig. 50).

The dataset contains 3 classes that can be viewed as three groups in the Fig. 51. Each group contains the anomalies (Fig. 51).

SCADA networks—anomaly-detection

The goal is to detect the anomalies in SCADA servers. The features are extracted from each server response: through-put (mb/s) -latency (ms). Mostly dataset are non-anomalous, but there is suspicion of getting anomalously in the dataset.

In order to detect the anomalous data the Gaussian distribution model is used. The data with low probability value are classified as anomalies.

Fig. 50 One class available data

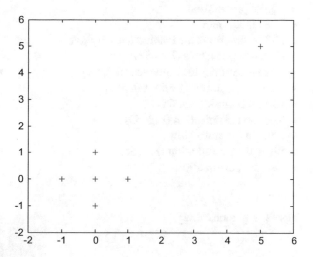

Fig. 51 The anomalies detected in three classes

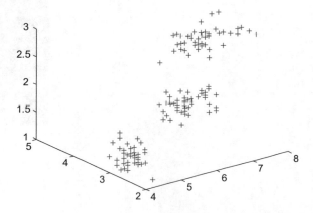

Case study: ShapeShifter Attack on the Faster R-CNN object detectors into an autonomous vehicle, connected to traffic network.

For one trained machine learning model E and an inoffensive instance $x \in X$ (correctly classified by E). The attacker has the goal to find different instance $x' \in X$, such that $E(x') \neq E(x)$ and distance metric $d(x, x') \leq \epsilon$ sufficiently small, with specified non-zero perturbation budget ϵ.

Faster R-CNN [71] object detector is attacked.

In this example a stop sign is choose to be perturbed and detected as the particular image, different from traffic signs.

The target class that will be perturbed stop sign (Fig. 52).

Stages:

- Model preparation
- Load label map
- Utility functions for loading and displaying images
- Create target labels (Fig. 53)
- Load the model for inference (Fig. 54)
- Define the plotting utility function
- Load the mask (Fig. 55)
- Run the optimization (Fig. 55)
- Load the perturbation
- Show the perturbation (Fig. 56)
- Test the perturbation

Fig. 52 The normal data

Fig. 53 Create target labels.
Load the model for inference

Fig. 54 Load the model for
inference

Fig. 55 Load the mask

Fig. 56 Test the anomalies
data

Conclusion: the vulnerability in already-learned object detectors is shown. Therefore, the security system should take into account the possible attack by the adversial to object detection systems.

6 Case Studies: Machine Learning Approach for Intrusion Detection

First attack on industrial controllers (Siemens 315, and 417) into a SCADA system was in 2010, according to [1]. It was an *autonomous attack*, without access to internet, through an USB pen drive into a local network from Natanz (Iran) fuel (uranium) enrichment plant (FEP). Stuxnet infected the Windows based computers through the communication link (Ethernet, Profibus, or MPI Siemens link). The SCADA system communicates with the controller through the programming software, therefore the DLL driver of the vendor was used to penetrate the fingerprinting process. The matched fingerprint enabled the Stuxnet code injection. There were used different kinds of attacks: on 325 controllers –Dead Foot—0xDEADF007 [2] and on the 417 controllers—*man-in the-middle*.

Despite of the computer controllers, the industrial controllers acts on the primary equipment (pump, valve so on)—physical one. Taking into consideration this, the intentional damage could take place.

Lagner provides a future solution to avoid the destruction cause by the Stuxnet: to monitor the industrial controllers to any change.

The first oriented cyber-attack on a Power Grid infrastructure was in the last month of the 2015 in Ukraine [3, 4]. The objective was to make a power outage of the utility grid. The remote control coordinated attack on distribution management system (DMS) of Ukrainian Kyivoblenergo energy [5] distribution company caused, for three hours, disconnection of 30 substations (23 of 35 kV, and seven of 110 kV). This intentional attack caused a black-out supported by 225,000 of customers [3]. Two SCADA Hijacking approaches were developed by the attackers [6].

Both the E-ISAC and SANS ICS (Industrial Control System) team analysed the cyber-attacks [4]. The attackers used a variety of techniques: e-mails phishing, advanced developed BlackEnergy 3 malware, the malware infection of the Microsoft

Office to access the Information Technology Network (ITN) of the electricity distribution company, theft of the credentials of the electric company, use of virtual private network tool to penetrate the network, manipulation of the remote access tools, serial to Ethernet communication Windows process was deleted by KillDisk, erasing the vital information on the master boot with a modified KillDisk, denial-of service attack toward the existing call center. In this way, the attackers used the control system, not the SCADA network environment [5]. As a solution, monitoring of network security was chosen as one of the active defence measures. In order to show the cyber-attack kill chain a Purdue model mapping was analysed in deep [6]. Two stages were identified: intrusion, and ICS attack. The phases of the cyber-attack kill chain are as follows: 1. reconnaissance; 2. weaponization and/or targeting; 3. delivery, 4. exploit, 5. install/modify- during this phase the malicious documents of the Microsoft Office package were sent to the users via e-mail (installation of the malicious software- KillDisk). The opening of the document enabled the Black-Energy 3 malware; 6.command and control—enabled the communication with the attackers through the control IP address; 7. Action and cyber-attack (execute the attack—through the Human Machine Interface of the SCADA system; the physical breakers were opened). Therefore, a cyber-attack can be viewed as a perturbation of the power system [6].

Based on these two different ways of attacks (stand-alone: Stuxnet, and remote control of DMS- Ukrainian attack), the authors intend to deliver the efficient solutions for intrusion protection on Industrial Control Systems and SCADA networks.

I. Robust estimation of the state space values

Case 1: the estimation of the real current value fails when the power system is attacked (Fig. 57)

The classical estimation theory fails under the attack conditions.

Case 2: the robust estimator can achieve the predicted normal value of the real system in conditions of attack (Fig. 58).

II. Stealthy Attack on Industrial Control System

By monitoring the time series from the sensors of the real ICS and applying stealthy-attack detection mechanism [67–70], [72] the changes into the process structure can be found (Figs. 59, 60, 61, 62, 63 and 64).

Significant changes appear between 4000 and 4800 values of index (Fig. 63).

III. Online Network Intrusion Detection System (NIDS)

The plug and play Network Intrusion Detection System, namely Kitsune, can detect attacks on-line, without being necessary supervision. The algorithm KitNET [73] is based on the ensemble of neural networks or autoencoders (Figs. 65 and 66).

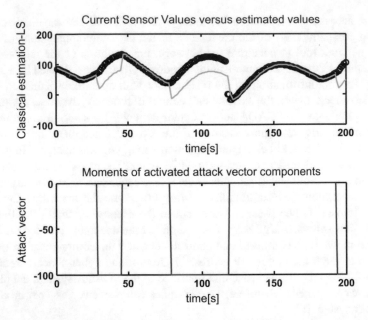

Fig. 57 The current sensor real value (black) and the estimated value (green) under attack (red)

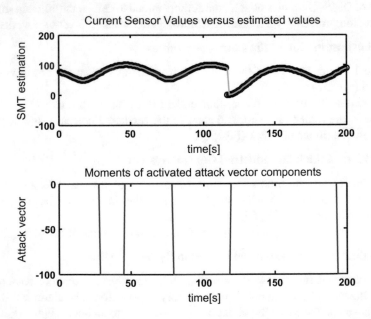

Fig. 58 The current sensor real value (black) and the estimated value (green) under attack (red)

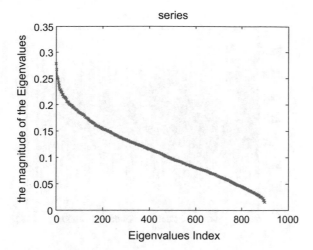

Fig. 59 The eigenvalue magnitudes

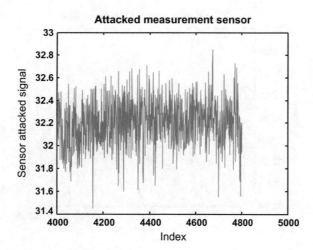

Fig. 60 The perturbation viwed as attacked vector

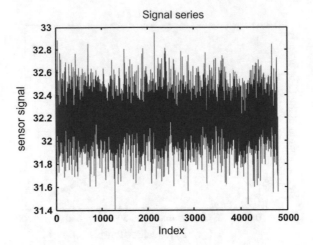

Fig. 61 The real data measurement series

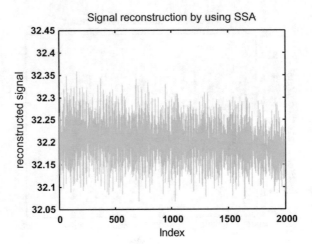

Fig. 62 The signal reconstruction with SA algorithm

Fig. 63 The amplitude scores: detection phase

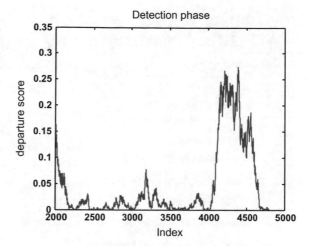

Fig. 64 The attack process on the sensor

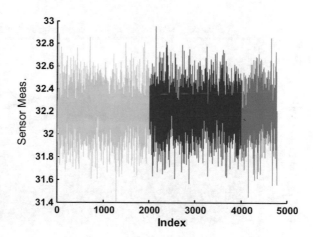

The anomalies detection by using KitNET [73]: the lower red line indicates the safety value for which no false positive alarms can occurs. The yellow line produces alert, the vulnerabilities tool detection check the devices; the upper blue line indicates an attack.

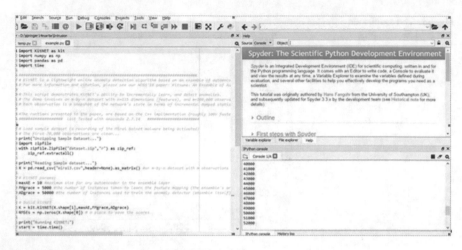

Fig. 65 Python implementation of KitNET

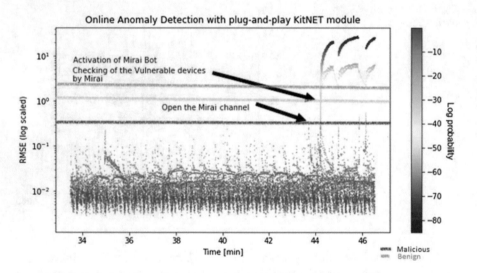

Fig. 66 The attack detection and system data protection under KITNET module

IV. Memetic Algorithm for Energy-Efficient Coverage Control in Cluster-Based Wireless Sensor Networks CoCMA

In Fig. 67 the uniform distribution of 100 sensor nodes (left) in a 50 m × 50 m sensing field area and the specific point of interest (POI) are shown.

In Fig. 68 the random distribution of the sensor nodes such that the energy efficiency is attained by applying CoCMA [74]. The yellow star-shaped points represent POIs, and the blue circles represent the sensing ranges for nodes (Figs. 69 and 70).

Fig. 67 Uniform
distribution of 100 sensor
nodes

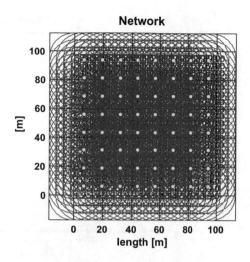

Fig. 68 Optimal efficiency
distributed sensor nodes

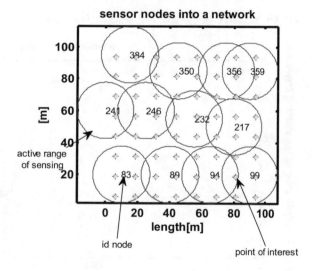

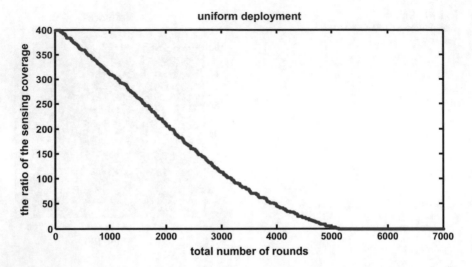

Fig. 69 Sensing coverage ratio versus the number of rounds in the uniform deployment scenario

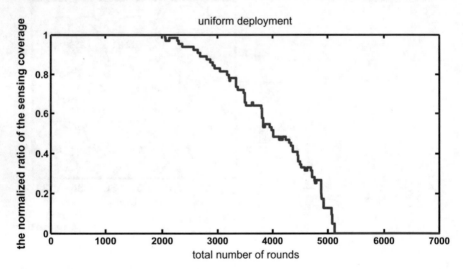

Fig. 70 The normalized ratio of the sensing coverage

7 Conclusion

The importance of the cyber-security on the ICS and SCADA is underlined through the examples. The comparisons between the conventional and modern industrial control systems, along with the security functions are highlighted. An overview of planning the adequate security solutions is delivered. The guideline to design the cyber-security for IDS and SCADA networks is highlighted through the specific standards. The simulation results of the Intrusion detection algorithms are delivered in this chapter. Different case studies approach for intrusion detection into ICS/SCADA systems are implemented, and the obtained numerical results are presented in this chapter.

References

1. Falco, J.: IT Security for Industrial Control Systems. NIST IR 6859. (2003). http://www.isd. mel.nist.gov/documents/falco/ITSecurityProcess.pdf
2. Irfan, N., Mahmud, A.: A novel secure SDN/LTE based architecture for smart grid security. In: Proceeding of IEEE International Conference on Computer and Information Technology (2016)
3. Machii, W., Kato, I., Koike, M., Matta, M., Aoyama, T., Naruoka, H., Koshima, I., Hashimoto, Y.: Dynamic zoning based on situational activitie for ICS security. In: IEEE 978-1-4799-7862-5/15 (2015)
4. Intel Corporation.: Reducing Cost and Complexity with Industrial System Consolidation. Retrieved March 2016 from: http://www.intel.com/content/www/us/en/industrial-automation/reducing-cost-complexity-industrial
5. Mix, S.: Supervisory control and data acquisition (SCADA) systems security guide. Electr. Power Res. Inst. (EPRI) (2003)
6. Duggan, D.: Penetration testing of industrial control systems. Sandia National Laboratories, Report No SAND2005-2846P (2005)
7. Stamp, J.: Common vulnerabilities in critical infrastructure control systems. Sandia National Laboratories. (2003). http://citeseerx.ist.psu.edu/viewdoc/download?doi=10.1.1.132.3264&rep=rep1&type=pdf
8. K. Scarfone, and P. Mell, NIST SP 800-94, "Guide to Intrusion Detection and Prevention Systems (IDPS)", Feb. 2007, http://csrc.nist.gov/publications/PubsSPs.htmlhttp://csrc.nist.gov/publications/PubsSPs.html#800-94
9. Rinaldi, S.: Identifying, understanding, and analyzing critical infrastructure interdependencies. IEEE Control Syst. Mag. **3**, 11–25 (2001)
10. Matthew, F.: Vulnerability testing of industrial network devices. In: Critical Infrastructure Assurance Group, Cisco Systems (2003). http://blogfranz.googlecode.com/files/franz-isa-device-testing-oct03.pdf
11. Peerenboom, J.: Infrastructure interdependencies: overview of concepts and terminology. Invited paper, NSF/OSTP Workshop on Critical Infrastructure: Needs in Interdisciplinary Research and Graduate Training, Washington, DC, 14–15 June 2001
12. Boyer, S.: SCADA: Supervisory Control and Data Acquisition, 4th edn. Research Triangle Park, North Carolina: International Society of Automation (2010)
13. Fraser, R.E.: Process Measurement and Control: Introduction to Sensors, Communication, Adjustment, and Control, Upper Saddle River. Prentice-Hall Inc, New Jersey (2001)
14. Knapp, E.: Industrial Network Security: Securing Critical Infrastructure Networks for Smart Grid, SCADA, and other Industrial Control Systems, Waltham. Syngress, Massachusetts (2011)

15. Bailey, D., Wright, E.: Practical SCADA for Industry. IDC Technologies, Vancouver (2003)
16. SCADA Security.: Advice for CEOs, IT Security Expert Advisory Group (ITSEAG)
17. American Gas Association.: AGA Report No. 12, Cryptographic Protection of SCADA Communications, Part 1: Background, Policies and Test Plan, 14 Sep, Mar 2006
18. Stanculescu, M., Badea, C.A., Marinescu, I., Andrei, P.C., Drosu, O., Andrei, H.: Vulnerability of SCADA and security solutions for a waste water treatment plant. In: Proceeding of IEEE-ATEE (2019)
19. Peterson, D.: Intrusion detection and cyber security monitoring of SCADA and DCS networks. ISA Automation West (AUTOWEST 2004), Long Beach, California, Apr 2004
20. https://ws680.nist.gov/publication/get_pdf.cfm?pub_id=902622
21. Duff, W.G.: Handbook Series on Electromagnetic Interference and Compatibility, vol. 7
22. Stouffer, K., Abrams, M.: Guide to Industrial Control Systems (ICS) Security, pp. 800–82. NIST Special Publication (2013)
23. Guide to Securing Legacy IEEE 802.11 Wireless Networks, NIST Special Publication 800–48 Rev. 1
24. https://nvlpubs.nist.gov/nistpubs/Legacy/SP/nistspecialpublication800-48r1.pdf
25. Establishing Wireless Robust Security Networks: A Guide to IEEE 802.11i
26. https://nvlpubs.nist.gov/nistpubs/legacy/sp/nistspecialpublication800-97.pdf]
27. http://etutorials.org/Networking/802.11+security.+wi-fi+protected+access+and+802.11i/Part+II+The+Design+of+Wi-Fi+Security/Chapter+7.+WPA+RSN+and+IEEE+802.11i/What+Is+IEEE+802.11i/
28. https://en.wikipedia.org/wiki/IEEE_802.11i-2004
29. http://etutorials.org/Networking/802.11+security.+wi-fi+protected+access+and+802.11i/Part+II+The+Design+of+Wi-Fi+Security/Chapter+7.+WPA+RSN+and+IEEE+802.11i/Differences+Between+RSN+and+WPA/
30. https://ec.europa.eu/eip/ageing/standards/ict-and-communication/interoperability/iec-61784_en)
31. Cybersecurity Interdisciplinary Systems Laboratory (CISL) Sloan School of Management, Room E 62-422 Massachusetts Institute of Technology Cambridge, MA 02142, https://cams.mit.edu/wp-content/uploads/2016-22.pdf
32. SP 800-48, Guide to Securing Legacy IEEE 802.11 Wireless Networks (2008)
33. The IAONA Handbook for Network Security Version 1.5—Magdeburg, June 6th 2006 http://www.ininet.ch/vpi-initiative/download/IAONA-Security-Guide-15-draft.pdf
34. Control Systems Cyber Security: Defense in Depth Strategies, David Kuipers Mark Fabro, May (2006). https://inldigitallibrary.inl.gov/sites/sti/sti/3375141.pdf
35. C37.1-2007-IEEE Standard for SCADA and Automation Systems, https://ieeexplore.ieee.org/document/4518930
36. https://www.isa.org/intech/201810standards
37. https://www.iec.ch/
38. https://www.iec.ch/cybersecurity/?ref=extfooter
39. Keith Stouffer Victoria Pillitteri Suzanne Lightman Marshall Abrams Adam Hahn, NIST Special Publication 800-82 Revision 2Guide to Industrial Control Systems (ICS) Security Supervisory Control and Data Acquisition (SCADA) Systems, Distributed Control Systems (DCS), and Other Control System Configurations such as Programmable Logic Controllers (PLC). (2015). https://nvlpubs.nist.gov/nistpubs/specialpublications/nist.sp.800-82r2.pdf
40. The North American Electric Reliability Council (NERC). http://www.nerc.com. Accessed 2019
41. https://www.nerc.com/pa/Stand/Cyber%20Security%20Permanent/Cyber_Security_Standards_Board_Approval_02May06.pdf
42. https://www.nist.gov/
43. Lemaire, L., Vossaert, J., De Decker, B., Naessens, V.: Extending FAST-CPS for the analysis of data flows in cyber-physical systems. In: Rak, J., Bay, J., Kotenko, I., Popyack, L., Skormin, V., Szczypiorski, K. (eds) Computer Network Security. MMM-ACNS 2017. Lecture Notes in Computer Science, vol. 10446. Springer, Cham (2017)

44. Lee, E.A.: Cyber physical systems: design challenges. In: Object Oriented Real-Time Distributed Computing (ISORC), 2008 11th IEEE International Symposium on, pp. 363–369. IEEE, 2008
45. Wang, E.K., Ye, Y., Xu, X., Yiu, S.M., Hui, L.C.K., Chow, KP.: Security issues and challenges for cyber physical system. In: Proceedings of the 2010 IEEE/ACM Int'l Conference on Green Computing and Communications and International Conference on Cyber, Physical and Social Computing, pp. 733–738. IEEE Computer Society (2010)
46. Chapman, J.P., Ofner, S., Pauksztelo, P.: Key factors in industrial control system security. In: Local Computer Networks (LCN), 2016 IEEE 41st Conference on, pp 551–554. IEEE (2016)
47. Sadeghi, A.R., Wachsmann, C., Waidner, M.: Security and privacy challenges in industrial internet of things. In: Proceedings of the 52nd Annual Design Automation Conference, 54p. ACM (2015)
48. Gligor, V.D., Pompiliu, D.: Block encryption method and schemes for data con_dentiality and integrity protection. US Patent 6,973,187. 6 Dec 2005
49. Agrawal, S., Boneh, D.: Homomorphic macs: Mac-based integrity for network coding. In: International Conference on Applied Cryptography and Network Security, pp. 292–305. Springer, 2009
50. Neuman, C.: Challenges in security for cyber-physical systems. In: DHS Workshop on Future Directions in Cyber-Physical Systems Security, pp. 22–24. Citeseer (2009)
51. McLaughlin, S., Konstantinou, C., Wang, X., Davi, L., Sadeghi, A.-R., Maniatakos, M., Karri, R.: The cybersecurity landscape in industrial control systems. Proc. IEEE **104**(5), 1039–1057 (2016)
52. Walker, M., Reiser, M.O., Tucci-Piergiovanni, S., Papadopoulos, Y., Lönn, H., Chokri, M., Parker, D., Chen, D., Servat, D.: Automatic optimization of system architectures using east-adl. J. Syst. Softw. **86** (10): 2467–2487 (2013)
53. Abrams, M., Weiss, J.: Malicious control system cyber security attack case study- maroochy water services, Australia (2008)
54. Lee, R.M.: The Industrial Cyber Threat Landscape, The Role of The Private Sector And Government in Addressing Cyber Threats to Energy Infrastructure. Dirksen Senate Office Building. 1 Mar 2018. https://www.energy.senate.gov/public/index.cfm/2018/3/full-committee-hearing-to-examine-cyber-security-in-our-nations-critical-energy-infrastructure-030118
55. Mohurle, S., Patil, M.: A brief study of wannacry threat: Ransomware attack 2017. Int. J. **8**(5) (2017)
56. Industrial Risk Assessment Map v2 (IRAM) HD. www.scadacs.org
57. http://ocularwarfare.com/
58. Laurens, L., Vossaert, J, De Decker, B., Naessens, V.: Assessing the Security of an Industrial. Hatchery using the FAST-CPS Framework, Report CW710, Dec 2017
59. http://ics-cert.us-cert.gov/Assessments Homeland Security. Cset: Cyber security evaluation tool (2014)
60. LeMay, E., Ford, M.D., Keefe, K., Sanders, W.H., Muehrcke, C.: Model-based security metrics using adversary view security evaluation (advise). In: Quantitative Evaluation of Systems (QEST), 2011 Eighth International Conference on, pp. 191–200. IEEE (2011)
61. Ford, M.D., Keefe, K., LeMay, E., Sanders, W.H., Muehrcke, C.: Implementing the advise security modeling formalism in mobius. In: Dependable Systems and Networks (DSN), 2013 43rd Annual IEEE/IFIP International Conference on, pp. 1–8. IEEE (2013)
62. Vu, A.H., Tippenhauer, N.O., Chen, B., Nicol, D.M,, Kalharczyk, Z.: Cybersage: a tool for automatic security assessment of cyber-physical systems. In: International Conference on Quantitative Evaluation of Systems, pp. 384–387. Springer (2014)
63. Sommestad, T., Ekstedt, M., Holm, H.: The cyber security modeling language: A tool for assessing the vulnerability of enterprise system architectures. Syst. J. IEEE **7**(3), 363–373 (2013)
64. https://www.nationalgrideso.com/document/61391/download
65. https://people.duke.edu/~rnau/411arim.htm

66. https://towardsdatascience.com/time-series-forecasting-arima-models-7f221e9eee06. https://www.kaggle.com/rihadv
67. https://machinelearningmastery.com/arima-for-time-series-forecasting-with-python/
68. Fanaee-T, H., Gama, J.: Event labeling combining ensemble detectors and background knowledge. In: Progress in Artificial Intelligence, pp. 1–15. Springer Berlin Heidelberg. (2013). http://link.springer.com/article/10.1007/s13748-013-0040-3
69. Lichman, M.: UCI Machine Learning Repository [http://archive.ics.uci.edu/ml]. Irvine, CA: University of California, School of Information and Computer Science (2013)
70. https://pk-shinies.shinyapps.io/ipto-ml/#section-loads-descriptives-statistics
71. Chen, S.T., Cornelius, C., Martin, J., Chau, D.H.P.: ShapeShifter: robust physical adversarial attack on faster R-CNN object detector. Georgia Institute of Technology, Atlanta, GA, USA
72. Aoudi, W., Iturbe, M., Almgren, M.: Truth will out: departure-based process-level detection of stealthy attacks on control systems. In: Proceedings of the ACM Conference on Computer and Communications Security: pp. 817–831. (2018). http://dx.doi.org/10.1145/3243734.3243781
73. Mirsky, Y., Doitshman, T., Elovici, Y., Shabtai, A.: Kitsune: An Ensemble of Autoencoders for Online Network Intrusion Detection. arXiv:1802.09089v2 [cs.CR] 27 May 2018
74. Jiang, J.A., Chen, C.P., Chuang, C.L., Lin, T.S., Tseng, C.L., Yang, E.C., Wang, Y.C.: CoCMA: energy-efficient coverage control in cluster-based wireless sensor networks using a memetic algorithm. Sensors. **9**, 4918–4940 (2009). https://doi.org/10.3390/s90604918, ISSN 1424-8220

Security Evaluation of Sensor Networks

Horia Andrei, Marian Gaiceanu, Marilena Stanculescu, Ioan Marinescu
and Paul Cristian Andrei

Abstract Traditional industrial control systems (ICS) were implemented especially
for isolated system and used specialized hardware and software control protocols.
Along with development of low-cost Internet Protocol (IP) devices, the ICS adopting
new information technologies (IT) solutions to promote systems connectivity and
data communication.

Keywords Control systems · Sensor networks · Security analysis · Defense
solutions

1 Introduction. Security Evaluation

Traditional industrial control systems (ICS) were implemented especially for isolated system and used specialized hardware and software control protocols. Along
with development of low-cost Internet Protocol (IP) devices, the ICS adopting new
information technologies (IT) solutions to promote systems connectivity and data
communication. ISC include Supervisory Control and Data Acquisition (SCADA)
systems, distributed control systems (DCS), and other control system configurations

H. Andrei (✉) · I. Marinescu
Doctoral School of Engineering Sciences, University Valahia Targoviste, Targoviste, Romania
e-mail: hr_andrei@yahoo.com

I. Marinescu
e-mail: ioan.marinescu@yahoo.com

M. Gaiceanu
Department of Control Systems and Electrical Engineering, Dunarea de Jos University of Galati,
Galati, Romania
e-mail: marian.gaiceanu@ugal.ro

M. Stanculescu · P. C. Andrei
Department of Electrical Engineering, University Politehnica Bucharest, Bucharest, Romania
e-mail: marilena.stanculescu@upb.ro

P. C. Andrei
e-mail: paul.andrei@upb.ro

© Springer Nature Switzerland AG 2020   263
E. Pricop et al. (eds.), *Recent Developments on Industrial Control Systems Resilience*,
Studies in Systems, Decision and Control 255,
https://doi.org/10.1007/978-3-030-31328-9_11

such as skid-mounted Programmable Logic Controllers (PLC) [1, 2]. The data processed by ICS are provided by sensors, as smart Wireless Sensor Network (WSN), which are responsible for monitoring, recording and remote communicating of real physical data to a central location [3].

ICS are used in important industry sectors such as energy, oil and natural gas, water and waste water treatment, transportation, chemical, pharmaceutical, automotive, aerospace, food and beverage etc. The data processed by ICS are provided by sensors, as smart Wireless Sensor Network (WSN), which are responsible for monitoring, recording and remote communicating of real physical data to a central location. The wireless sensors include sensors of electrical parameters, temperature, humidity, gravity, pressure, and lighting [4, 5].

Besides the advantages of using new IT technologies such as a fast process control, an increase of connectivity to other systems, a great flexibility and reliability, they reduce the isolation from external disturbing factors. These vulnerabilities of modern ICS are all the more important and can cause great damage as they relate to the real parameters of systems that control vital industries [6].

The difficulty and complexity of the ICS transition from the traditional to the intelligent state consists of the fact that this must be done in order to solve the present problems and to prevent those which the system might face in the future by allowing the new technological innovations. All of these must take place while the system is in operation, without affecting the consumers. At the same time, the security attacks associated with the traditional industrial system, based on a small number of challenges, are increasing with particular emphasis on network sensors data and communicating components. About these threats and potential attacks, the authors refer in the second section of this chapter.

Considering the technical requirements of the ICS, cyber-security solutions will be presented. General issues and development trends in this area such as cryptography, attack detections and preventions, secure routing and location security, are mentioned in the next section.

In order to provide an example of potential attacks and solutions for restoring the data provided by network sensors, the authors propose an implementation of a system under sensors cyber-attack which will detect the attack and remove it, restoring the normal operation of the system. Imagining a cybernetic attack on the values of electrical parameters provided by the network sensors and by the data acquisition system, the authors suggest some cryptography solutions to protect and to keep the data unchanged. The implementation methods and advantages of these algorithms are presented in the fourth paragraph of this chapter. In order to implement a structural cyber-system, the authors developed a mathematical model of the cyber-attacks. One problem that comes with the detection of anomalies in an industrial cybernetic system is the formal verification of software and hardware through the SMT (Satisfiability Modulo Theories) approach. Constraint Satisfaction Problems (CSP) can be solved with programming tools called constraint programming (CP). In CSP relationships between variables are declared as constraints. Constraints are specified by domains: integer numbers, real, rational, logical (boolean). Solving the CSP problem is finding an allocation to variables that satisfy all constraints. To solve CSP problems, search algorithms based on CP are used: Boolean satisfiability problem (SAT solver), genetic algorithms, mixed integrated linear programming (MILP), so on. Satisfiability Modulo Theories (SMT) is a decision problem for logical formulas that determine whether a first-order formula is certifiable in terms of a logical theory. Modern, SMT is applied in establishing the necessary and sufficient conditions to estimate the states of a discrete system based on a number of attacked sensors. This estimator is robust to noise and can provide the states of the original system under cyber-attack conditions.

The chapter ends with conclusions and a large number of references in the field of security evaluation and solutions.

The chapter is organized as follows: Sect. 2 presents an overview of the threats and attacks on sensor networks of ICS. The cyber-security solutions and cryptography methods are described in Sect. 3. In Sect. 4 is analyzed an implementation example of a system under sensors cyber-attack will detect the attack and remove it, restoring the normal operation of the system. The conclusions are drawn in Sect. 5. At the end of the chapter the list of bibliographic references is presented.

2 Threats and Attacks

Over time, the dependence of the society (domestic and industrial consumers, government institutions, the business environment, education, public or private medical services and other critical infrastructure operators) on energy infrastructure has increased, implying the need to implement reliability and safety policies to prevent black-outs or dysfunctions, regardless of the causes that could be the origin of these unpleasant events.

Physical security means the security of the power system against physical attacks by individuals or organizations, with the intention of destroying its key points, disrupting its operations (sabotage, terrorist attack) or removing valuable parts as a theft (for example, stealing copper conductors).

Cyber security refers to protecting the power system against threats of theft/destruction/manipulation of company data or customer information databases or manipulating sensors and equipment to interrupt activity.

Although different, the two domains interconnect, the vulnerabilities being complex, and a possible attack aimed at the malfunction of the power system will be carried out on both planes simultaneously.

Although it is a major development in power systems, SCADA adds cyber vulnerabilities that can be exploited by an expanding range of attackers (hostile states, competing companies, profit hackers, terrorist organizations, etc.).

As far as the beginning of this system is concerned, nowadays SCADA solutions are currently being implemented with standardized electronics (usually communicating over the Internet) and component information is easily accessible on-line, for any individual [7].

The quoted article models the consequences of a cyber-attack on the Automatic Generation Control (AGC) which is an automatic adjustment of power produced by power plants generators depending on the power grid load in order to maintain the system's frequency constant. Since maintaining frequency at the reference value of the power system is performed automatically in SCADA, the authors consider the scenario with high probability of occurrence in practice and significant damage as follows: the attacker intercepts the read signal on the way to the command center and returns a false value, causing the generator to change its speed which will modify the system's frequency [8].

A list of interruptions of power supply services in the world is compiled by the Nation Master and is accessible at https://www.nationmaster.com/country-info/stats/Energy/Electrical-outages/Days and another list classified by the number of affected users is available on the Wikipedia platform and permanently updated at https://en.wikipedia.org/wiki/List_of_major_power_outages.

Further, we present a series of undesirable events, resulting in the interruption of electricity supply, as follows:

1. **Romania**

Shortly after the earthquake that devastated Romania on March 4, 1977 an energy black-out due to a human error was propagated in the national power system, with important economic consequences assessed with an impact on the economy more pronounced than the effects of the aforementioned earthquake. So far, no such event has occurred, but in the event of its occurrence nowadays it would result in loss of lives and billions of material damage, also affecting Romania's energy export capabilities as well.

2. **Western Europe**

   - In 2005s winter when a combination of massive snow, frost, and winds led to a five-day power outage interruption, affecting more than 80,000 consumers in Germany, Belgium and the Netherlands, causing a loss of 20 million Euros [9].
   - In 2006, when an interruption of power supply in Germany produced a cascading effect propagating through eleven European countries and affecting 15 million people for three days [10].

3. **North America**

   - On Aug. 14, 2003, the most significant power failure occurred in North America, 50,000,000 people were out of power, the return to normal lasted between 2 days and 2 weeks. Following the analysis of the causes that led to this collapse of power, the US authorities concluded the following:
     - poor implementation of vegetation management that has led to the cable branches contact;
     - the inability of operators to view events in the system;
     - failure to operate between safety margins, inefficient operational communication and coordination;

- inefficient training of operators in order to recognize and respond accordingly in case of emergency;
- inappropriate power resources [11].

The power grid is impossible to protect against every physical threat given its size as well as the fact that certain sites are located in isolated or less accessible areas, at least at certain times of the year. Thus, as in the case of combating other forms of crime, the most effective method of protection is the proactive collection of information by the intelligence structures, corroborated with the action of the police forces on the prevention, deterrence and annihilation of hostile actions on the power system before the hostile actions are effectively enforced.

- Such a scenario has already taken place in the US on April 16, 2013. Although the attack, commonly called "the sniper attack Metcalf" has not reached its goal and the subject has not been brought to public attention by mass-media, it is of particular gravity. Attackers have previously cut optical fiber to make telecommunications impossible between Metcalf transformation plant (owned by Pacific gas and electricity company, Coyote, California) and authorities. Subsequently, they opened fire on the 17 transformers in the station, who lost the cooling oil, overheating and going out of function. No one was injured or killed in this sabotage attempt, most likely executed by professionals because the authorities did not find evidence at the scene of the attack, no claim was published (which is why it is considered a sabotage action, not a terrorist attack) and no motivation was presented to public opinion. The investigation set the course of events according to Fig. 2, and stopped without identifying the guilty, while the repairs to the affected elements lasted for a month. The attack did not reach its obvious goal (disrupting power) because the company managed to redirect electricity through other distribution lines to the region, avoiding a black-out [12].

4. **Ukraine**

A country waging undeclared war with the Russian Federation has been subjected to several attacks on the power system, as follows:

- In 2015, a cyber-attack took off around 30 electrical stations by taking control of system control and data acquisition systems, disabling operators from control centers, simultaneous with rewriting component software to block any attempts to restore the power supply and then using software (Kill Disk) to erase data inside control center computers. The operation was successfully carried out through the following steps: company employees received infected emails through which attackers managed to gain access to the networks of targeted companies, then in order to gain access to SCADA they harvested employee access log-in data in order to inject malicious code into the system's closed loop. Following the initial attack on the power grid, a denial of service attack blocked the customer support center in order to keep the problems hidden from the operators and increase the time of the black-out, also targeting the customers confidence as from their point of view the companies were doing nothing to remediate the situation [13–15].

- In 2016, a cyber-attack with malware (known as Industroyer or Crash Override) especially created for an attack on power systems has partially interrupted electricity in Kiev. ESET (antivirus service developer) and Dragos (an industry-focused cyber security company) identified the abovementioned malware as the most advanced software framework targeting power grid infrastructures. It's developed for a wide area of critical infrastructures sabotage, being adaptable, scalable, and automated with potential to be used against any network in the world with minimal human effort and involvement. The above-mentioned specialists consider the attack to be a beta test of the malware in real life operation, as the magnitude of the attack and its consequences could have easily been much more severe [16–18].

5. Venezuela

The country is undergoing a humanitarian crisis, its president is not recognized by a part of the population and most of the world's states, and has closed its borders even for humanitarian aid (even though the population suffers from food and medication access shortages) under the guise of an invasion of foreign powers who condemns his regime.

In addition to these problems, in early March, the capital faced a generalized power outage due to a combined cyber-attack targeting the Simon Bolivar hydroelectric power plant generating electricity (powering almost three-quarters of the country), an attack targeting telecommunications and subsequently a series of physical attacks on power stations [19, 20].

We have previously mentioned the attack targeting the Energetic Dispatcher controlling frequency, now we present the effects of fire at a transformer station. In the documents taken from the archive of former Romanian Autonomous Electricity Registry (the company that had responsibilities in generation, transport and distribution of electricity before division and privatization) regarding the event produced at the 400/110/20 kV at Smârdan power station in Galați County on July 17, 1997 is investigated the explosion followed by fire at transformer no. 2 (250 MVA, 400/110 kV) [7]. Regarding the location and nature of the defect, the transformer's protections functioned correctly and quickly isolated the defect, corroborated with the fixed fire-extinguishing system which deployed properly.

The event investigation reveals the cause of the event: failure of the insulating passage insulation—the R 110 kV phase over the current transformer and the priming of the electric arc between the phase and the cuvette (the R phase carrier). As a result of the internal defect, a strong gas evolution resulted in an overpressure in the vessel, which resulted in the dismantling of the 400 kV insulating passages R and S with the support pans, the shearing of the bolts and the firing of the fire. The high temperature caused by the fire led to the damage of the transformer no. 2, the 400 kV and 110 kV bar fields, the tread and the fire extinguishing system.

Figure 1 highlights the damage caused by fire to a transformer, the most complex and costly component in a transformation station.

Fig. 1 Destroyed remains of a large transformer (250 MVA, 400/110 kV) from the 400/110/20 kV Smârdan station following 10 h of fire fighting

3 Defense Issues

Cyber-security solutions and cryptography methods are consequences of the fact that may people's life depends, mostly, on the information correctness and security. This derives from using, for example, on-line payment instruments, mobile phones, credit cards, bank transfers, etc. In the last few years, an increased attention is payed to the security in the strategically domain—represented by the energy system. The traditional ICS moved to the new era by quickly adapting to the new advances. For example, to ensure the connectivity and data communication, ICS encapsulates the Internet Protocols and adopted the IT solutions. A lot of key industrial sectors, such as energy, oil and natural gas, transportation, etc., are using ICS on large scale. ICS includes first, among other important features, the SCADA systems. Data processed by ICS are collected via sensors which have as main tasks the monitoring and the transmission of physical data to the place where they will be further processed, and critical decisions will be taken. These decisions can greatly affect our every day life.

One of the problems we face nowadays due to the great technological advances is those related to ensure the cyber-security. For example, a potential attack on the ICS can be represented by altering the data provided by the network sensors. We can imagine an attack against the electric parameters provided by the networks sensors and by the SCADA system. Therefore, we propose an *information protection analysis approach*. The information analysis approach leads to the identification of three

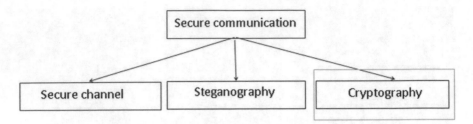

Fig. 2 Major methods for a secure communication between data transmitter and receiver

major methods categories used to realize secure communication between at least two correspondents, as presented in Fig. 2.

A secure channel is a channel to which only the respective correspondents can have access. Nowadays, in real life, this solution for ensuring the information confidentiality is no longer plausible. It has been practiced in domains such as military and diplomacy domains. Steganography is a method used for hiding the transmitted information by means of apparently inoffensive information, named support information which can be text, image etc. This method makes the information to be invisible for the unadvised. Cryptography is a method used to transmit information in an unintelligible form, but which can be recomposed by the message receiver.

In terms of cyber-security solutions for protecting the data collected from the sensors networks of ISC, an appropriate approach is to use cryptographic methods. The implementation of cryptographic solutions implies to consider the so-called cryptographic resistance. The cryptographic resistance of a system is its capacity to oppose deciphering.

Cryptographic systems offer different levels of cryptographic resistance, determined by the degree of resisting deciphering. The main indicators used in establishing the cryptographic resistance level are [21]:

– *The costs* determined by human and material resources allocated to realize the deciphering process. If these costs are bigger than the value of the obtained information, then probably the system is sure.
– *The time necessary to decipher the system*. If the necessary time exceeds the information validity period, then, probably the system is sure.

If the lengths of the messages are smaller than the unique distance of the system, then, probably the system is sure.

When stating these indicators, one used the formula *probably the system is surer* because the estimation of the costs and of the time necessary to decipher the system cannot be done precisely, not knowing exactly the potential of the cryptanalyst. More, there are no sufficient conditions to ensure a full information protection, because the results obtained in cryptanalysis experience a fast and continuous development. In addition, for any cryptographic system, for which the keys' space is finite, the system is theoretically indecipherable. This statement can be proved by deciphering the cryptograms using all possible keys and resulting a unique plaintext, if the length of the cryptogram is bigger than the unicity distance.

The sure system is the system One Time Pad. To maintain it in the indecipherable class there should be fulfilled some requirements regarding the exploitation and management of the keys.

Although there are many types of cryptographic resistance, the most important ones are the following two:

1. *Unconditioned resistance* assumes the impossibility of deciphering the system, no matter how many human and material resources are allocated. The deciphering time allocated is unlimited. The unconditioned resistance is equivalent to the perfect secret, specific to perfect systems. The only system having this property is the One Time Pad system.
2. *The computational resistance* assumes the impossibility of deciphering with available current and future human and material resources, detained by an imaginary opponent. The collocation *available resources* are interpretable and cannot be precisely estimated.

Whenever *information protection analysis approach* is under discussion, both coding (ciphering) and decoding (deciphering) process should be considered. In literature there are *four general methods*, unanimously accepted, used in deciphering process, presented below and in Fig. 3:

1. Attack based on knowing the cryptograms (*cipher text-only attack*)
2. Attack based on knowing the plaintext (*known plaintext attack*)

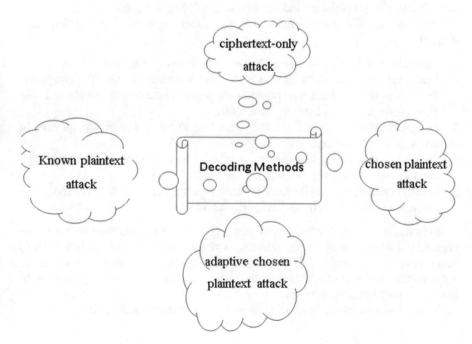

Fig. 3 General methods used in decoding process

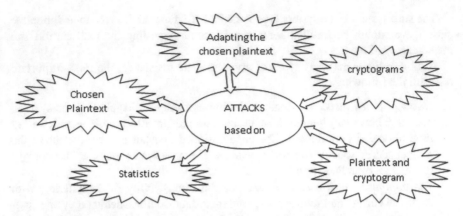

Fig. 4 Main procedures to be used for an attack

3. Attack based on a chosen plaintext (*chosen plaintext attack*)
4. Attack based on adapting the chosen plaintext (*adaptive chosen plaintext attack*).

This type of attack is a special case of the attack from point (3)—attack based on a chosen plaintext. The cryptanalyst has the possibility to adapt the plaintext during the deciphering process, function of the obtained results. If for *known plaintext attack* one starts from a very long plaintext, for the *adaptive chosen plaintext attack* one starts from a shorter plaintext and this is adapted during the process.

For each type of attack, the main procedures that can be used are the following (Fig. 4):

1. General procedures used for attacks based only on cryptograms
2. Statistical methods: correlation tests, coincidence index, dictionary-type procedures, genetic algorithm type procedures, genetic, tabu-search, divide and conquer, algebraic attack, Birthday attack etc.
3. General procedures used for attack based on knowing the plaintexts and the cryptograms (linear cryptanalysis, slide attack etc.).
4. General procedures used for attacks based on chosen plaintext (differential cryptanalysis, boomerang-type attack, rectangle attack etc.)
5. General procedures used for attacks based on adapting the chosen plaintext (differential cryptanalysis, boomerang-type attack).

To evaluate a cryptographic system important human and material resources are necessary. Human resources should have a serious mathematical and cryptology background. The evaluation of cryptographic resistance of a system assumes the cryptanalytic attack methods specific to the system under evaluation process to be known, as well as the general ones.

Among the general methods, the following can be mentioned [22]:

1. *Statistical analysis*, which assumes that the cryptograms are subjected to a set of tests which should underline the cryptographic problems of a system. This fact implies the interpretation, from cryptanalytic point of view, of the system's particularities and the elaboration of some statistics that should emphasize these problems.
2. *General cryptographic analysis method*, according to the four methods of cryptanalytic attacks.

Classical cryptographic systems are divided into two big classes:

(A) **Substitutions class**: the elementary ciphering units can be characters, bigrams, words etc., which are being replaced by uniform, not-uniform, ciphering representations, with multiple representations proportional or equal for each elementary ciphering unit. For the substitution systems case the quantitative aspects of a language are totally transferred in the cryptogram.

In classical cryptography there are *four categories of substitution systems*:

1. Simple substitutions (mono-alphabetic), which should consist of nature, number and structure of the ciphering representations.
2. Non-uniform substitutions.
3. Polygram substitutions.
4. Polyalphabetic substitutions.

(B) **The transposition class** (elementary ciphering units are keeping their identities but they lose their positions).

The transpositions consist of the following types of systems:

1. Transpositions with complete mesh.
2. Transpositions with incomplete mesh.
3. Transpositions of grid-type mesh (square or after certain predefined structure).

Cyber-security solutions and cryptography methods, when used to secure ICS, should consider both the systems requirements and the costs for implementing the appropriate solution. In these key domains and not only, the use of cryptographic methods for sensitive data security is a must and the majority of the network designs include also the information protection analysis.

4   Case Studies: Security Analysis and Defense Solutions of Public Grid Electrical Parameters

In order to implement a structural cyber-system, the authors developed a mathematical model of the cyber-attacks. One problem that comes with the detection of anomalies in an industrial cybernetic system is the formal verification of software and hardware through the SMT (Satisfiability Modulo Theories) approach. Constraint Satisfaction Problems (CSP) can be solved with programming tools called constraint programming (CP). In CSP relationships between variables are declared as constraints. Constraints are specified by domains: integer numbers, real, rational, logical (boolean). Solving the CSP problem is finding an allocation to variables that satisfy all constraints. To solve CSP problems, search algorithms based on CP are used: Boolean satisfiability problem (SAT solver), genetic algorithms, mixed integrated linear programming (MILP), so on.

SMT is a decision problem for logical formulas that determine whether a first-order formula is certifiable in terms of a logical theory.

Modern, SMT is applied in establishing the necessary and sufficient conditions to estimate the states of a discrete system based on a number of attacked sensors. This estimator is robust to noise and can provide the states of the original system under cyber-attack conditions.

Both solvers, SAT and SMT, can be used to solve a large equations system. The SAT solvers are fundamental to computer science. They are applied only to conjunctive normal form of the boolean equations [23]. At the same time, the fast developing of the digital technology makes accessible the use of both SAT and SMT solvers. The dependency between the SAT and SMT solvers are as follows: the SAT solvers make the input expressions of the SMT to be converted into CNF. Therefore, the SMT is used as frontend to SAT. CryptoMiniSAT is used as external SAT solver, i.e. as backend of SMT solvers. There is also SMT solvers including the own SAT solver (e.g. Z3 [24, 25]).

Due to the diversity of mathematical models used in cyber-physical systems, the problem of designing and analyzing these systems is very difficult. Cyber-physical systems have continuous and discrete (combinatorial) dynamics being subjected to non-linear constraints. A case study of an electric power generation network subjected to cyber-attack on sensors is presented. The problem is to maintain the features of the original system under attack conditions (a large part of the information taken from the sensors is compromised). The solution to the problem is to use a cyber-state-specific estimator. Thus, in the case study the results obtained by using two types of estimators: classical (least squares method) and cyber-physical (based on SMT/SAT) method.

Based on the mathematical model in the state space of the IEEE14bus network [26–29], the system, command, and output matrices (A, B, C) are known. The considered mathematical model contains a number of 10 states, while the number of sensors that provide the information required to control the system is 35. It will be considered a number of 16 sensors as being attacked (the results provided are compromised). Under these conditions, the cyber estimator will be able to provide the real states of the original system (the original system means the system without attacked sensors, i.e. normal operation conditions).

The graphical representation of such a system is shown in Fig. 5.

The considered electrical network includes the generation, transmission and distribution of energy. The electrical network includes loads, reactive compensation elements, electric generators (Fig. 5). Each synchronous generator is represented as a voltage source.

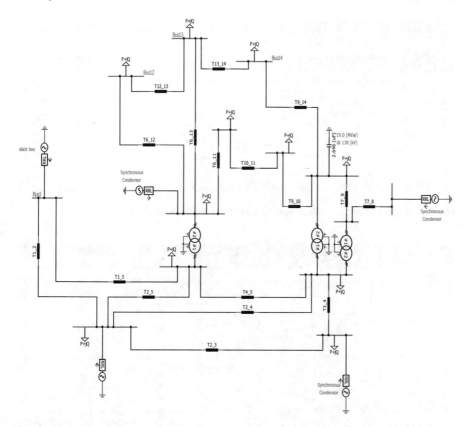

Fig. 5 The IEEE 14-bus system [26]

The sensors measure the actual injected power at each bus (14 sensors), the real power circulated in each branch (20), and a sensor providing the rotor angle for the synchronous generator (total 35 sensors). The sensor of maximum importance to be secured is the utmost sensor that of the rotor angle measurement. It contains vital information for monitoring and controlling the power generation system. The numerical example provided took into account compromising the information (sensor attack) of a number of 16 sensors.

In this case study, the authors provide a solution based on cyber-physical system (CPS) to obtain the accurate state information of an IEEE14bus [29] system subjected to sensor attacks. The mathematical model of the system can be represented in the state space by the system matrix **A**, the control matrix **B**, and the output matrix **C**, shown in Figs. 6 and 7.

It starts from the mathematical model in the time domain, in the standard state space form, provided in the technical literature derived from the laws of energy conservation and expressed as follows:

Variables - Ac

Ac

10x10 double

	1	2	3	4	5	6	7	8	9	10
1	0	0	0	0	0	1	0	0	0	0
2	0	0	0	0	0	0	1	0	0	0
3	0	0	0	0	0	0	0	1	0	0
4	0	0	0	0	0	0	0	0	1	0
5	0	0	0	0	0	0	0	0	0	1
6	-4.4881	2.3938	0.9402	0.6652	0.4889	-1	0	0	0	0
7	2.3938	-4.5737	1.0717	0.6230	0.4852	0	-1	0	0	0
8	0.9402	1.0717	-2.7437	0.3900	0.3418	0	0	-1	0	0
9	0.6652	0.6230	0.3900	-2.1746	0.4963	0	0	0	-1	0
10	0.4889	0.4852	0.3418	0.4963	-1.8122	0	0	0	0	-1

Fig. 6 The system matrix A (10 × 10 double) of the continuous system

Variables - Cc

Ac | vector_senzori_atacati | Cc

35x10 double

	1	2	3	4	5	6	7	8	9	10
2	-2.6805	5.0215	-1.1135	-0.6911	-0.5364	0	0	0	0	0
3	-1.0266	-1.1527	3.0247	-0.4485	-0.3968	0	0	0	0	0
4	-0.0415	-0.0534	-0.1113	0.0498	0.1564	0	0	0	0	0
5	-0.0150	-0.0603	-0.0395	0.1261	-0.0112	0	0	0	0	0
6	-0.6847	-0.6472	-0.4134	2.3562	-0.6110	0	0	0	0	0
7	0	-2.1063e-16	0	4.8934e-16	6.3027e-16	0	0	0	0	0
8	-0.4889	-0.4852	-0.3418	-0.4963	1.8122	0	0	0	0	0
9	0.0141	0.0175	0.0169	-0.1317	0.0832	0	0	0	0	0
10	0.0020	0.0025	0.0025	-0.0191	0.0120	0	0	0	0	0
11	0.0023	0.0028	0.0027	-0.0213	0.0134	0	0	0	0	0
12	-0.0056	-0.0069	-0.0067	0.0521	-0.0329	0	0	0	0	0
13	0.0060	0.0074	0.0071	-0.0556	0.0351	0	0	0	0	0
14	6.4297e-04	7.9593e-04	7.6946e-04	-0.0060	0.0038	0	0	0	0	0
15	3.7890	-2.5647	-0.7569	-0.2344	-0.2329	0	0	0	0	0
16	1.1287	-0.0821	-0.2717	-0.4801	-0.2949	0	0	0	0	0
17	0.6292	0.9077	-1.5240	0.0147	-0.0276	0	0	0	0	0
18	0.3208	0.7816	-0.2554	-0.4042	-0.4427	0	0	0	0	0
19	0.1585	0.7675	-0.0909	-0.5360	-0.2990	0	0	0	0	0
20	-0.3975	-0.2450	1.5007	-0.4338	-0.4245	0	0	0	0	0
21	-0.6891	-0.1035	0.6940	-0.5206	0.6192	0	0	0	0	0

Fig. 7 The output matrix **C** (35 × 10 double) of the continuous system

The control vector is set to 0 values in order to underline the attack power and the moment of attack.

In Fig. 8 the components of the attack vector are depicted and also graphically represented in Fig. 9.

By applying both type of estimators, SMT solver shown in Fig. 10 and least squares shown in Fig. 11, the following numerical results (Fig. 12) are obtained [30].

From the Fig. 13 the compromising results obtained by using LS estimator are shown.

	7	8	9	10	11	12	13	14	15	16
1 15	27	30	26	2	29	22	16	19	10	33
2										

Fig. 8 The attack vector

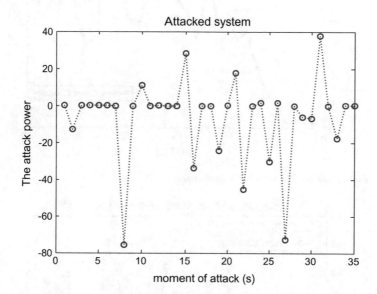

Fig. 9 The output of the system under attack (35 × 1 double)

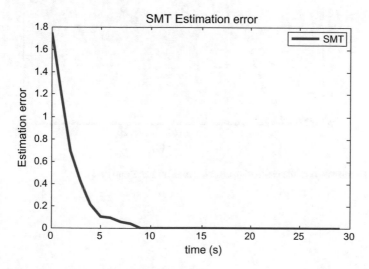

Fig. 10 The estimation error of the SMT estimator

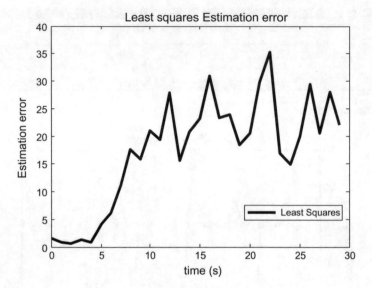

Fig. 11 The estimation error based on LS estimator

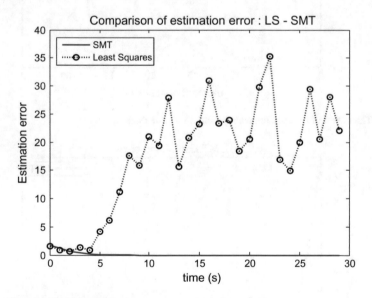

Fig. 12 Comparison of the estimation errors: based on SMT and on LS

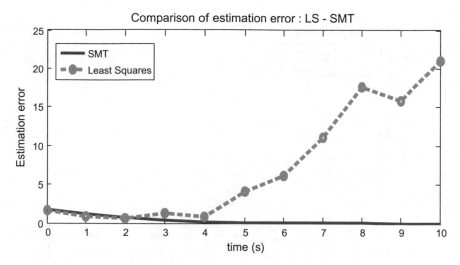

Fig. 13 The magnified first 10 s time interval

It could be noted that the least square estimator follows approximately the SMT estimator until the attack moment ($t = 2$ s). At the moment of attack the traditional estimation method fails, conducting further to wrong results. From the Figs. 9, 10, 11, 12 and 13 the first attack component takes place at 2 s. Starting with the first attack moment the obtained results based on the LS estimator are compromised. Despite of the LS estimator, the SMT estimator works well under the attack components (the error between the original state vector and the estimated vector based on SMT is vanish, according to Fig. 13).

Figure 14 presents the numerical results of the state-space for three types of the systems: the original—x, the cyber-physical system xSMT, and the estimated system based on least squares method xLS.

Several conclusions can be drawn:

– The error of estimation is very large for the estimated states based on LS method and decreases significantly (5×10^{-18}) based on CPS (Fig. 14). Therefore, the numerical results validate the methodology of attack protecting of the IEEE14 bus;
– This case study has demonstrated that new theories are being considered in cyber-attacks;
– CPS-based estimator based on SMT solvers manages to isolate corrupted sensors and rebuild the normal state of the electrical network.

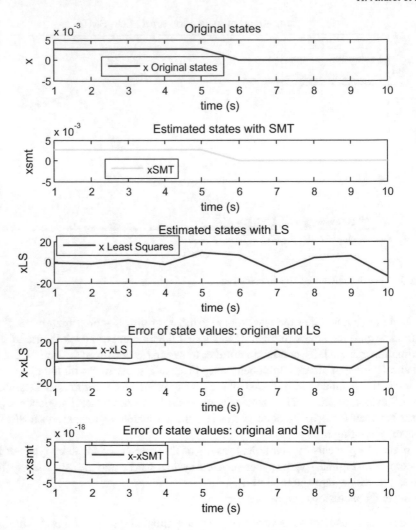

Fig. 14 The states comparison: x—the original state vector, xLS—the estimated state vector by using Least Squares method, xSMT—the estimated state vector based on Satisfiability Modulo Theories (SMT)

5 Conclusion

This chapter provides insight into the most commonly used protection methods, but also into the attacks to which they are subjected. It presents cryptography and cyber-security solutions for network sensors and data acquisition systems of industrial consumers. The example show how to obtain the accurate state information of an IEEE14bus system subjected to sensor attacks by using SMT solvers that isolate corrupted sensors and rebuild the normal operation state of the public grid.

References

1. Stouffer, K., Falco, J., Scarfone, K.: Guide to Industrial Control Systems (ICS) Security, 2015, NIST Special Publication 800-82 (2015)
2. Peterson, D.: Intrusion Detection and Cyber Security Monitoring of SCADA and DCS Networks, ISA (2004)
3. Diaz, A., Sanchez, P.: Simulation of attacks for security in wireless sensor network. Sensors **16**, 2–27 (2016)
4. Shukla, J., Babli Kumari, K.: Security threats and defense approaches in wireless sensor networks: an overview. IJAIEM **2**, 165–175 (2013)
5. Jinghua, Z.: Wireless sensor network technology based on security trust evaluation model. iJOE **14**(4), 211–226 (2018)
6. Singh, J.: Security issues in wireless sensor networks (wsn). Lect. Notes Eng. Comput. Sci. **2170**(1), 40–45 (2015)
7. Marinescu, I., Botea, B., Andrei, H.: Critical infrastructure risk assessment of Romanian power systems. In: Proceedings of IEEE-ISEEE Conference (2017)
8. Esfahani, P.M., Vrakopuolou, M., Margellos, K., Lygeros, J., Andersson, G.: Cyber attack in a two-area power system: impact identification using reachability. In: Proceedings of American Control Conference (2010)
9. https://en.wikipedia.org/wiki/List_of_major_power_outages
10. Final Report System Disturbance on 4 November 2006, UCTE. https://www.entsoe.eu/fileadmin/user_upload/_library/publications/ce/otherreports/Final-Report-20070130.pdf
11. "North American Electric Reliability Council, Technical Analysis of the August 14, 2003, Blackout: What Happened, Why, and What Did We Learn?", North American Electric Reliability Council (2004)
12. Smith, R.: Assault on California power station raises alarm on potential for terrorism. Wall Street J. (2014)
13. Zetter, K.: Inside the cunning, unprecedented hack of Ukraine's Power Grid (2016). https://www.wired.com/2016/03/inside-cunning-unprecedented-hack-ukraines-power-grid/
14. Medairy, B.: Lessons from Ukraine's energy grid cyber-attack (2017). https://www.boozallen.com/s/insight/thought-leadership/lessons-from-ukranians-energy-grid-cyber-attack.html
15. TLP: White Analysis of the Cyber Attack on the Ukrainian Power Grid (2016). https://ics.sans.org/media/E-ISAC_SANS_Ukraine_DUC_5.pdf
16. Greenberg, A.: Crash override: the malware that took down a power grid (2017). https://www.wired.com/story/crash-override-malware/
17. Lee, R.M.: Crashoverride: Analyzing the Malware that Attacks Power Grids. https://dragos.com/resource/crashoverride-analyzing-the-malware-that-attacks-power-grids/
18. Osborne, C.: Industroyer: An in-Depth look at the Culprit Behind Ukraine's Power Grid Blackout (2018). https://www.zdnet.com/article/industroyer-an-in-depth-look-at-the-culprit-behind-ukraines-power-grid-blackout/
19. Maduro, N.: Presents Details of Attacks Against Electrical System in Venezuela (2019). https://www.telesurenglish.net/news/Maduro-Presents-Details-of-Attacks-Against-Electrical-System-20190311-0019.html
20. Teruggi, M.: New Sabotage and Power Outage in Venezuela March 25–26 (2019). https://www.workers.org/2019/03/29/new-sabotage-and-power-outage-in-venezuela-march-25-26/
21. Andrei, A., Stănculescu, M.: Cryptography versus Cryptanalysis (in Romanian: Criptografie versus criptanaliza). Ed. Printech (2014)
22. Deaconu, I.D., Stănculescu, M., Chirilă, A. I., Năvrăpescu, V., Andrei, H.: On automatic transfer switch system security. In: Proceedings of IEEE-International Conference on Applied and Theoretical Electricity (ICATE) (2018)
23. De Moura, L., Bjorner, N.: Z3: An efficient SMT solver. In: Proceedings of International Conference on Tools and Algorithms for the Construction and Analysis of Systems, TACAS'08/ETAPS'08 (2008)

24. https://www.microsoft.com/en-us/research/people/nbjorner/
25. https://uofi.app.box.com/s/o4c20a1y3onih9yq8woe/file/10768277893
26. https://www.google.ro/url?sa=t&rct=j&q=&esrc=s&source=web&cd=1&ved=2ahUKEwjbm
    6qClffgAhWB5KQKHZfBAvUQFjAAegQICRAC&url=https%3A%2F%2Fwww.researchgate.
    net%2Fprofile%2FHarikrishnan_Nair2%2Fpost%2FCould_anyone_provide_me_the_PSCAD
    _model_of_IEEE14-bus_system_or_any_bigger_size%2Fattachment%2F59d644c6c49f478072
    ead71a%2FAS%253A273819317538816%25401442295020231%2Fdownload%2FIEEE_14-
    bus_technical_note.docx&usg=AOvVaw12ywIiAkKH2BIEkFxl-Le2
27. Pasqualetti, F., Dorfler, F., Bullo, F.: Attack detection and identification in cyberphysical sys-
    tems. IEEE Trans. Autom. Control **58**(11), 2715–2729 (2013)
28. Pasqualetti, F., Dörfler, F., Bullo, F.: Cyber-physical attacks in power networks: models, funda-
    mental limitations and monitor design. In: Proceedings of IEEE-Conference on Decision and
    Control, and European Control Conference (2011)
29. Mahdi, M., El-Arini, M., Fathy, A.: Identification of coherent generators using fuzzy C-means
    clustering algorithm and construction of dynamic equivalent of power system. J. Electr. Syst.
    6(2), 2–18 (2010)
30. Shoukry, Y., Chong, M., Wakaiki, M., Nuzzo, P., Puggelli, A., Sangiovanni-Vincentelli, A.,
    Seshia, S., Hespanha, J., Tabuado, P.: SMT-Based observer design for cyber-physical systems
    under sensor attacks. ACM Trans. Cyber-Phys Syst **2**(1), 36–50 (2018)

Innovative Hardware-Based Cybersecurity Solutions

Octavian Ionescu, Viorel Dumitru, Emil Pricop and Stefan Pircalabu

Abstract Cyber threats are currently targeting not only the individual computers, but the industrial infrastructures managed by industrial control systems, that are often interconnected in large networks. The recent developments in the field of electronic devices, namely sensors and wearable equipment, and the evolution of communication technologies permitted the adaptation of Internet of Things (IoT) devices on a large scale. These IoT devices contains computing power, good connectivity, but are prone to a large number of cybersecurity vulnerabilities and attacks. This chapter is focused on describing some innovative hardware-based cybersecurity solutions designed especially for Internet of Things devices. The proposed solutions are based on physical unclonable functions (PUF) encryption methods. The first method is based on unique features which are the results of variable entries in the process of manufacturing semiconductor devices. An example of an encryption system already developed and available on the market is presented. The second method is based on the proper-ties of memristors, mainly the nonlinearity of the I–V characteristics and the particularities in their production process, which allow generating unique material states. By using this method, it is possible to generate very large random encryption keys, using few electrical components. This solution could be incorporated in various security systems that uses parallel communication channels to send

O. Ionescu (✉)
National Institute for Research and Development in Microtechnologies, IMT Bucharest, Bucharest, Romania
e-mail: octavian.ionescu@imt.ro

V. Dumitru
National Institute of Materials Physics, Magurele, Romania
e-mail: viorel.dumitru@infim.ro

E. Pricop
Petroleum-Gas University of Ploiesti, Ploiesti, Romania
e-mail: emil.pricop@upg-ploiesti.ro

S. Pircalabu
Cyberswarm Inc., San Mateo, CA, USA
e-mail: stefan@cyber-swarm.net

© Springer Nature Switzerland AG 2020
E. Pricop et al. (eds.), *Recent Developments on Industrial Control Systems Resilience*, Studies in Systems, Decision and Control 255,
https://doi.org/10.1007/978-3-030-31328-9_12

283

and receive the voltage level to be used in reading the memristors, as well as to physically control the connection to the main data transfer channel, being it Ethernet, Wi-Fi or another broad-band connection.

Keywords Cybersecurity · Memristor · Physical unclonable functions · Data encryption · Key generating · Physical link control

1   Introduction

The recent developments in the field of electronic devices, namely sensors and wearable equipment, and the evolution of communication technologies permitted the adaptation of Internet of Things (IoT) devices on a large scale. These IoT devices contains computing power, good connectivity, but are prone to a large number of cyber-security vulnerabilities and attacks. The complexity and number of cyber security threats have increased along with the need for new security solutions that can be used to protect devices against them.

Protecting IoT devices is a challenging task due to their reduced processing power and due to energy saving requirements. Data encryption may solve some of the security problems, as it can be used for both data privacy and also for checking the integrity of the data sent to or received from any connected device, since it can prevent any malicious party to modify or read the data. However, encryption on various IoT devices is difficult to achieve as most specific algorithms require high processing power for encoding data. The majority of the algorithms that can be used on low processing power devices are not very strong and that data might be decrypted with more or less efforts. Another vulnerability of the current encryption systems is the possibility for an attacker to steal the encryption key which is stored in an improper way.

A significant number of strong security solutions were developed, but since they use complex algorithms and multi-layered encryption methods, they are not designed for usage in industrial or in IoT environments. It has to be stated that sensors and actuators in the majority of existing control systems do not possess high computing power and do not implement authentication or any other security methods. Taking into consideration these facts, the need for reliable, robust and cost-effective solutions for protection of low-processing power devices (IoT devices, sensors, etc.) is obvious. The problem is stringent since the world is now pushing towards the introduction of Internet of Tings (IoT) in every aspect of our lives, as some consider the next industrial revolution will come from using IoT devices alongside artificial intelligence. Researchers are speaking frequently about *Smart Cities, Intelligent House, Intelligent Agriculture, Smart* Grid and *Autonomous cars* as aspects of the 4th Industrial Revolution—Industry 4.0. All these concepts are based on IoT devices, sensors and data exchanges.

According to Ref. [1], there is a network of about 6 billion devices which need to be connected in the near future that, unfortunately, to this moment, don't have a reliable solution for the main unsolved problem of protecting the devices in this enormous network against malicious attacks. Some may consider that there isn't a real problem if someone is watching the electrical power consumption of a user; however, the real problem is if the same person would be able to open/close the water electrical vane or, even worse, the electrical vane of natural gas in someone else's house.

Another security issue of IoT devices is the possibility of being hacked and introduced in botnets. One relevant example is Mirai that was used for DDoS (Distributed Denial of Service) attacks on the servers of services like Twitter, Netflix and PayPal in 2016. Countless IoT devices that power everyday technology like surveillance cameras and smart-home devices were hijacked by the malware and used in botnets for launching DDoS attacks against various servers [2, 3].

Since the control systems are critical for the functioning of various infrastructures and the adoption of IoT devices is increasing, it is necessary to develop powerful methods for increasing their cyber security. The software implementations are complex and difficult in these cases. In this chapter a hardware-based security solution based on physical unclonable functions (PUF) is described. This approach can ensure not only the data security, but it can uniquely identify any of the connected devices, permitting mutual device authentication.

In the second section of the chapter semiconductors specific physical unclonable functions as well as some recent applications are described. The memristors I–V characteristic is presented and a possible application of these functions in encryption key generators. The last part of the chapter is focused on the design a Vernam cipher based on the memristor PUF generator.

2 Semiconductors Specific Physical Unclonable Functions (PUFs)

Initially observed and proposed to be used for anti-counterfeiting currency in 1983 by Bauder [4], PUFs are unique features which are the results of variable entries in the process of manufacturing semiconductor devices. Although the variations are not affecting the functionality of the semiconductor device, they are generating some unique differences physically embedded into the semiconductor device and can be used in the process of identifying a circuit or a device [5]. Another important application is of PUF is generating unique encryption keys, application which will be described in this section.

There are several types of semiconductor peculiarities which could be exploited in or-der to generate a PUF. One of the most known PUF is the one generated by the random properties of the gate oxide inhomogeneous properties occurring in semiconductors manufacturing process [6], this being considered as an intrinsic device function. In addition to the randomly generated fractures which are the ideal source for physical unclonable function the IC design laboratories could enhance security level by implementing controlled oxide rupture PUFs. There it was demonstrated that oxide rupture PUF can extract uniformly-distributed binary bits through amplification and self-feedback mechanism [6]. The random bits generated are providing the users with a large flexibility to choose their own key-generation method and system management approaches. The notable weaknesses of the PUFs generated by oxide fractures are created by the fact that they are subject to environmental variations such as temperature, supply voltage and electromagnetic interference, which are affecting their stability. It is impossible to predict the temperature of the device when the process of inter-rogation occurs, thus random errors will make difficult to use them as key generators.

Another category of PUFs are explicitly induced in the process of IC production by slightly modifying the production process parameters and by using various coatings with different properties. Although these PUFs are much more stable, and less influenced by environmental factors, the increased production costs make them less attractive.

Taking into account the increased need for cyber security solutions, the route from the concept [4] to the implementation of PUFs embedded on ICs was quite short and now there are several companies which are producing such circuits. Xilinx has advertised already its new PUF IP, supplied by Verayo, which, as declared, is producing a unique device "fingerprint" that provides a cryptographically strong Key Encryption Key (KEK) known only to the device [7]. The PUF is based on CMOS manufacturing process variations such as threshold voltage, oxide thickness, metal shape, resistances, and capacitances which are producing uniqueness between devices. Another example is the Altera Stratix 10 FPGAs with high-performance FPGA security capabilities enabled through an innovative Secure Device Manager (SDM). This SDM features a Physically Unclonable Function for Key Material and Identity based on technology by Intrinsic-ID [8]. Recently, Maxim Integrated has advertised his DS28E50, the first Deep Cover® secure authenticator with the SHA-3 algorithm, combined with the Chip DNA physically unclonable function (PUF) [9].

3 Methods of Using Semiconductor Generated PUF in Encryption

The way that encryption works on the devices provided with PUF capabilities is quite complex. Instead of embodying a single cryptographic key, IC provided with PUF capabilities are working by using challenge–response authentication method in order to analyze its microstructure. Due to the random modifications introduced during the manufacturing processes when there it is applied a physical stimulus to the structure, this reacts in an unpredictable (but repeatable) manner [10].

The system works based on a *challenge* sent by the user which wants to access the protected device, however, the sender cannot get a Challenge-Response pair directly, since the PUF never releases responses directly, in order to prevent a third party

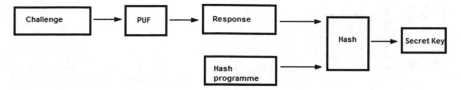

Fig. 1 Block diagram of the method to be used for restraining the access of a third party to PUF direct response

to find its response. Consequently, there it will be necessary to use a hash function which could return the response needed. A hash function is defined as a function which could be used to map data of arbitrary size onto data of a fixed size. There it should be mentioned that not all the functions are suitable for this operation. The cryptographic hash functions are the one that allows the user to verify if the input data map onto an established hash value, however, when the input data is unknown consequently is difficult to reconstruct it knowing only the stored hash value. The hash generating program is based on mathematical algorithm that maps data of arbitrary size ("Response") to a bit string of a fixed size (the "Hash") actually implementing a one-way function impossible to invert [11]. The block diagram of this method used for restraining the access of a third party to PUF direct response is presented in Fig. 1.

The inclusion of hash function is not enough to obtain the pair Challenge-Response from an IC provided with PUF, an extra function respectively a Pre-Challenge is needed as presented in Fig. 2.

The complete algorithm for obtaining the pair Challenge-Secret Key for an IC provided with PUF is presented in Fig. 3.

After an IC with PUF implemented capability is produced, the user would have to generate a challenge-response pair. The user will provide the pre-challenge and the hash function, and the IC will produce the Response. The user obtains the Challenge by computing the hash function and the Pre-Challenge. Thus, the user will have the Challenge-Response pair, the Hash function which are public; however, the Response will not be known to anybody since the Pre-Challenge is not known. An example of circuit developed for this is the one produced by Maxim Integrated [9] is presented in Fig. 4.

The *IC 2* block in Fig. 4 is an I2C-to-1-WireM bridge device (DS2484) with the role to interface directly to standard (100 kHz/or fast 400 kHz max) I2C masters performing protocol conversion between the I2C master and any downstream 1-Wire

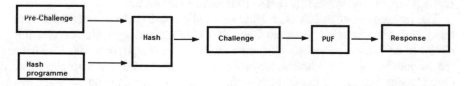

Fig. 2 Block diagram of the method used to obtain the response form an IC with PUF capabilities

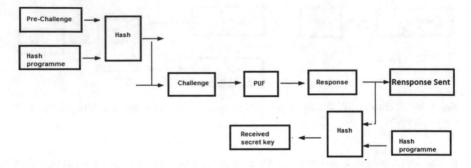

Fig. 3 The algorithm to be implemented for obtaining the pair Challenge-Secret Key

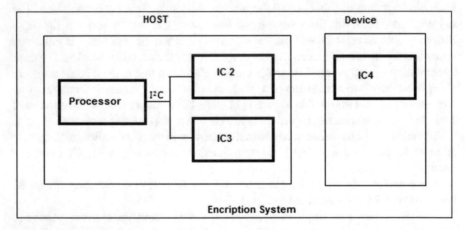

Fig. 4 Secure authentication with ARM processors

slave devices characteristics, thus avoiding the discovery of the unique value used by the chip crypto-graphic functions [9].

The integrated circuit IC3 in Fig. 4 is the DS2476 and according to the catalogue description [9] is a DeepCover secure ECDSA (Elliptic Curve Digital Signature Algorithm) and HMAC (keyed-Hash Message Authentication Code) SHA-256 coprocessor for the DS28C36. This coprocessor is able to compute the required HMACs or ECDSA signatures and to perform necessary operations on the DS28C36. The DS2476 provides also a set of cryptographic tools derived from integrated asymmetric (ECC-P256) and symmetric (SHA-256) security functions.

The integrated circuit IC4 is an ECDSA public key-based secure authenticator that incorporates Maxim's patented ChipDNA PUF technology which enables this circuit to deliver cost-effective protection against invasive physical attacks. Using the random variation of semiconductor device characteristics that naturally occur during wafer fabrication, the ChipDNA circuit generates a unique output value that is repeatable over time, temperature, and operating voltage [9].

There are several concerns related to this approach one of the most important being the price. It would be quite expensive to provide the basic elements of an IoT network such as remote sensors and actuators with the ICs with PUF capabilities. Since the previous method is not entirely usable in low cost IoT applications researchers were looking for some other hardware devices which are provided with PUF.

4 Memristors Specific Physical Unclonable Functions

Memristors are devices which presents a unique characteristic resulted by composing the properties of resistors, capacitors, and inductors. The name of memristor was awarded for its capability of memorizing the resistance, the term "memristor" being an abbreviation of memory and resistor. According to the theory proposed by Chua [12], the memristor was described as the fourth passive device together with resistors, capacitors, and inductors. In a very simple approach memristors could be also described as passive circuit elements that maintain a relationship between the time integrals of current and voltage across a two-terminal element. The symbol for memristor as an electronic device is presented in Fig. 5.

Consequently, the resistance of a memristors is varying according to a devices memristance function, thus by using small read charges, the memristor could offer in-formation related to the previous applied voltages. This theory nowadays had a series of critics, one of the most remarkable being the article of Isaac Abraham where the author is contradicting Chua theory stating and demonstrating [13] that the ideal memristor is an unphysical active device and any physically realizable memristor is a nonlinear composition of resistors with active hysteresis. This statement changes somehow the vision on memristors and emphasizes that there are still only three fundamental passive circuit elements. However, the discussions about being a passive circuit or an active circuit is irrelevant since the effect of interest for cryptography is related to one of the most important memristor properties, the fact that it has an unclonable behavior.

There are several types of memristors, which all are presenting the same properties although the mechanisms behind this behavior are different and a brief description of different classes of memristors is presented in Fig. 6.

Within these types there are two classes of memristors:

- charge controlled memristors;
- flux-controlled memristors.

Fig. 5 Symbol for memristor

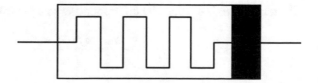

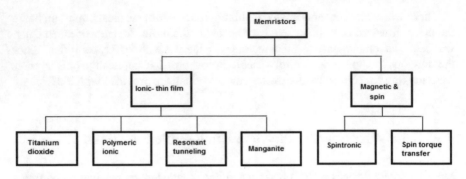

Fig. 6 Types of memristors

The equations which are describing the charge controlled memristors are presented hereunder and starts from describing the relationship between the magnetic flux φ and charge q.

$$\varphi = f(q) \tag{1}$$

By differentiating in (1) results:

$$\frac{d\varphi}{dt} = \frac{df(q)}{dq} \cdot \frac{dq}{dt} \tag{2}$$

where it could be identified that the voltage is $v(t) = \frac{d\varphi}{dt}$ and the current $i(t) = \frac{dq}{dt}$ thus the Eq. (2) could be written as in (3).

$$v(t) = M(q) * i(t) \tag{3}$$

$$M(q) = \frac{df(q)}{dq} \tag{4}$$

The notation M(q) in (3) and (4) represents the memristance and it is since it is defined similar to a resistor M(q) is measured in Ohms.

Similar to the equations above there could be written the equations for a flux controlled memristor (5):

$$q = f(\varphi) \tag{5}$$

The relation (6) is obtained by differentiating relation (5).

$$\frac{dq}{dt} = \frac{df(\varphi)}{d\varphi} \cdot \frac{d\varphi}{dt} \tag{6}$$

Identified that the voltage is $v(t) = \frac{d\varphi}{dt}$ and the current is $i(t) = \frac{dq}{dt}$, thus the Eq. (6) could be written like in (7).

$$i(t) = W(q) * v(t) \tag{7}$$

where:

$$W(q) = \frac{df(\varphi)}{d\varphi} \tag{8}$$

$W(q)$ was named memductance and it is similar with the conductance in the resistor case. Its measurement unit is Siemens.

One of the most interesting characteristics of these memristors is their I–V characteristic which presents a pinched hysteresis loop, always crossing the origin. This could be explained easily since when applying a signal to a memristor, if the voltage is zero, the current will be also zero. Another interesting feature of the memristors is that the shape of the pinched hysteresis loop will change with frequency [14], while increasing the frequency the hysteresis loop is shrinking as presented in Fig. 7.

Applying a voltage or a current to a memristor, results in a significant change in its resistance. Waser and Aono [15] presented two switching methods in memristors used in resistive memories:

- Cation migration—this mechanism is based on phenomenon of electrochemical growth and dissolution of metallic filaments.

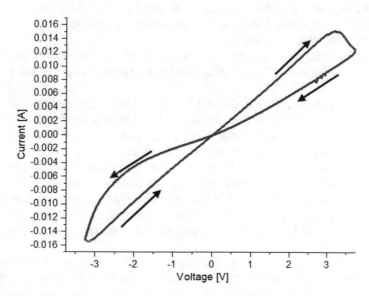

Fig. 7 I–V characteristic of the memristor

- Anion migration—this mechanism is based on phenomenon of forming/suppressing sub-oxide conductive paths through local redox processes.

Based on the electrical polarity, the switching could be differentiated in:

- Unipolar—when the polarity of voltage and current does not affect the switching process.
- Bipolar—when the switching process depend on the voltage and current polarity.

There should be mentioned that in both unipolar and bipolar switching mechanisms if the reading of the state is done at a small voltage, the state remains unchanged while reading process.

5 Methods of Using Memristor PUF in Encoding Systems

Recent studies demonstrated that there it could be achieved much more than two states (On/Off) from memristors. Actually, as presented in the US Patent Application 62/844,936 [16] an IGZO memristor employing thin film transistor (TFT) technology was developed and this one presents the capability of being set in a large number of resistive states. For instance, by applying subsequent positive voltage sweeps with increasing amplitude between 4 and 6.5 V as depicted in Fig. 8a, the resistance of the memristor could be increased in small steps as depicted in Fig. 8b. It is interesting that the I–V characteristics are nonlinear ones. Therefore, that means that by reading one resistive state with different voltages, different resistance values are obtained. The reading voltages could be lower than the setting voltages, in order not to further modify the state of the memristor.

Therefore, using only one memristor by setting the resistance in various stats and reading each state with different voltages a large pool of resistance values will be available.

In this way, an interval from 0.1 to 4 V will be used and the sweep voltage will have a difference of 0.1 V from one read to other results a number of 40 different resistance values for each used memristor; these values may be used for generating an encryption key. In addition, considering the fact that each memristor could be set in at least 20 different states (as presented in Fig. 6b), the result will be a large variety of random values to be used for issuing a random encryption key. Taking into consideration that the value of 0.1 V was selected only as an example and to make the process easily quantifiable, it could be foreseen that much small values of reading voltage values could be used (0.01 V as an example), thus the number of characters becomes much larger.

One way in which the values resulted after reading the memristor at different voltages may be interpreted is by using intervals of resistance to transform the resistance value into an ASCII (American Standard Code for Information Interchange) character or any other type of character encoding. For example, it could be established that an electrical resistance value read between 500 and 600 Ω, could be associated

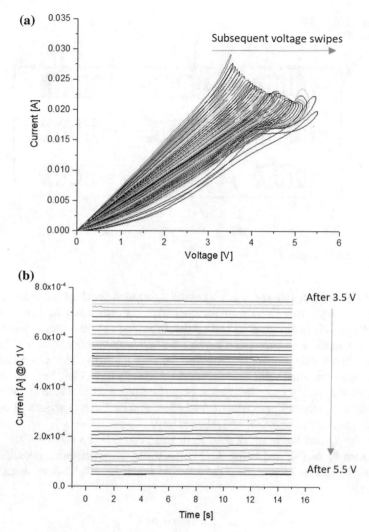

Fig. 8 a I–V Characteristics of one memristor. **b** Multiple resistive states obtained with the IGZO memristor

with character "A", or an electrical resistance between 1000 and 1100 Ω, could be associated with a string of characters which may represent a partial encryption key "WXYZ". By using this method, we could generate keys of arbitrary size by using very few memristors.

In addition to this depending on the required strength of the encryption/decryption key multiple memristors can be arranged to form a matrix, or a vector as shown in Fig. 9. These memristors can be used to generate encryption keys, as it was explained previously, by setting them in a particular state and then reading them using different voltage values.

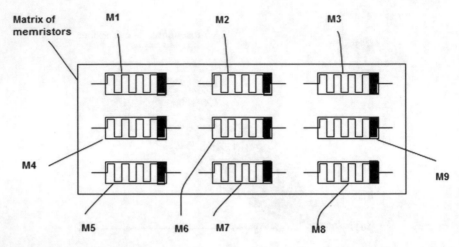

Fig. 9 Matrix of memristors

A similar voltage value (e.g. 2.5 V) could be applied on all of the memristors of the matrix; considering that the memristors were set previously in different states with different values the result will be a string or a matrix of nine completely random values to be used as an encryption key; by reading with another voltage value (e.g. 0.5 V) another 9 values could be generated (or even a longer string of characters, if the electrical resistance value is interpreted not as only one value, but as a string of characters of arbitrary size), this process could be repeated to create an encryption key of any size. Depending on the method of encryption the key generated could be used for symmetrical or asymmetrical encoding.

Considering that the result of memristor matrix read is the matrix of resistances values, namely M (9), and being A (11) the string of characters representing the message that has to be sent, then the encoded message will be R as described in relation (10).

$$M = \begin{bmatrix} m_{11} & m_{12} & m_{13} \\ m_{21} & m_{22} & m_{23} \\ m_{31} & m_{32} & m_{33} \end{bmatrix} \tag{9}$$

$$R = A * M \tag{10}$$

Consequently, the receiver should decrypt the message A by using the inverse of the matrix M, namely M^{-1}, as presented by Eq. (11).

$$A = R * M^{-1} \tag{11}$$

Another advantage of this method for generating encryption keys is that the key generation can happen on multiple devices, without actually sending the encryption key; instead, the voltage values used to generate the key on the device that sends data can be sent to the device that receives the data, that device being able to apply those voltage values on its own matrix of memristors and generate the matching symmetrical key (for symmetrical encryption) or the inverse function for an asymmetrical encryption. Because of the fact that the key is generated using a physical device, the memristor matrix, where each memristor can be set in a unique and different state, even if a malicious user intercepts the voltage values used to generate the key used for decrypting the data, it will be impossible to generate the decryption key, as he would need the physical memristor matrix to apply the voltages on.

Here we have to introduce another feature of memristors: due to the specific steps in the production process there are quite significant differences from batch to batch of memristors produced even more there are differences between memristors located in different places of the wafer. To exemplify this feature, we can take one technological step in IGZO memristor production: magnetron sputtering deposition/coating. There are inherent variations in the process from batch since the target cathode coating material used in PVD (Physical Vapor Deposition) presents different characteristic erosion patterns on the target face after each batch.

In comparison, the classic method of encoding requires the key to be sent to or stored on the encoding/decoding devices. When using the memristors, the key could be generated based on a short message containing the level of voltage which should be applied to read the memristors (this way no key would be sent) on any device that has a memristor matrix attached to it.

As stated above, the key generation on multiple devices can happen not only by reading memristors values at different, low voltage pulses, but also by changing the memristor state during the key generation (by applying high voltage sweeps or voltage pulses), and then reading it with a low voltage. This modification of the memristor state can happen in various ways, by choosing to apply a voltage pulse or a voltage sweep for a period of time (those sweeps, or pulses may have negative or positive value, to increase the electrical resistance value or to decrease it). We can choose the intensity of the voltage sweeps or pulses (to be 4 V or 6 V or 10 V etc.) the duration while the voltage is applied on the memristor (0.5, 1, 2 s etc.) and various combinations of those parameters, which will result in different resistance values, and different encryption keys. The same process must be applied on the sender and the receiver of data, to generate the key needed to encrypt and decrypt data.

In order to further increase the safety of the system and thus to secure the information sent/received is to use a separate communication channel to send the information concerning the level of voltage used for writing/reading the memristors and eventually the structure of the encryption/decryption key. This method using the GSM/GPRS channel as presented in the block diagram from Fig. 10 (proposed in US Patent application US 2019/0132340 A1) proved to be reliable and useful for applications requiring a high level of security (actuators, upgrade of software for machine tools, etc.).

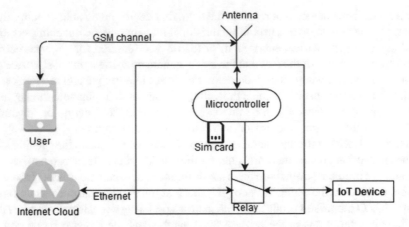

Fig. 10 Additional channel used for sending the reading voltage levels

6 Vernam Cipher Design Based on Memristor PUF Generator

A Vernam cipher was mathematically proven to be unbreakable if all the following conditions are met:

(1) the key must be truly random;
(2) the key must be at least the same size as the text that needs to be encrypted;
(3) the key is never reused in whole or in part;
(4) the key is kept completely secure [17].

The main reasons why the Vernam cipher was not used in digital devices is that there is a high risk that an attacke could gain access to the device and steal the encryption key or identify the key generation algorithm used for generating the keys. Another reason is that the key generation algorithms weren't able to generate completely random keys.

In Fig. 11 is presented the proposed implementation of a Vernam cipher based on a memristor matrix for generating the keys. The problems presented in the previous paragraph are fixed by the fact that we can isolate the device doing the encryption physically from the data transmission channel (for example the Ethernet channel) and that we can generate completely random encryption key, of any size, using the memristors. Also, because of the fact that, by using this method, we never send the actual key, but just the voltages used to generate the key, and by using the secure GSM channel to send the voltages, we solved the problem of a malicious party intercepting the keys.

The first condition of the Vernam cipher is met by using the property of memristor to generate completely random values, this being used to generate completely random voltage sweep values that are applied to the memristor matrix to generate the encryption key.

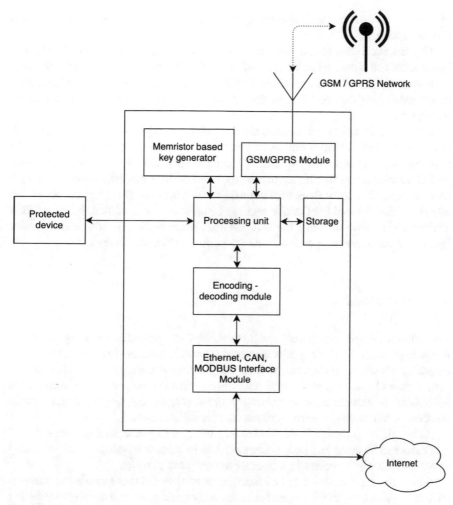

Fig. 11 A cyber security system with memristor based generating key device

The second condition of the Vernam cipher is met by simply generating as many voltage sweep values are needed to generate an encryption key the same size, or bigger than the plaintext that needs to be encrypted.

The third condition of the Vernam cipher is met by deleting the key immediately after being used to encrypt data, and generating another completely random key the next time when a message needs to be encrypted, by using the property of memristors to generate completely random values; also, during the encryption and the sending of voltage values from the sender device to the receiver device, the system is isolated from the Ethernet channel by opening the contacts of two relays (one placed between the protected device and the system, and another placed between the system and the Ethernet cloud) of another possible embodiment of the system described in

Fig. 10, so no one can get inside the system during the encryption process to steal the encryption key.

The last condition of the Vernam cipher is met, firstly by destroying the key immediately after the encryption, and by delivering the voltage values (to the receiver of the data transferred over the Ethernet) that can only be used on the memristor matrix of the receiver of data to generate the unique encryption key that must be used to decrypt data.

For IoT devices, another method, similar to the one explained above can be used, where instead of having a matrix of memristors placed on all the devices of the network, we may have only one memristor for each IoT device. This memristor would match another memristor placed on a central location, like a server where it must transmit data to, or from where it may receive updates. By using an encryption algorithm that is capable of data integrity check, we can ensure that the data that is sent won't be modified along the way, even if the encryption itself would be easily broken, as the important process is the integrity check, not data privacy.

7 Conclusions

The solutions based on physical unclonable functions generated either by peculiarities of manufacturing process of ICs or the one specific to some classes of memristors are viable solutions for the new cyber security challenges. Requiring less processing power than classic software-based encryption methods and providing unclonable identification features these methods are more suitable for sensors and actuators protection in industrial control systems and for IoT devices.

Although the progress is significant and utility was proven, more research and work have to be done and both systems has to be further developed and somehow simplified in order to be easily implemented and cost effective.

Based on the previous described facts, it could be stated that memristors presents PUF embedded characteristics, and consequently could offer an excellent competing solution with the one presented previously (PUF based on defects in gate silica oxide insulator). The big advantages of memristors is that they are more stable to the environment variations as well as the fact that they could be embedded easily into any circuitry and could be write/read with a lower number of additional components.

A big advantage of the proposed memristor based key generation device is that, by using the additional GSM/GPRS channel instead of sending the decryption key, there could be sent voltage values, rendering anyone intercepting these values incapable of generating the decryption key, as they would need the physical device containing the memristors (which are each set in unique states).

References

1. IHS: Internet of Things (IoT) connected devices installed base worldwide from 2015 to 2025 (in billions). https://www.statista.com/statistics/471264/iot-number-of-connected-devices-worldwide/
2. Sathwara, S., Dutta, N., Pricop, E.: IoT Forensic A digital investigation framework for IoT systems. In: Proceedings of 10th International Conference Electronic Computer Artificial Intelligence. ECAI 2018. https://doi.org/10.1109/ecai.2018.8679017 (2019)
3. Spognardi, A., De Donno, M., Dragoni, N., Giaretta, A.: Analysis of DDoS-Capable IoT Malwares. In: Proceedings 2017 Federated Conference on Computer Science and Information Systems vol. 11, pp. 807–816. https://doi.org/10.15439/2017f288 (2017)
4. Bauder, D.: An Anti-Counterfeiting Concept for Currency Systems. Sandia Natl. Labs, Albuquerque, NM, Technical Report (1983)
5. Pappu, R., Recht, B., Taylor, J., Gershenfeld, N.: Physical one-way functions. Science 297(80–), 2026–2030 (2002). https://doi.org/10.1126/science.1074376
6. Wu, M.Y., Yang, T.H., Chen, L.C., Lin, C.C., Hu, H.C., Su, F.Y., Wang, C.M., Huang, J.P.H., Chen, H.M., Lu, C.C.H., Yang, E.C.S., Shen, R.S.J.: A PUF scheme using competing oxide rupture with bit error rate approaching zero. In: Digital Technical Paper—IEEE International Solid-State Circuits Conference, vol. 61, pp. 130–132. https://doi.org/10.1109/isscc.2018.8310218 (2018)
7. Verayo PUF: IP on Xilinx Zynq UltraScale + MPSoC Devices Addresses Security Demands. http://www.prnewswire.com/news-releases/verayo-puf-ip-on-xilinx-zynq-ultrascale-mpsoc-devices-addresses-security-demands-300357805.html
8. Intrinsic-ID: Altera Reveals Stratix 10 with Intrinsic-ID's PUF technology. https://www.intrinsic-id.com/altera-reveals-stratix-10-with-intrinsic-ids-puf-technology/
9. Maxim-Integrated: DS28E38 DeepCover® Secure ECDSA Authenticator with ChipDNA PUF Protection. https://www.maximintegrated.com/en/products/embedded-security/DS28E50.html
10. McGrath, T., Bagci, I.E., Wang, Z.M., Roedig, U., Young, R.J.: A PUF taxonomy. Appl. Phys. Rev. 6 (2019). https://doi.org/10.1063/1.5079407
11. Halevi, S., Krawczyk, H.: Strengthening Digital Signatures Via Randomized Hashing, pp. 41–59. https://doi.org/10.1007/11818175_3 (2006)
12. Chua, L.: Memristor-the missing circuit element. IEEE Trans. Circuit Theory 18, 507–519 (1971). https://doi.org/10.1109/TCT.1971.1083337
13. Abraham, I.: The case for rejecting the memristor as a fundamental circuit element. Sci. Rep. 8 (2018). https://doi.org/10.1038/s41598-018-29394-7
14. Chua, L.: Resistance switching memories are memristors. Appl. Phys. A Mater. Sci. Process. 102, 765–783 (2011). https://doi.org/10.1007/s00339-011-6264-9
15. Rainer, W., Masakazu, A.: Nanoionics-based resistive switching memories. Nat. Mater. 6, 833–840 (2007)
16. Dumitru, V.-G., Besleaga Stan, C., Botea, I.-M.: Memristive device, US Patent Application, US Provisional Application Serial No 62/844,936 (2019)
17. Vernam, G.S.: Automatic telegraph switching system plan 55-A. Trans. Am. Inst. Electr. Eng. Part I Commun. Electron. 77, 239–247 (2013). https://doi.org/10.1109/tce.1958.6372793

Legal Issues of Deception Systems in the Industrial Control Systems

Pavol Sokol, Radoslav Benko and Laura Rózenfeldová

Abstract Deception systems, and within them deception industrial control systems, present a newly emerging type of defence in cybersecurity, providing for the detection, analysis and defence against cyber-attacks. The deception technology focuses on the attackers, their point of view and methodology used to exploit and navigate networks to identify and exfiltrate data. The chapter discusses the nature of the deception Industrial Control Systems and the legal issues encompassed with their use. It provides the legal framework of the fundamental right to privacy and the fundamental right to personal data protection, as well as the legal framework of the liability, predominantly in the area of tort law, applicable to the use of the deception Industrial Control Systems, the provider of these systems must be aware of.

Keywords Deception system · Honeypot · ICS · EU law · Privacy · Liability

1 Introduction

The rapid growth of information and transfer of messages has led the cyber security to become an important part of the Industrial control systems (ICS). Traditional security tools, methods and techniques applied in cyber security are becoming ineffective and insufficient, as the attackers are able to easily circumvent commonly applied security mechanisms (firewalls, detection and prevention systems etc.). To counteract this

P. Sokol (✉)
Faculty of Science Institute of Computer Science, Pavol Jozef Šafárik University in Košice, Košice, Slovakia
e-mail: pavol.sokol@upjs.sk

R. Benko
Faculty of Law Institute of International Law and European Law, Pavol Jozef Šafárik University in Košice, Košice, Slovakia
e-mail: radoslav.benko@upjs.sk

L. Rózenfeldová
Faculty of Law, Department of Commercial Law and Business Law, Pavol Jozef Šafárik University in Košice, Košice, Slovakia
e-mail: laura.rozenfeldova@student.upjs.sk

© Springer Nature Switzerland AG 2020
E. Pricop et al. (eds.), *Recent Developments on Industrial Control Systems Resilience*,
Studies in Systems, Decision and Control 255,
https://doi.org/10.1007/978-3-030-31328-9_13

301

development, it is necessary to gather and investigate in depth as much information on the cyber-attacks and their perpetrators as possible.

The use of **deception systems** may be a suitable tool in this regard. A deception system provides false, delayed or incomplete information, and misleads attackers into a course controlled by the operator of the deception systems [1]. In general, deception is "*a kind of persuasion. Persuasion means trying to get someone or something to help you achieve goals that would be costly or impossible for you to achieve on your own*" [2].

Rowe in [2] provides a taxonomy of the deception methods. Qassrawi et al. in [3] shifts the idea of these taxonomies towards cyber security and presents examples of the deception systems. According to [2, 3] the deception systems include:

- a system hiding things in the background (**masking**)—e.g. monitoring of users by modifying the operating system to hide its traces;
- a system hiding something as something else (**repackaging**)—e.g. embedding of attack-thwarting software within otherwise innocent utilities;
- a system hiding something by having it overshadowed by something else (**dazzling**)—e.g. sending a lot of error messages to attackers when they perform harmful activities;
- a system imitating aspects of something else (**mimicking**)—e.g. fake filesystem or directory, which looks like the real file system or directory;
- a system, which creates new and often "fake" objects that may interest the deceivee (**inventing**)—e.g. a piece of software left for attackers to download, such as honeyfiles, which are bait files intended for hackers to access [4]. A specific example of honeyfiles is honeysheet [5];
- a system, which uses diversions unrelated to the object of interest (**decoying**)—e.g. snooping login credentials to attackers to encourage them to log in (Honeytoken [6]).

The most used deception system is a **honeypot.** The term honeypot was defined by the public forum of over 5000 security professionals in 2003 *as "an information system resource whose value lies in being probed, attacked, or compromised"* [7]. Honeypot "*provides a defense mechanism in which they deceive attackers into believing that they are compromising a real production system*" [3]. Honeypots may be categorized by the level of interaction, purpose, role and deployment. This paper considers the categorization of honeypots according to the level of interaction and purpose of usage.

As regards **the level of interaction,** it presents the maximum range of attack possibilities allowed by a honeypot to an intruder to have. **Low-interaction honeypots,** that are dominant among the honeypots used, implement targets to attract or detect attackers by the emulation of characteristic elements of a particular operating system or network services. In contrast, **high-interaction honeypots** use complete operating systems, allowing the attackers to access all services on the operating system and platform in question, with the purpose of gathering more realistic information about the attackers, their methods and attacks. Honeypots providing the attackers with more ability to interact than do the low-interaction honeypots, but supporting

Table 1 Comparison of honeypots based on interaction [8, 9]

Comparative aspects	Low interaction honeypot	Medium interaction honeypot	High interaction honeypot
Interaction	Services	Emulated environment	Full system
Risk measure	Low	Medium	High
Amount of data	Low	Medium	High
Level of risk	Low	Medium	High
Control needed	No	No	Yes

less functionality than high-interaction solutions, present **medium-interaction honeypots**. The properties of each type of honeypots based on the interaction relevant for the legal analysis are shown in Table 1.

The second classification of honeypots relevant to the legal analysis is the classification based on their **purpose**, where research honeypots and production honeypots can be distinguished. A **research honeypot** is used to obtain information about the blackhat community, without any direct value to the organization, which must protect its information. In comparison, a **production honeypot** is used within an environment of organization to protect the organization and help mitigate the risk [5].

Honeypots can also be classified according to their **role**. On the one hand, honeypots for the client-side attacks are called **client side honeypots**; on the other hand, **server side honeypots** are used for the detection of new exploits, collection of malware and for further research of the threat analysis.

Network of honeypots is defined as a **honeynet**. A successful deployment of a honeynet means the successful deployment of its architecture. There are some core elements of the honeynet architecture [10]:

- **data control**—the first core element, enabling the control of and containment of the attacker's activity;
- **data capture**—the second core element, allowing the monitoring and logging of all of the attacker's activities within the honeynet;
- **data collection**—the third core element, that ensures all captured data to be stored in one central location, if more honeynets are used by the organization;
- **data analysis**—optional element, which analyses the data collected from the honeynet.

Another example of the deception technology is based on the digital bait approach. **Digital bait** "*is a false digital entity created by the administrators for discovering the adversary*" [3]. The idea of **honeytoken** can be traced back to Spitzner. He defines honeytoken as "*an artificial digital data item (e.g. a credit card number, a database entry or login credentials) planted into a genuine system resource (e.g. databases, file systems or e-mail inboxes)*" [6]. Similar definition can be found in [11]. According to Bercovitch, it is "*an artificial data item that is so similar to real tokens such that even an expert in the relevant domain will not be able to distinguish between real tokens and the honeytoken*" [11].

1.1 Deception Industrial Control Systems

An interesting field for the application of the deception technologies is an industrial environment. Industrial honeypots, honeynets and honeywords are examples of the deception systems under the protection of the ICS.

ICS honeypots support a large number of industrial protocols, especially Modbus, SNMP, http, S7comm etc. The most known industrial honeypot is Conpot [12]. It is a low-interaction honeypot, which collects data about attacks and attackers targeting industrial control systems. It supports a vast number of industrial communication protocols such as Modbus, IPMI, SNMP, S7 and Bacnet. Other examples of ICS low-interaction honeypots simulate electricity power grids (e.g. Gridpot [13]), a measuring device commonly used in the oil and gas industry for tanks at petrol stations (e.g. GasPot [14]) or an air conditioning system (e.g. HoneyPhy [15]). An example of a medium-interactive honeypot is XPOT honeypot [16]. It simulates a Siemens S7-300/400 models. Siemens S7-300 can be simulated by high-interactive honeypot CryPLH [17].

In the field of industrial control systems, we can also find ICS **honeynets**. The first project in this field was SCADA Honeynet Project [18]. This project started in 2004 and was aimed at the creation of a framework for the simulation of industrial networks. It was able to simulate TCP/IP stack of a device, specific industrial protocols (e.g. Modbus), SCADA applications etc. Another example is the Digital Bond's SCADA Honeynets [19]. This honeynet uses at least two machines. One machine monitors the network activity and the other machines simulate a programmable logic controller with several services (e.g. Modbus, FTP, Telnet, and SNMP). A newer example of a network of honeypots is the distributed industrial honeypot system called DiPot [20]. This system monitors scanning and attacking behaviors against industrial control systems. It uses basic network protocols (e.g. HTTP, FTP) and most open industrial protocol (e.g. Modbus, S7comm, SNMP, IPMI, Kamstrup).

An example of **honeyword** in ICS word is HoneyPoint [21]. It is a deception system emulated device in ICS, which is honeypot (simulated network services) and honeyword. It provides manipulated documents, which allow tracking an attacker every time when the honeyword (document) is opened after its download.

1.2 Legal Issues

In general, the development and operation (usage) of the deception systems to provide safe industrial environment is accompanied by technical and legal circumstances.

Several legal and ethical aspects must be taking into account, including privacy, entrapment and liability, by the operator of the deception system. As most of the industrial control deception systems are honeypots, respectively honeynets, our legal analysis focuses primarily on these systems.

Within the legal framework, we examine the relevant legislation affecting the member states of the European Union (EU). This chapter discusses the EU law like regulations (e.g. General Data Protection Regulation), directives (e.g. NIS directives), but as well international agreements (e.g. The Council of Europe's Cybercrime Treaty). The EU law is either an integral part of the member states' law (EU regulations, certain international agreements) or it is supposed to be implemented into the member states' law (EU directives). In this regard, we must mainly distinguish between regulations which are directly applicable in all member states and directives which require the implementation into the national law). Moreover, the relevant case law of the Court of Justice (CJEU) and the European Court of Human Rights (ECHR) is also considered in this analysis.

The objective of this contribution is to address the issue of the legal regulation of the deception systems' operation. The paper focuses on the issues of privacy and personal data protection, liability issues, ethical aspects, and research issues. The entrapment issue is, however, omitted, as it is not regulated by the EU law, but in the national law of the individual Member States.

Several contributions of this paper can be defined. Firstly, the review of the research literature regarding the legal aspects of the deception systems. Secondly, the legal analysis of the deployment and usage of industrial control deception systems from the EU law perspective.

This chapter will be organised into five sections. Introduction to the deception systems will be outlined in Sect. 1. Related works will be discussed in Sect. 2, considering the previous papers analysing the legal aspects of deception systems, specifically issues of privacy and liability. Section 3 will focus on the privacy and the personal data protection law analysis. We will consider specifically the legal framework established by the EU law. Moreover, the basic concepts of personal data protection in the EU and data collected by the deception systems will be discussed. Section 4 outlines the liability issue, mainly by the provision of an overview of tort law in the operation of deception systems. The Sect. 5 concludes this chapter.

2 Related Research

In this section the related works on legal issues of deception systems will be presented. Deception systems and the relevant legal issues are discussed by Fraunholz et al. In [22] the state of art regarding the legal issues of deception systems is outlined. This issue is closely examined in [1], where the authors focus on the general aspects (privacy, entrapment and liability) and domain specific law that is applicable, e.g. to research or government. German and European law is considered.

Other works focus on the individual types of deception systems, specifically regarding honeypots and honeynets, while examining these issues particularly from the perspective of the law of the US and the EU. However, certain works consider other legal orders and provide a comparison with the US law. To illustrate, Warren et al. address Australian law as regards the same issue [23].

The most discussed issues in this regard are privacy, liability and entrapment. These issues were stated by Spitzner who provided a definition of a honeypot for the first time [24]. More issues later emerged, e.g. regarding the data publication, ethical issues, research questions, the copyright protection etc. In the following text we provide an overview of the relevant works that provided a basis for the implementation and usage of the most known deception systems.

Spitzner outlined the legal issues of honeypots from the US law perspective for the first time. As regards privacy, he concluded that the use of honeypots significantly impacts the right to privacy. Moreover, if honeypots are used to harm third parties, he expects the possibility of litigation considering the establishment of liability. As to the entrapment issue, he concludes that "*honeypots are not a form of an entrapment*" [9].

Salgado formulates the relevant legal issues to take into account when honeypots are used within the US, specifically focusing on the need to limit rights to monitor and record all activities in a computer network. Moreover, the operator's liability for the abuse of honeypots to harm other servers and networks is outlined. As regards entrapment, Salgado considers this issue as overstated [24].

In [25] Mokube outlined the aspects to be considered when honeypots and honeynets are deployed and used within the US. Similarly, three relevant issues are examined: privacy, liability and entrapment. Despite of the operators' responsibility to secure their networks, their right to monitor the activities in these networks is limited by the relevant legislation. As regards the liability of operators as well as attackers, the authors implied that honeypots may be misused to harm other networks and computers.

In contrast, Scottberg et al. recognized honeypots as an entrapment [26]. The privacy issues are also studied. The attacker's files are not protected, as attackers cannot be considered as having legitimate accounts or interests. As regards liability, in line with the previous authors, it is discussed that compromised honeypots may be used to attack other computer networks and systems. This may lead to the establishment of the operator's liability in due diligence of assets.

All of the above-mentioned authors discuss the legal issues of honeypots and honeynets from the US law perspective. The first authors, who focus on honeypots and honeynets from the EU law perspective was Dornseif et al. In the paper [27], legal issues of honeypots from the perspective of the EU law were examined, focusing more on the civil and criminal liability as regards honeypots, and an overview of the relevant issues arising from the national law of Germany was provided.

Other works can be found by Sokol et al. [28] where the legal issues of honeypots considering their generations from the perspective of the EU law were formulated. This paper considered the core elements of honeynets (data control, data capture and data collection), as well as issues in connection with privacy and liability. In the paper [29], Sokol et al. expanded the analysis of the privacy issues of honeypots and honeynets with respect to their technical aspects, specifically focusing on the legal framework of privacy, legal grounds for data processing, and the IP address aspects. In another paper [30], authors discuss the civil and criminal liability aspects of honeypots and honeynets, as well as cybercrime and liability of attackers. This chapter is established on these papers.

3 Privacy Issues of the Use of Deception ICS

As was mentioned above, the use of deception ICS encompasses several legal issues, mainly the issues of legal liability and privacy. The issue of privacy seems to play less of a role within the use of the ICS device, since the system it aims to protect, does not operate primarily with personal data, but relates to industrial processes. The ICS devices control, capture, collect and analyse data, but the data relates to the production, and does not present personal data of the legitimate users.

The provider of deception ICS operates with the data relating to the attack, its technique, proceeding, type of operating system and web browser used in a computer from which the attack originated, or which applications run in the computer as well as IP address of the computer. In that context the question arises, whether data processed by the provider of deception ICS can be considered as personal data and thus whether the required protection is to be provided.

In most cases, the computer from which the attack on the ICS originated is not the attacker's computer, but a computer of a third party abused by the intruder of the attack. Irrespective of the fact whether the computer from which the attack originated belongs to the intruder (attacker) or to someone else, all of the data received from the computer, which should be considered as personal data, must be protected by the operator of the ICS device.

However, the literature on understanding the legal aspects, including that of the privacy, of deception ICSs' use is absenting, despite of the fact that such an understanding is highly desirable. Application of deception ICSs in practice faces three major legal challenges relating to the right of privacy, resp. right to protection of personal data. Firstly, to identify the data collected and processed by deception ICS's provider, which should be considered as personal data, secondly, to determine the legal basis for the processing of such data and, thirdly, to find a fair balance between the legitimate aim of processing of such data and the relevant rights of the person concerned (data subject rights) stemming from the fundamental right to privacy, resp. right to personal data protection. The third legal challenge relates to the way how the interference with the personal data protection rights by using deception ICSs should be legally regulated.

In the following text we will outline the legal framework of privacy issues the provider of deception ICS must be aware of when collecting and processing personal data of the data subject. Firstly, we will outline the law of the EU and of the Council of Europe as regards the right to privacy and the right to personal data protection.

Later, we will consider the application of the EU's legal framework on the right to personal data protection as a part of the right to privacy to the use of deception ICS.

3.1 The Law of the EU and of the Council of Europe on the Right to Privacy and the Right to Personal Data Protection

The right to privacy is protected under numerous international agreements.[1] In Europe, within the Council of Europe, the right to privacy is guaranteed by Article 8 of the Convention on the protection of human rights and fundamental freedoms (also known as the European Convention on Human Rights, hereinafter as "ECHR") from 1953. With regard to automatic processing of personal data, the Council of Europe's Convention n. 108 from 1981 [31] was adopted and ratified together with the ECHR by all of the Member States of the EU. The Convention n. 108 introduced the definition of such terms as personal data, automatic processing or controller of the file, and provided the basic principles for data processing and rules on the transborder flow of data (Article 2, Chapters 2, 3 and 4 of Convention n. 108). The additional protocol adopted in 2001 stipulated provisions on the establishment of independent supervisory bodies and rules on the transfer of data to third countries [32]. The basic source of EU primary law which recognizes the fundamental right to respect for private and family life and the fundamental right to protection of personal data is the Charter of Fundamental Rights of the European Union (hereinafter as "Charter") [33].

Unlike the Charter, the ECHR does not recognize the fundamental right to the protection of personal data as a separate fundamental right. Article 8 of the ECHR states that "*everyone has the right to respect for his private and family life, his home and his correspondence*". The article specifies four spheres of the privacy protection: private life, family life, a person's home and correspondence. Personal data protection is considered as an integral part of the right to respect for private life, as the private life of an individual may be affected when their personal data are collected, recorded or analysed. The notion of private life covers individual communication with others. The communications, via phone or internet, be they private or business, are not only protected under the notion of "correspondence", but also under the notion of

---

[1]To illustrate, by Article 12 of the Universal Declaration of Human Rights from 1948, Article 17 of the International Covenant on Civil and Political Rights from 1966, both adopted within the United Nations Organisation. Within the Organisation for Economic Cooperation and Development (OECD) the Guidelines on the protection of privacy and transborder flows of personal data were adopted on the 23rd of September 1980.

"private life".[2] The notion of private life also encompasses public information, if it is systematically collected and stored in files held by authorities.[3]

3.2 Right to Privacy Versus Right to Personal Data Protection

The roots of the right to personal data protection resides in the right to privacy [34]. A link between these rights can be established through the appearance of the notion of informational self-determination, which encompasses control over one's own personal information. Westin states: *"[p]rivacy is the claim of individuals ... to determine for themselves when, how, and to what extent information about them is communicated to others"* [35]. Some considers that the data protection enshrines informational self-determination and that the right to privacy is in fact the right to informational self-determination. The notion of informational self-determination has German roots (informationelle Selbstbestimmung) [36]. It considers the consent to data processing as the key element for lawful data processing. The general approach of the EU and the Council of Europe to the data protection is not based on the notion of informational self-determination. Within the law of the EU and the Council of Europe the notion of consent plays an important role, but it is not the only legal ground for legitimate data processing.[4] The EU data protection rules create a system of checks and balances which enable lawful processing even in the absence of the consent of the person involved [34]. It can be deduced that the right to data protection as introduced by the Charter is based on the system of checks and balances. According to Article 8 (2) of the Charter *"personal data must be processed on the basis of the consent of person concerned, or on some other legitimate basis laid down by law"*. The rights to privacy and the right to data protection are closely linked, but should not be perceived as the same right. The incorporation of the right to data protection in the Charter reflects the elaborated and conclusive system of checks and balances which secures lawful processing of personal data [34].

---

[2]Judgement of the European Court of Human Rights (hereinafter as "ECtHR") from the 6th of September 1978, Klass a. o. v. GER, Complaint No. 5029/71, § 41; Judgment of the ECtHR from the 2nd of August 1984, Malone v. UK, Complaint No. 8691/79, § 64; Judgement of the ECtHR from the 25th of June 1997, Halford v. UK, Complaint No. 20605/92, § 44.

[3]See: Judgment of the ECtHR from the 4th of May 2000 (GC), Rotaru v. ROM, Complaint No. 28341/95, § 43.

[4]To the lawfulness of processing personal data of data subject (to the legitimate ground for data processing) within the EU, see: Article 6 para 1 of regulation (EU) 2016/679 of the European Parliament and of the Council of 27 April 2016 on the protection of natural persons with regard to the processing of personal data and on the free movement of such data, and repealing Directive 95/46/EC (General Data Protection Regulation). OJ L 119, 04/05/2016, pp. 1–88. Compare to Article 15 para 3 of the Council of Europe Convention n. 108, which requires the express consent of the person concerned for data processing.

3.3 EU Law on the Protection of Personal Data

Contrary to other international agreements on human rights the Charter expressly contains in its Article 8 the right to the protection of personal data.

According to the Explanations to the Charter, Article 8 of the Charter is based on Article 16 of the Treaty on the Functioning of the EU,[5] on Article 39 of the Treaty on the EU and on regulation (EU) 2016/679 on the protection of natural persons with regard to the processing of personal data and on the free movement of such data (General Data Protection Regulation, hereinafter as "GDPR") [37], which replaced the directive 95/46/EC on the protection of individuals with regard to the processing of personal data and on the free movement of such data. It is also based on the Article 8 of the ECHR and on the Council of Europe Convention n. 108 for the protection of individuals with regard to automatic processing of personal data as well as on the regulation (EU) 2018/1725 on the protection of natural persons with regard to the processing of personal data by the Union institutions, bodies, offices and agencies and on the free movement of such data [38], which repealed the former regulation (EC) 45/2001 on the protection of individuals with regard to the processing of personal data by the Community institutions and bodies and on the free movement of such data.

The above-mentioned regulations of the EU contain conditions and limitations for the exercise of the personal data protection right. The limitations for this right should also be based on the provision of Article 8 (2) and Article 52 (1) of the Charter. As to the application and interpretation of the rights and freedoms contained in the Charter, according to Article 52 (3) of the Charter in so far as the Charter contains rights which correspond to the rights guaranteed by the ECHR, the meaning and scope of those rights shall be the same as those laid down by the said Convention. This provision shall not prevent Union law from providing more extensive protection. Since Article 8 of the Charter has its roots in the law of the Council of Europe, ECHR and the Convention n. 108, it shall be interpreted with respect to them. With respect to the processing of personal data in the electronic communication's sector and to ensure the free movement of such data and of electronic communication equipment and services in the EU, the so called e-Privacy directive was adopted at the level of the EU [39]. The e-Privacy directive harmonises the provisions of the Member States as regards the requirement to ensure an equivalent level of protection of fundamental rights and freedoms, in particular the right to privacy, in the electronic communication's sector. The provisions of the e-Privacy directive particularise and complement GDPR for the purposes mentioned above. This directive focuses on the obligations of the providers of publicly available electronic communication services or of public communication networks.

GDPR lays down *"rules relating to the protection of natural persons with regard to the processing of personal data and rules relating to the free movement of personal data"* (Article 1 (1) GDPR). It protects *"fundamental rights and freedoms of natural*

---

[5]Former Article 286 of the Treaty establishing the European Community.

persons and unifies the level of protection of the right to personal data protection within the EU".[6]

GDPR applies to *"the processing of personal data wholly or partly by automated means and to the processing other than by automated means of personal data which form part of a filing system or are intended to form part of a filing system".*[7] Therefore we can conclude that GDPR should be applicable to the use of deception ICS systems, except the case the Article 2 (2) letter d) of GDPR applies.[8]

3.4   Legal Definition of the Deception ICS Under the EU Law

According to the resources available to us, there is no explicit legal definition of deception ICS in the international or EU law. However, a deception system can be considered, according to the EU law, as an electronic communication service with the electronic communication network according to the Directive 2002/21/EC on a common regulatory framework for electronic communications networks and services (hereinafter as "Framework Directive") [40]. The **electronic communications service,** according to Article 2 letter c) of the Framework Directive means *"a service normally provided for remuneration which consists wholly or mainly in the conveyance of signals on electronic communications networks, including telecommunications services and transmission services in networks used for broadcasting, but exclude services providing, or exercising editorial control over, content transmitted using electronic communications networks and services"*. The **electronic communications network**, according to Article 2 letter a) of Framework Directive, means *"transmission systems and, where applicable, switching or routing equipment and other resources which permit the conveyance of signals by wire, by radio, by optical or by other electromagnetic means, including satellite networks, fixed (circuit- and packet-switched, including Internet) and mobile terrestrial networks, electricity cable systems, to the extent that they are used for the purpose of transmitting signals, networks used for radio and television broadcasting, and cable television networks, irrespective of the type of information conveyed"*.

Article 4 (1) of e-Privacy directive states that *"the provider of a publicly available electronic communications service must take appropriate*

---

[6] According to Article 1 (3) of GDPR, *"the free movement of personal data within the Union shall be neither restricted nor prohibited for reasons connected with the protection of natural persons with regard to the processing of personal data".*

[7] See: Article 2 of GDPR, which states in a positive manner what is covered by GDPR and in a negative manner what is not covered by it.

[8] According to Article 2 para 2 letter d) of the GDPR, *"it does not apply to the processing of personal data by competent authorities for the purposes of the prevention, investigation, detection or prosecution of criminal offences or the execution of criminal penalties, including the safeguarding against and the prevention of threats to public security".*

technical and organisational measures to safeguard security of its services, if necessary in conjunction with the provider of the public communications network[9] with respect to network security".

3.5 Which Data, the Provider of ICS Device Operate with, Should Be Considered as Personal Data?

According to GDPR **personal data** means *"any information relating to an identified or identifiable natural person"* ("data subject"). **Data subject** is thus considered to be *"every identified or identifiable natural person to whom information relates"*. An **identifiable natural person** is *"one who can be identified, directly or indirectly, in particular by reference to an identifier such as a name, an identification number, location data, an online identifier or to one or more factors specific to the physical, physiological, genetic, mental, economic, cultural or social identity of that natural person"* (Article 4 (1) GDPR). Despite the wide definition of personal data, in some cases it may be not clear, whether certain information in a specific situation qualifies as personal data, for example IP addresses. The e-privacy Directive distinguishes three types of data in electronic communication, of which it provides the definitions—traffic data, location data and communication data (Article 2 letters b), c) and d) of the e-Privacy directive). **Traffic data** means *"any data processed for the purpose of the conveyance of a communication on an electronic communications network or for the billing thereof"*. **Location data** means *"any data processed in an electronic communications network or by an electronic communications service, indicating the geographic position of the terminal equipment of a user of a publicly available electronic communications service"*. **Communication** means *"any information exchanged or conveyed between a finite number of parties by means of a publicly available electronic communications service"*. Information conveyed as a part of a broadcasting service to the public over an electronic communications network is excluded, except to the extent that it can be related to an identifiable subscriber or user receiving the information.

The data which provider of the ICS device retains enables to trace and identify the source of a communication and its destination, the date, time, duration and type of a communication, as well as users' communication equipment, and to establish the location of a mobile communication equipment. That data includes, inter alia, the name and address of the subscriber or registered user and an IP address for internet services. *"That data makes it possible, in particular, to identify the person with whom a subscriber or registered user has communicated and by what means, and to identify the time of the communication as well as the place from which that communication*

---

[9] According to Article 2 letter (d) of e-Directive, public communications network means *"an electronic communications network used wholly or mainly for the provision of publicly available electronic communications services"*.

took place. Further, that data makes it possible to know how often the subscriber or registered user communicated with certain persons in a given period".[10]

While within the use of deception systems devices all three types of data (traffic, location and communication data) may be relevant, within the use of deception ICSs only the traffic data and the location data shall be relevant. A specific type of traffic data is an Internet Protocol (IP) address.

The purpose of the attacks' prevention and the later prosecution of the ICS's intruders is ensured through the storage of information regarding all access operations of the ICS by the provider of the deception ICS. Information obtained after the attack of the ICS by the intruder includes the name of the web page or file to which access was sought, the terms entered in the search fields, the time of access, the quantity of data transferred, an indication of whether access was successful, and the IP address of the computer from which access was sought.

The Court of Justice of the EU (CJEU) in its judgment in the case Scarlet Extended held that "*IP addresses of internet users were protected personal data because they allow users to be precisely identified*" [41]. The IP address should be therefore considered as personal data, but only if it may be allocated to a certain specific person. In such a case, the practice of storing the IP addresses of the users without their consent or without any other legitimate ground for such storage thus qualifies as an infringement of the right to the protection of personal data. However in most cases, the attack of the ICS originates not directly from the intruder's computer, but from the computer of a third party, to which the IP address was allocated. The intruder of the ICS misuses the computer of another person in order not to be revealed and to avoid the legal consequences of his/her conduct. The IP address of the computer from which the attack of ICS originated can be allocated to this third party, and not to the intruder. But no matter to whom, whether to the intruder him/her-self or to another person, the IP address can be allocated, if the allocation of the IP address to a specific person is possible, and thus the IP address must be considered as personal data relating to this person. In such a case, the provider of the ICS device should respect all of the legal duties relating to the protection of personal data, such as to base the processing of the IP address on a legitimate ground and to ensure that the rights of the concerned person stemming from the fundamental right to the protection of personal data will be observed. As regards the use of the ICS device, one type of person (different from the intruder and the third party whose computer was abused to execute the attack) potentially concerned by the processing of IP addresses by the provider of the ICS device, might be excluded, namely, the legitimate user of the ICS, since the ICS do not have any legitimate users. The legitimate user might be concerned as data subject in the case of using different deception system devices from the ICS devices.

---

[10]Compare: Judgment of the Court of Justice, Tele2 Sverige AB v Post- och telestyrelsen and Secretary of State for the Home Department v Tom Watson and Others, joined cases C–203/15 and C–698/15, EU:C:2016:970, para 98. See also, by analogy, with respect to Directive 2006/24, judgment of the Court of Justice, Digital Rights Ireland Ltd v Minister for Communications, Marine and Natural Resources and Others and Kärntner Landesregierung and Others, joined cases C-293/12 and C-594/12, EU:C:2014:238, para 26.

In considering the IP address as a personal data, the distinction between a static and dynamic IP address has to be considered. *"Internet service providers allocate to computers of internet users either a static IP address or a dynamic IP address, the latter to be an IP address which changes each time there is a new connection to the internet. Unlike static IP addresses, dynamic IP addresses do not enable a link to be established, through files accessible to the public, between a given computer and the physical connection to the network used by the internet service provider"*.[11]
A **dynamic IP address** *"does not constitute information relating to an identified natural person, since such an address does not directly reveal the identity of the natural person who owns the computer from which a website was accessed, or that of another person who might use that computer"*.[12] However, it must be determined whether the possibility to combine a dynamic IP address with the additional data held by the provider of ICS device constitutes means likely reasonably to be used to identify the data subject.[13]

The CJEU stated *"that an identifiable person is one who can be identified, directly or indirectly. The use by the EU legislature of the word indirectly suggests that, in order to treat information as personal data, it is not necessary that that information alone allows the data subject to be identified"*. Furthermore, recital 26 of GDPR states that *"to determine whether a person is identifiable, account should be taken of all the means likely reasonably to be used either by the controller or by any other person to identify the said person"*. In so far as that recital refers to the means **likely reasonably** *"to be used by both the controller and by any other person, its wording suggests that, for information to be treated as personal data, it is not required that all the information enabling the identification of the data subject must be in the hands of one person"*. The fact that the additional data necessary to identify the computer from which the ICS was attacked are held not by the provider of the deception ICS, but by e.g. the user's internet service provider or potentially the other subject, does not appear, according to the CJEU, *"to be such as to exclude that dynamic IP addresses registered by the provider of the deception ICS constitute personal data within the meaning of GDPR"*.[14] We may conclude, that even the dynamic IP address registered by the provider of the deception ICS at the time of the access of the deception ICS made accessible to the public by the ICS provider may constitute personal data within the meaning of GDPR, in relation to that provider, if the latter has the legal means which enable it to identify the data subject with additional data possessed by a third party, e.g. the internet service provider, relating to that person.

According to the CJEU, *"the personal data protection rules of the EU precludes from excluding, categorically and in general, the possibility of processing certain*

---

[11] See: Judgment of the Court of Justice, Patrick Breyer v Bundesrepublik Deutschland, C–582/14, EU:C:2016:779, para 16.

[12] *Ibid*, para 38.

[13] Compare to: Judgment of the Court of Justice, Patrick Breyer v Bundesrepublik Deutschland, C–582/14, EU:C:2016:779, para 45.

[14] Compare to: Judgment of the Court of Justice, Patrick Breyer v Bundesrepublik Deutschland, C–582/14, EU:C:2016:779, paras 40–44.

categories of personal data without allowing the opposing rights and interests at issue to be balanced against each other in a particular case.[15] Therefore, it cannot be definitively prescribed by the legislation, *"for certain categories of personal data, the result of the balancing of the opposing rights and interests, without allowing a different result by virtue of the particular circumstances of an individual case"*.[16] It is up to the provider of the deception ICS to balance the objective (legitimate aim), which the provider pursues by processing the personal data of the user of a computer from which the attack of the deception ICS originated, against the interests or fundamental rights and freedoms of the user. The provider of the deception ICS may collect and use personal data relating to the user of the computer from which the attack of the ICS was done, without the consent of the user, *"only in so far as the collection and use of that information are necessary to reach the legitimate aim pursued by the provider of the deception ICS"*.[17]

3.6 Data Subject Rights Stemming from Personal Data Protection According to EU Law

Every data subject, identified or identifiable natural person to whom information relates, can invoke his or her rights under GDPR—right to information (right to provide the data subject with certain information), right to access to personal data, right to rectification, right to erasure ("right to be forgotten"), right to restriction of processing, right to data portability, right to object, right not to be subject to a decision based solely on automated processing, including profiling (Articles 12–23 of GDPR).

According to the EU law, the provider of the deception ICS when processing the personal data of the data subjects (intruder of deception ICS or the user of the computer used by the intruder to attack) should respect the rights of the data subject enshrined in the fundamental right to the protection of personal data.

Another key actor defined by GDPR is **controller**, which is *"any natural or legal person, public authority, agency or other body which, alone or jointly with others, determines the purposes and means of the processing of personal data; where the purposes and means of such processing are determined by Union or Member State law"*.[18] **Processing**, according to GDPR, means *"any operation or set of operations which is performed on personal data or on sets of personal data, whether or not by*

---

[15]Ibid, para 62.

[16]See, to that effect, also: Judgment of the Court of Justice, Asociación Nacional de Establecimientos Financieros de Crédito (ASNEF) and Federación de Comercio Electrónico y Marketing Directo (FECEMD) v Administración del Estado, joined cases C–468/10 and C–469/10, EU:C:2011:777, paras 47 and 48.

[17]Compare: Judgment of the Court of Justice, Patrick Breyer v Bundesrepublik Deutschland, C–582/14, EU:C:2016:779, paras 63-64.

[18]The controller or the specific criteria for its nomination may be provided for by Union or Member State law.

automated means, such as collection, recording, organisation, structuring, storage, adaptation or alteration, retrieval, consultation, use, disclosure by transmission, dissemination or otherwise making available, alignment or combination, restriction, erasure or destruction". The provider of the deception ICS can be considered as the controller for the purposes of the application of the GDPR provisions.

The provider of the deception ICS may perform an operation with the personal data of a specific kind defined by the GDPR as profiling. According to GDPR, **profiling** means *"any form of automated processing of personal data consisting of the use of personal data to evaluate certain personal aspects relating to a natural person, in particular to analyse or predict aspects concerning that natural person's performance at work, economic situation, health, personal preferences, interests, reliability, behaviour, location or movements"* (Article 4 (4) of GDPR). According to Article 22 of GDPR, *"the data subject shall have the right not to be subject to a decision based solely on automated processing, including profiling, which produces legal effects concerning him or her or similarly significantly affects him or her"*. Data subject can be the subject of such a decision only *"if the decision*:

- *is necessary for entering into, or performance of, a contract between the data subject and a data controller*;
- *is authorised by Union or Member State law to which the controller is subject and which also lays down suitable measures to safeguard the data subject's rights and freedoms and legitimate interests*; *or*
- *is based on the data subject's explicit consent"*.

In these cases, the data controller shall implement suitable measures to safeguard the data subject's rights and freedoms and legitimate interests, at least the right to obtain human intervention on the part of the controller, to express his or her point of view and to contest the decision.

3.7 Data Quality Requirements Under the EU Law

Article 5 of GDPR contains several principles relating to the processing of personal data. The principles mirror the "data quality". According to Article 5 of GDPR personal data shall be:

- *"processed lawfully, fairly and in a transparent manner in relation to the data subject"* (**"lawfulness, fairness and transparency"**),
- *"collected for specified, explicit and legitimate purposes and not further processed in a manner that is incompatible with those purposes"* (**"purpose limitation"**),[19]
- *"adequate, relevant and limited to what is necessary in relation to the purposes for which they are processed"* (**"data minimisation"**),

---

[19]Further processing for archiving purposes in the public interest, scientific or historical research purposes or statistical purposes shall, in accordance with Article 89(1) of GDPR, not be considered to be incompatible with the initial purposes.

- *"accurate and, where necessary, kept up to date"* (**"accuracy"**),[20]
- *"kept in a form which permits identification of data subjects for no longer than is necessary for the purposes for which the personal data are processed"* (**"storage limitation"**),[21]
- *"processed in a manner that ensures appropriate security of the personal data, including protection against unauthorised or unlawful processing and against accidental loss, destruction or damage, using appropriate technical or organisational measures"* (**"integrity and confidentiality"**).

A controller is the subject required to be able to demonstrate compliance of personal data processing with the above-listed principles.

One of the principles relating to the quality of data processing is the lawfulness of personal data processing. Processing of personal data is lawful only if and to the extent that it is based on the legal ground laid down in Article 6 (1) of GDPR. Article 6 of GDPR provides several legal grounds for lawful processing of personal data. The provider of deception ICS could base his/her processing of data subject's personal data, except on the basis that the data subject consented to the processing of their personal data for one or more specific purposes (referred to in point (a) of Article 6 (1) of GDPR), on the basis referred to in point (c) of Article 6 (1) of GDPR, according to which *"processing is lawful only if and to the extent that it is necessary for the compliance with a legal obligation to which the controller is subject"*. The legal obligation to which the provider of deception ICS is subject and which could legitimize his/her processing of data subject's personal data can be found in Article 4 (1) of e-Directive, according to which *"the provider of a publicly available electronic communications service must take appropriate technical and organisational measures to safeguard security of its services, if necessary in conjunction with the provider of the public communications network with respect to network security"*. Another possible legal basis on which the provider of deception ICS could base his/her processing of data subject's personal data could be found in point (f) of Article 6 (1) of GDPR, according to which *"processing is lawful only if and to the extent that it is necessary for the purposes of legitimate interests pursued by the controller or by a third party, except where such interests are overridden by the interests or fundamental rights and freedoms of the data subject which require protection of personal data, in particular where the data subject is a child"*. The legitimate interest pursued by the provider of deception ICS could rest on securing the safety of a computer network and network services of a certain organisation. The deception ICS might be used as the Intrusion Deception System (IDS). The deception ICS can increase the security of the organisation. In practice, the consent of data subject as a legitimate ground for processing

---

[20]Every reasonable step must be taken to ensure that personal data that are inaccurate, having regard to the purposes for which they are processed, are erased or rectified without delay.

[21]Personal data may be stored for longer periods insofar as the personal data will be processed solely for archiving purposes in the public interest, scientific or historical research purposes or statistical purposes in accordance with Article 89(1) of GDPR subject to implementation of the appropriate technical and organisational measures required by GDPR in order to safeguard the rights and freedoms of the data subject.

his/her personal data by the provider of deception ICS might be excluded, since it is impossible for the provider of the deception ICS to obtain it.

The legitimate ground for the provider of deception ICS to processing activities may be in principle covered by:

- the compliance with legal obligation and
- the performing a task for the purposes of the legitimate interests pursued by the controller or by a third party.

Another possible legitimate ground for the provider of deception ICS to process the personal data, in the case when the provider is of public authority character, can be found in the provision of Article 6 (1) point (e) of GDPR, according to which "*the processing is lawful if it is necessary for the performance of a task carried out in the public interest or in the exercise of official authority vested in the controller*".

The provider of deception ICS may be of public authority character for example in the case of being a computer security incident response team ("CSIRT") according to directive (EU) 2016/1148 concerning measures for a high common level of security of network and information systems across the Union ("NIS directive") [42]. According to NIS directive, Member States are obligated to "*designate national competent authorities, single points of contact and CSIRTs with tasks related to the security of network and information systems*". According to Annex 1 to NIS directive, "*the requirements and tasks of CSIRTs shall be adequately and clearly defined and supported by national policy and/or regulation*". CSIRTs' tasks shall include at least the following:

- monitoring security incidents;
- providing early warning and dissemination of information to relevant stakeholders about security risks and security incidents;
- responding to security incidents;
- providing security risk analysis and situational awareness;
- participating in the CSIRTs network.

The CSIRT may operate with deception system (for example honeynet), which simulate ICS in order to monitor the actual state of security threats menacing ICS including critical infrastructure. The critical infrastructure includes for example, energy (electricity, oil, or gas), transport, drinking water supply and distribution infrastructures. Deception systems simulating ICS contribute to the increase of security of these systems either as reactive (as detecting systems) or proactive (as systems of testing the vulnerability of these systems). According to Article 14 (1) of the NIS directive, "*Member States shall ensure that operators of essential services take appropriate and proportionate technical and organisational measures to manage the risks posed to the security of network and information systems which they use in their operations. Having regard to the state of the art, those measures shall ensure a level of security of network and information systems appropriate to the risk posed*".

Regardless of the legal ground, on which the processing of data subject's personal data by the provider of deception ICS is based, another important duty is imbedded by the GDPR to the provider of deception ICS if the provider processes the personal data. According to Article 32 of GDPR, *"the controller has a legal obligation to implement appropriate technical and organisational measures to ensure a level of security appropriate to the risk of varying likelihood and severity for the rights and freedoms of natural persons"*. In fulfilling the obligation, except of the risk, the state of the art, the costs of implementation and the nature, scope, context and purposes of processing have to be considered. The measures the controller (provider of ICS device) shall take may include inter alia as appropriate:

- the ability to ensure the integrity, confidentiality, availability and resilience of information systems and their services;
- the ability to restore the availability and access to personal data;
- a process for testing, and evaluating the effectiveness of technical and organisational measures;
- the pseudonymisation and encryption of personal data.

3.8 Possible Limitations and Derogations on the Exercise of the Data Subject's Rights by the Provider of Deception ICS Under the EU Law

Article 52 of the Charter regulates the possible limitations on the exercise of the rights and freedoms recognised by the Charter. According to Article 52 (1) of the Charter, *"any limitation on the exercise of the rights and freedoms recognised by this Charter must be provided for by law and respect the essence of those rights and freedoms"*. The fact that any limitation should be provided for by law is a repetition of the Article 8 (2) of the Charter which states that *"the data processing, if not based on the consent of the person concerned, should have a legitimate basis laid down by law"*. Any processing of personal data without the consent of the data subject should be justified in accordance with the Article 52 (1) of the Charter, which also requires to respect the principle of proportionality. It requires the limitations to be adopted only if necessary and to genuinely meet objectives of general interest recognised by the Union or the need to protect the rights and freedoms of others. Similarly, according to the ECHR, any interference with the rights under Article 8 (1) of the ECHR, which enshrines the right to respect for private life is justified according to Article 8 (2) of the ECHR *"if it is in accordance with the law and is necessary in a democratic society in the interests of national security, public safety or the economic well-being of the country, for the prevention of disorder or crime, for the protection of health or morals, or for the protection of the rights and freedoms of others"*. In the individual case the Court (ECtHR) examines whether one or more aims listed in Article 8 (1) have been pursued. To collect and release personal information without the individual being able to foresee the measure or having access to the data, *"under exceptional conditions, may be necessary in a democratic society in the interests of national security and/or for the prevention of disorder or crime".*[22]

Acquiring, storing as well as saving or using data constitute the interference with the fundamental right to data protection. The restriction of the fundamental right to data protection is constituted in a case when certain tools are used for the acquisition of information. In particular, secret surveillance measures, interception of communications or wiretapping interfere with the right to data protection.[23] The mere fact that some private documents are confiscated constitute an interference with one's privacy.[24] Also, the mere storing of personal information as such and the release of the information establish an interference with the right to respect for private life.[25] All the interferences with the fundamental right to data protection require justification. Processing of personal data by the provider of deception ICS without the explicit consent of the data subject requires the justification as well.

The requirement of any interference with fundamental right to personal data protection to be prescribed by law relates to legal ground for the processing of personal data, the issue which was discussed above in the paper.

According to the case law of the ECtHR, as to the view of the requirement of foreseeability, required by the ECHR, the requirement of the foreseeability is not the same in the special context of the interception of the communications for the purpose of police investigations. It cannot mean that an individual should be able to foresee when the authorities are likely to intercept their communications so that they can adapt their conduct accordingly. However, *"the law must be sufficiently clear in its terms to give citizens an adequate indication as to the circumstances in which and the conditions on which public authorities are empowered to resort to this secret interference with the respect for correspondence/to restore to any such secret measures. Powers of secret surveillance of citizens are tolerable under the ECHR only in so far as strictly necessary for safeguarding the democratic institutions".*[26] In the case of secret surveillance or interception of electronic communications by the public authorities, due to the severity of such interference with one's private life and correspondence, *"the domestic law must be sufficiently clear in its terms to give citizens an adequate indication as to the circumstances in and conditions*

---

[22] See: Judgment of the ECtHR from the 6th of September 1978, Klass a. o. v GER, Complaint No. 5029/71, § 48.

[23] See: Judgment of the ECtHR from the 6th of September 1978, Klass a. o. v GER, Complaint No. 5029/71, § 41; Judgment of the ECtHR from the 2nd of August 1984, Malone v UK, Complaint No. 8691/79, § 64; Judgment of the ECtHR from the 25th of March 1998, Kopp v SUI, Complaint No. 23224/94, § 53; Judgment of the ECtHR, Amann v SUI, Complaint No. 27798/95, § 44 at seq.

[24] See: Decision of the European Commission of Human Rights from the 10th of December 1975, X., Complaint No. 6794/74.

[25] See: Judgment of the ECtHR from the 26th of March 1987, Leander v SWE, Complaint No. 9248/81, § 48; Judgment of the ECtHR from the 4th of May 2000, Rotaru v ROM, Complaint No. 28341/95, § 46; Judgment of the ECtHR from the 4th of December 2008, Marper v UK, Complaint No. 30562/04, § 67 (storage of fingerprints, DNA-profiles and cellular samples).

[26] See: Judgment of the ECtHR from the 6th of September 1978, Klass a. o. v GER, Complaint No. 5029/71, § 42.

on which public authorities are empowered to any such secret measures".[27] The collection, storage and transfer of personal data are often exercised secretly, without the knowledge of the person concerned. Therefore, in such cases the legal basis has to meet special requirements as to how precise a regulation has to be and safeguards from arbitrary use by an authority [43].

To examine the proportionality of the provider of the deception ICS's interference with the rights of the data subject by processing his/her personal data requires a fair balance to be struck between the interest of the data subject to use its rights and the legitimate aim pursued by the provider of the deception ICS. In evaluating whether or not an interference with the data subject's rights, relating to the fundamental right to personal data protection, is necessary, the deception ICS provider's legitimate interest in processing certain data has to be balanced against the data subject's interests in protecting his/her rights. Within the evaluation, the character of personal data and its importance to a person's enjoyment of his or her data subject rights are to be considered. Thus, even though

- to comply with legal obligation or
- to perform a task for the purposes of the legitimate interests pursued by the controller or by a third party or eventually
- to perform a task carried out in the public interest or in the exercise of official authority vested in the controller, may in principle be a legitimate aim pursued by the provider of deception ICS.

The evaluation, whether or not an interference with the data subject's rights is necessary, "*will depend on such factors as the nature and seriousness of the interests at stake of the data subject and the gravity of the interference*".[28]

4 Liability Issues of the Use of Deception ICS

The deployment and use of deception systems may prove problematic in practice due to the possible abuse of these systems that can be misused or used to attack other systems (e.g. server or computer networks). In this regard, the relevant liability issues must be considered. The term **liability** refers to a legal duty or obligation [44] and is defined as "*a legal responsibility for one's acts or omissions*". "*If a person (entity) fails to meet that responsibility, it leaves him open to an action from court such as a lawsuit or a court order*" [45]. This paper focuses specifically on the civil liability. As regards honeynets, liability issues are closely linked to the core element of a honeynet—data control. Therefore, in relation to the deception systems, liability issues can be distinguished in several cases:

---

[27] Judgment of the ECtHR from the 9th of June 2009, Kvasnica v SVK, Complaint No. 72094/01, § 79.

[28] Compare: Judgment of the ECtHR from the 26th of February 1997, Z. v FIN, Complaint No. 22009/93, § 99.

- damage of third parties caused by the attacker's interaction with the deception system (e.g. honeypot),
- damage resulting from the publication of information collected by the deception system (e.g. honeypot, honeytoken).

4.1   The Approaches to Civil Liability in EU Law

In the area of liability relations there is no legislation explicitly regulating the liability as regards the use of deception systems. This chapter outlines the existing EU legislation on the liability, specifically focusing on the issues relevant to the liability relations in connection with the deception systems, including deception ICS systems. The legal framework on liability relations is characterised by its fragmentation within the EU. In general, it can be classified into:

- **primary law of EU**—regulated in the primary documents of the EU, e.g. regarding the liability of the EU;
- **secondary law of EU**—especially regulations and directives;
- **case law**—judgements of the CJEU and the European Court of Human Rights;
- **soft law regulation**.

Soft law regulation's purpose is mainly to provide recommendations. Soft law is contained in numerous initiatives that can serve as a model for the future legal relations' regulation. To illustrate, such initiatives include documents of the International Institute for the Unification of Private Law (UNIDROIT), Principles of European Contract Law (PECL), Common Frame of Reference (CFR), as well as the Principles of European Tort Law (PETL) [46].

In the following chapters we will outline the applicable legal framework as regards the secondary EU law, including a closer examination of the Principles of European Tort Law.

4.2   The Secondary Law of EU

The secondary law of EU contains legislation on numerous areas regulated on the EU level, particularly by regulations and directives that are further supplemented by the national law of the Member States. This includes such areas as transportation (Regulation of the European Parliament and of the Council (EU) 261/2004), competition (Directive of the European Parliament and of the Council 2014/104/EU) or consumer law (Council Directive 85/374/EEC concerning liability for defective products [47]).

One of the important areas as regards liability includes the legislation regarding the information society services in the internal market (electronic commerce) contained in the Directive 2000/31/EC on certain legal aspects of information society services, in particular electronic commerce, in the Internal Market (hereinafter as "**Directive on electronic commerce**"). This directive "*seeks to contribute to the proper functioning of the internal market by ensuring the free movement of information society services between the Member States*" [48]. Moreover, this directive also regulates the legal relations established when information society services are provided. This directive will be applicable as regards the deception systems only in the case that a deception system will be considered as an information society service according to its legal definition. Within the meaning of Article 1(2) of EU Directive 98/34/EC as amended by EU Directive 98/48/EC **the information society service** is defined as "*any service normally provided for remuneration, at a distance, by electronic means and at the individual request of a recipient of services*". Examples of this service include web services, email services, database services, internet security services etc.

Section 4 of the EU Directive on electronic commerce contains several provisions on the liability of intermediaries. According to these provisions, a service provider is not liable in three cases:

- **mere conduit**—service provider "*does not*:

  - *initiate the transmission,*
  - *select the receiver of the transmission, and*
  - *select or modify the information contained in the transmission*";

- **caching**—service provider "*does not modify the information, complies with conditions on access to the information, complies with rules regarding the updating of the information, specified in a manner widely recognised and used by industry, does not interfere with the lawful use of technology, widely recognised and used by industry, to obtain data on the use of the information*";
- **hosting**—service provider "*does not have actual knowledge of illegal activity or information and, as regards claims for damages, is not aware of facts or circumstances from which the illegal activity or information is apparent*".

A deception system is an information society service when a **deception system is provided as a service (DSaS)**. The purpose of this service is to provide the recipient with the overview of the security of their computer network or an information system.

In the present, numerous examples of the provision of honeypots as a service can be identified (HaaS) [49]. These do not provide ICS honeypots primarily as a service, but the nature of the service does not exclude it. As regards HaaS, two situations must be distinguished. Firstly, the HaaS provider provides a honeypot as a whole, including its administration. In this situation the safe harbour rules are not applicable and the HaaS provider's liability is identical to the situation when honeypots are administered independently. This situation will be examined in more detail in the following chapter.

Secondly, the situation when the service provider only creates an environment for the creation and administration of honeypots, specifically ensuring data control and data capture, and where the service recipient further choses from the pre-defined honeypots. In this case the relevant provisions on safe harbours will be applicable, specifically those regarding the hosting liability exemption. To illustrate, such a situation would include one where the recipient can use a honeypot that enables the hosting of files (e.g. Dionaea [50], DiPot [20]). The HaaS provider will not be liable for the damage caused by the recipient (user) by their use and administration of a honeypot.

In relation to the use of deception systems, the liability for damage caused in the area of robotics and artificial intelligence is also of relevance. In this area the European Parliament adopted a resolution containing recommendations for the European Commission [51] and the European Commission adopted at the end of 2018 a communication [52]. As regards the artificial intelligence, the area of machine learning is also mentioned, which will in a certain way determine the deception systems as well. These documents outline the legislation applicable in relation to the liability for damage in the area of robotics and artificial intelligence (AI), including the above-mentioned directive concerning the liability for defective products. According to resolution: whereas according to the current legal framework product liability (above mentioned directive on liability for defective products [47])—"*where the producer of a product is liable for a malfunction - and rules governing liability for harmful actions -where the user of a product is liable for a behaviour that leads to harm - apply to damages caused by robots or artificial intelligence (AI)*".

4.3   The Principles of European Tort Law

In this paper we consider the soft law approach to the liability issues regarding the deployment and usage of deceptions systems as regulated in the Principles of European Tort Law (hereinafter only as "PETL"). PETL are the result of several years of efforts of a group of leading civilians working under the auspices of the European Group on Tort Law, with institutional support from the Austrian European Centre for Torture European Law of Insurance and Insurance Law. They were presented to a wider legal community together with a commentary at the conference in Vienna in 2005 [53].

The basic norm of liability gives a rough overview of the main reasons of liability. It is defined in PETL as "*a person (legal and natural) to whom damage to another is legally attributed and who is liable to compensate that damage*" (Art. 1:101. (1) PETL). Damage may be attributed in particular to the person (Art. 1:101. (2) PETL):

- "*whose conduct constituting fault has caused it*; *or*
- *whose abnormally dangerous activity has caused it*; *or*
- *whose auxiliary has caused it within the scope of his functions*".

4.4 General Conditions of Liability

PETL specifies requirements that need to be fulfilled for the liability to be established, specifically:

1. the **preconditions of liability** as stipulated in the II. part of PETL:

   - material or immaterial harm to a legally protected interest,
   - damage, and
   - causation.

2. the **bases of liability** as stipulated in the III. part of PETL:

   - fault,
   - abnormally dangerous activities, and
   - liability for others.

As regards the preconditions of liability as stipulated in the II. part of PETL, these must be fulfilled cumulatively. In contrast, as regards the bases of liability stipulated in the III. part of PETL, it is only required that one of these bases is fulfilled.

The specific nature of PETL is determined by the fact that they substitute illegality with the concept of a protected interest. The difference between these concepts is that the illegality requires the infringement of a specific legal norm, while for the infringement of a protected interest to occur a direct contradiction with a specific legal norm is not a prerequisite. This is of particular relevance as regards the deceptions systems, as it is not necessary to identify a specific legal norm regulating the liability relations in the area of deception systems, but the relevant protected interests and their infringement must be considered.

Numerous criteria are established to consider whether a protected interest was infringed. According to Art. 2:102 (1) of PETL, "*the protection scope of an interest depends on its nature; the higher its value, the precision of its definition and its obviousness, the more extensive protection is granted*". From the perspective of ICS deception systems, legally protected interests may include human dignity (e.g. privacy, personal data), economic interests or contractual relationships (e.g. functional computer networks, ICS device). On one hand, "*extensive protection is granted to property rights, including those in intangible property*" (Art. 2:102 (3) PETL). On the other hand, "*protection of pure economic interests or contractual relationships may be more limited in scope*" (Art. 2:102 (4) PETL).

Another basic condition for the establishment of liability is damage. According to Art. 2:101 of PETL, "*damage requires material or immaterial harm to a legally protected interest*". Art 1:101 PETL refers to the "damage" as such, but Art. 2:101 shows that only compensable damage is of relevance and it further defines it. The Article that completes Art. 2:101 is Art. 2:103 of PETL, according to which "*losses relating to activities or sources which are regarded as illegitimate cannot be recovered*". In other words, the damage caused to the attackers cannot be recovered (e.g. in the case of an active defense).

The protected interest and the existence of damage are completed by the third precondition as defined in PETL in the form of causation. Causation is recognized by all legal systems as a requirement of tortious liability [54]. This concept is systematically included in the third Chapter of PETL which is divided into:

- condition sine qua non, and
- scope of liability.

In the case of the sine qua non condition, "*an activity or conduct (hereinafter only as 'activity') is a cause of the victim's damage if, in the absence of the activity, the damage would not have occurred*" (Art. 3:101 PETL). Examples of such activities include inadequately secured deception systems (particularly in the case of high interaction honeypots or honeynets) or insufficient monitoring of deception systems (e.g. in the case of a save and further distribution of illegal content). Under Article 3:102 of the PETL, in the case of several acts, each of which would itself cause damage at the same time, each act is deemed to be the cause of the injured party's damage. As regards the operation of deception systems, any of the above-mentioned activities is sufficient to cause damage, meaning every such an activity is to be considered as the cause of damage.

Article 3:104 of PETL differentiates between two potential causes. The first category is stipulated in Article 3:104 (1) of PETL, according to which "*if an activity has definitely and irreversibly led the victim to suffer damage, a subsequent activity which alone would have caused the same damage is to be disregarded*". If a deception system was used within a botnet to attack a network service (e.g. web server) and subsequently this service lost its functions (web server is down), further activity (abuse) of the deception system is to be disregarded. Another example presents a situation when subsequent activities caused greater damage than one that was caused by the previous activity. According to Article 3:104 (2) of PETL, "*subsequent activity is to be considered if it has led to additional or aggravated damage*". This would include a situation, when an operator of a deception system enables illegal content to be available, which may lead to additional or aggravated damage.

Another relevant provision in this regard is Article 3:106 of PETL which considers uncertain causes within the victim's sphere. According to this Article, "*the victim has to bear his/her loss to the extent corresponding to the likelihood that it may have been caused by an activity, occurrence or other circumstance within his/her own sphere*". This could be illustrated on an example of a damage or complete loss of functions caused to a network service (e.g. web server by the abuse of a detection system by the attacker and by the later lasting power outage). In this case it is not clear, which of the causes led to the permanent damage of data. In this situation, the operator of a deception system will be liable only in the scope of ½. In the remaining part the victim will bear the consequences, as the power outage can be considered as an occurrence within its own sphere.

Where an activity is a cause, whether and to what extent damage may be attributed to a person depends on factors such as (Scope of liability, Art. 3:201 PETL):

- *"the foreseeability of the damage to a reasonable person at the time of the activity, taking into account in particular the closeness in time or space between the damaging activity and its consequence, or the magnitude of the damage in relation to the normal consequences of such an activity;*
- *the nature and the value of the protected interest;*
- *the basis of liability;*
- *the extent of the ordinary risks of life; and*
- *the protective purpose of the rule that has been violated".*

4.5 Bases of Liability

As was mentioned above, PETL distinguishes 3 bases of liability. This Chapter only examines the strict liability and liability based on fault. Within the concept of strict liability, the authors briefly consider abnormally dangerous activities (Art. 5:101 of PETL).

"*A person who carries on an abnormally dangerous activity is strictly liable for damage characteristic to the risk presented by the activity and resulting from it*" (Art. 5:101 (1) PETL). An activity is **abnormally dangerous** if (Art. 5:101 (2) PETL):

- *"it creates a foreseeable and highly significant risk of damage even when all due care is exercised in its management and*
- *it is not a matter of common usage".*

Deception ICS are not a matter of common usage. The second condition is fulfilled if foreseeable and highly significant risk of damage exists. There are two factors which should be considered when weighing the risk such as: gravity of harm and the likelihood of the damage [55]. It is our view that to establish strict liability only one of the above mentioned factors needs to be fulfilled. The likelihood of a deception system's abuse becomes higher with the scope of interaction of the system used. To illustrate, high-interactive honeypots or honeynets are more likely to be abused than low-interactive honeypots or honeytokens. The gravity of damage will not only be determined by the scope of interaction, but also by activities that the system in question pretends. For example, if a deception system provides a fake filesystem or directory, which looks like the real file system or directory of a legitimate user, it allows an attacker to store and share illegal content in deception systems. In this case, the gravity of harm will be higher than in the situation when a deception system simulates environment for the attacker, but does not allow them to use this environment to attack other networks or systems.

On the other hand, within liability based on fault, a person (operator of a deception system—natural or private person) is liable on the basis of fault for intentional or negligent violation of the required standard of conduct. The required standard of conduct (Art. 4:102 (1) PETL) is "*that of the reasonable person in the circumstances, and depends, in particular, on the nature and value of the protected interest involved, the dangerousness of the activity, the expertise to be expected of a person carrying it on, the foreseeability of the damage, the relationship of proximity or special reliance between those involved, as well as the availability and the costs of precautionary or alternative methods"*.

The security of the deception system must be ensured—this is the duty of the deception system's operator. This obligation may be achieved through the reaction care or proactive care. In the case of the reaction care, the duty of the operator to respond to the statement is a consequence of the obligation to intervene in a particular case (e.g. in a situation when the attacker hacks the server and attacks the no-deception systems). In such a case, the operator has a duty to respond to this attack (e.g. attack redirecting, connections blocking). On the other hand, the proactive care consists of ensuring a certain level of security of the deception system usage in the future. In this case the operator is required to adopt security measures to prevent successful attacks. Important question is what measures is the operator required to adopt. In our opinion, the operator is obligated to adopt minimal measures against attacks already carried out.

The possible interpretation of what presents the maximum of proactive care can be found in Article 15 of the EU Directive on electronic commerce. Despite of the fact that this provision relates to information society services, it can be used to determine the maximum limit of proactive care in the EU law. Within the meaning of Article 15 of the EU Directive on electronic commerce, no general obligation to monitor the information which is transmitted or stored, nor a general obligation to actively seek facts or circumstances indicating illegal activity can be imposed. This provision implies the limit between what is required and possible in proactive care and which requirements exceed the operators' liability. To illustrate a case exceeding the possible limits in proactive care, the zero-day vulnerabilities can be considered, as it presents the attacks or threats that exploit a previously unknown vulnerability in computer applications. Another example in this regard are specific vulnerabilities unknown to the professional community (e.g. honeynet.org).

An interesting issue in this regard is the impact of the degree of interaction on the level of risk associated with the operation of honeypots. In the case of a high-interaction honeypot (e.g. software network router), its operator must apply stricter technical and organisational measures to safeguard security, in contrast to an operator of a medium or a low interaction honeypot.

The above-stated standard *"may be adjusted when due to age, mental or physical disability or due to extraordinary circumstances the person cannot be expected to conform to it"* (Art. 4:102 (2) PETL). Extraordinary circumstances in the case of a deception system can be the above-mentioned zero-day or zero-hour vulnerability.

Equally interesting is the interpretation of the relevant provisions: in consideration of the state of art and the cost of their implementation, these measures shall ensure a level of security appropriate to the risk. We believe that clarification and interpretation of this formulation is left to the transposition of that provision to the national law of the EU Member States.

One of the operator´s duty is to act positively protect others from damage. This duty max exist (Art. 4:103 PETL):

- *"if law so provides, or*
- *if the actor creates or controls a dangerous situation, or*
- *when there is a special relationship between parties or*
- *when the seriousness of the harm on the one side and the ease of avoiding the damage on the other side point towards such a duty".*

For the operator of a deception system the first two cases may be relevant. There are several examples of the duty to act positively to protect others from damage that may exist if law so provides. The first example is contained in Article 13a of EU directive 2009/140/EC that outlines the security and integrity of networks and services [56]. According to this article, the operators of deception systems *"take appropriate technical and organisational measures to appropriately manage the risks posed to the security of networks and services. Having regard to the state of the art, these measures shall ensure a level of security appropriate to the risk presented"*. In particular, measures shall be taken to prevent and minimise the impact of security incidents on users and interconnected networks. Another example can be found in Article 4(2) of E-privacy directive [39]. According to this article, in the case of a particular risk of a breach of the security of the network, the operator *"must inform the production network users concerning such risk and, where the risk lies outside the scope of the measures to be taken by the operator of deception systems, of any possible remedies, including an indication of the likely costs involved"*.

The second example is the duty to act positively to protect others from damage that may exist if the actor creates or controls a dangerous situation. It is the most relevant case for the operator of a deception system. The deployment of deception systems (e.g. honeypot, honeynet) presents a dangerous situation to the production network and to the no-deception systems and the operator of deception systems controls this situation. Based on this, the operator of a deception system has a duty to act positively to protect others from damage.

Lastly, the liability for omissions to protect others is closely related to the culpable breach of operator's duty. In a number of national legal orders, fault is presumed. The possible forms of fault are direct intention, indirect intention, conscious negligence and unconscious negligence. In the case that operator of a deception system does not respond to the attack and his omission reinforces the damage, the operator is usually solidary liable for the entire implementation of harm.

5 Conclusion

In this chapter, we outline the basic legal framework of the deception ICS. As we have mentioned, these tools enable effective protection of information systems and networks of organisations.

The growing importance of the fundamental right to personal data protection and the need for adequate harmonised rules in this area has been reflected by the EU in the Lisbon Treaty with the introduction of specific provisions on data protection in

Article 16 TFEU and Article 39 TEU. The EU has the competence to adopt rules on data protection in all EU policy areas. The fundamental right to data protection as enshrined in Article 8 of the Charter constitutes the heading of a set of rights and obligations and limitations to these which are established as an elaborated system of checks and balances [34]. This system of checks and balances differentiates the fundamental right to personal data protection from the right to privacy. It provides the right to data protection with a feature that goes beyond the notion of informational self-determination.

The provider of a deception ICS shall be aware of all his/her legal obligations stemming from the EU's rules on personal data protection. The rules relate mainly to the identification of which data should be considered as personal data and thus be protected, the legal ground for legitimate processing of subject's personal data as well as other legal principles relating to the processing of personal data, which mirror the "data quality" requirement and the legal framework within which the exercise of the data subject's rights stemming from the fundamental right to personal data protection may be restricted.

Whether liability for the infringement of the relevant provisions in this regard can be established presents another issue entirely. In practice, the interpretation of the legislation and its applicability to a specific provider of a deception ICS may prove problematic. The existing legislation or the lack thereof do not ease the process of considering whether all liability preconditions are present. One solution present soft law instruments, the applicability of which, however, is questionable and must be established in each individual case separately. As the existing legal framework does not provide exact answers, numerous issues will arise, possibly discouraging the use of deception systems in practice due to fears of possible repercussions.

Acknowledgements We thank our colleagues from the Czech chapter of The Honeynet Project for their valuable inputs and comments. This paper is funded by the Slovak APVV projects under contract No. APVV-14-0598 and No. APVV- APVV-17-0561.

References

1. Fraunholz, D., Lipps, C., Zimmermann, M., Duque Antón, S., Mueller, J.K.M., Schotten, H.D.: Deception in information security: Legal considerations in the context of German and European law. In: Lecture Notes in Computer Science (including subseries Lecture Notes in Artificial Intelligence and Lecture Notes in Bioinformatics). pp. 259–274 (2018)
2. Rowe, N.C., Rrushi, J.: Introduction to cyberdeception. Springer International Publishing, Berlin (2016)
3. Qassrawi, M.T., Hongli, Z.: Deception methodology in virtual honeypots. In: NSWCTC 2010—The 2nd International Conference on Networks Security, Wireless Communications and Trusted Computing (2010)
4. Yuill, J., Zappe, M., Denning, D., Feer, F.: Honeyfiles: deceptive files for intrusion detection. In: Proceedings from the Fifth Annual IEEE SMC Information Assurance Workshop. pp. 116–122. IEEE (2004)

5. Martin Lazarov and Jeremiah Onaolapo and Gianluca Stringhini: Honey sheets: What happens to leaked google spreadsheets? In: 9th Workshop on Cyber Security Experimentation and Test (CSET) 2016. USENIX Association, Austin, TX (2016)
6. Graves, R., Stingley, M.: Honeytokens and honeypots for web ID and IH. (2015)
7. Spitzner, L.: Honeypots: Catching the insider threat. In: The 19th Annual Conference on Computer Security Application (ACSAC). pp. 304–313 (2003)
8. Krishnaveni, S., Prabakaran, S., Sivamohan, S.: A survey on honeypot and honeynet systems for intrusion detection in cloud environment. J. Comput. Theor. Nanosci. **15**, 2949–2953 (2018)
9. Spitzner, L.: Honeypots: tracking hackers. Reading: Addison-Wesley, Boston (2003)
10. Abbasi, F.H., Harris, R.J.: Experiences with a generation III virtual honeynet. In: 2009 Australasian Telecommunication Networks and Applications Conference, ATNAC 2009—Proceedings (2009)
11. Bercovitch, M., Renford, M., Hasson, L., Shabtai, A., Rokach, L., Elovici, Y.: HoneyGen: An automated honeytokens generator. In: Proceedings of 2011 IEEE International Conference on Intelligence and Security Informatics, ISI 2011 (2011)
12. Conpot—ICS/SCADA Honeypot, http://conpot.org/
13. Gridpot: Open source tools for realistic-behaving electric grid honeynets, https://github.com/sk4ld/gridpot
14. GasPot, https://github.com/sjhilt/GasPot
15. Litchfield, S.: HoneyPhy: A physics-aware CPS honeypot framework, https://smartech.gatech.edu/handle/1853/58329, (2017)
16. Lau, S., Klick, J., Arndt, S., Roth, V.: POSTER: Towards Highly Interactive Honeypots for Industrial Control Systems. In: Proceedings of the 2016 ACM SIGSAC Conference on Computer and Communications Security. (2016). https://doi.org/10.1145/2976749.2989063
17. Buza, D.I., Juhàsz, F., Miru, G., Fèlegyhàzi, M., Holczer, T.: CryPLH: Protecting smart energy systems from targeted attacks with a PLC honeypot. In: Lecture Notes in Computer Science (including subseries Lecture Notes in Artificial Intelligence and Lecture Notes in Bioinformatics) (2014)
18. Pothamsetty, V., Franz, M.: SCADA HoneyNet Project: Building Honeypots for Industrial Networks, http://scadahoneynet.sourceforge.net/
19. Peterson, D.G.: Siemens S7 Honeynet? | Digital Bond. http://www.digitalbond.com/blog/2011/07/27/siemens-s7-honeynet/
20. Cao, J., Li, W., Li, J., Li, B.: DiPot: A distributed industrial honeypot system. In: Lecture Notes in Computer Science (including subseries Lecture Notes in Artificial Intelligence and Lecture Notes in Bioinformatics) (2018)
21. Brent Huston: HoneyPoint Security Server ICS/SCADA Deployment Example—MSI :: State of SecurityMSI :: State of Security. https://stateofsecurity.com/honeypoint-security-server-icsscada-deployment-example/
22. Fraunholz, D., Anton, S.D., Lipps, C., Reti, D., Krohmer, D., Pohl, F., Tammen, M., Schotten, H.D.: Demystifying Deception Technology: A Survey. arxiv.org. (2018)
23. Warren, M.J., Hutchinson, W.: Australian hackers and ethics. Australas. J. Inf. Syst. (2015). https://doi.org/10.3127/ajis.v10i2.163
24. Spitzner, L.: The honeynet project: Trapping the hackers, (2003)
25. Mokube, I., Adams, M.: Honeypots: Concepts, Approaches, and Challenges. In: ACM-SE 45 Proceedings of the 45th annual southeast regional conference (2007)
26. Scottberg, B., Yurcik, W., Doss, D.: Internet honeypots: protection or entrapment? Presented at the (2003)
27. Dornseif, M., Gärtner, F.C., Holz, T.: Vulnerability Assessment using Honeypots. PIK—Prax. der Informationsverarbeitung und Kommun. **27**, 195–201 (2007). https://doi.org/10.1515/piko.2004.195
28. Sokol, P., Host, J.: Evolution of legal issues of honeynets. Studies in Systems. Decision and Control, pp. 179–200. Springer, Cham (2016)
29. Sokol, P., Míšek, J., Husák, M.: Honeypots and honeynets: issues of privacy. Eurasip J. Inf. Secur. (2017). https://doi.org/10.1186/s13635-017-0057-4

30. Sokol, P., Andrejko, M.: Deploying honeypots and honeynets: Issues of liability. In: Communications in Computer and Information Science. pp. 92–101 (2015)
31. Convention n. 108 of 18. January 1981 of the Council of Europe for the protection of individuals with regard to automatic processing of personal data. (1981)
32. Additional Protocol to Convention 108 of 8. November 2001 regarding supervisory authorities and transborder data flows, ETS 181. (2001)
33. Charter of Fundamental Rights of the European Union. OJ C 202, 7.6.2016. (2016)
34. Peers, S., Hervey, T., Kenner, J., Ward, A.: The EU Charter of fundamental rights: a commentary. (2014)
35. Westin, A.F.: Privacy and freedom. Wash. Lee Law Rev. **25**, 166 (1968)
36. Albers, M.: Informationelle Selbstbestimmung, Baden-Baden, 2005, zugl. Habil. (2005)
37. Regulation (EU) 2016/679 of the European Parliament and of the Council of 27 April 2016 on the protection of natural persons with regard to the processing of personal data and on the free movement of such data, and repealing Directive 95/46/EC (General Da. (2016)
38. Regulation (EU) 2018/1725 of the European Parliament and of the Council of 23 October 2018 on the protection of natural persons with regard to the processing of personal data by the Union institutions, bodies, offices and agencies and on the free movement. (2018)
39. Directive 2002/58/EC of the European Parliament and of the Council of 12 July 2002 concerning the processing of personal data and the protection of privacy in the electronic communications sector (Directive on privacy and electronic communications). OJ L. (2002)
40. Directive 2002/21/EC of the European Parliament and of the Council of 7 March 2002 on a common regulatory framework for electronic communications networks and services (Framework Directive). (2002)
41. Judgment of the Court of Justice, Scarlet Extended SA v Société belge des auteurs, compositeurs et éditeurs SCRL (SABAM), C 70/10, EU:C:2011:771. (2011)
42. Directive (EU) 2016/1148 of the European Parliament and of the Council of 6 July 2016 concerning measures for a high common level of security of network and information systems across the Union. OJ L 194, 19.7.2016, pp. 1–30
43. Glas, L.R.: European convention on human rights. Netherlands Q. Hum. Rights. **31**, 505–510 (2014). https://doi.org/10.5771/9783845258942
44. Black, H.C.: A dictionary of law. Yale Law J. **1**, 88 (2006). https://doi.org/10.2307/783720
45. Garner, B.A.: Black's Law Dictionary: Shield law. (2014)
46. Koch, B.A.: The European group on tort law and its principles of European tort law. Am. J. Comp. Law. **53**, 189–205 (2005)
47. Council Directive 85/374/EEC of 25 July 1985 on the approximation of the laws, regulations and administrative provisions of the Member States concerning liability for defective products OJ L 210, 7.8.1985, pp. 29–33. (1985)
48. Directive 2000/31/EC of the European Parliament and of the Council of 8 June 2000 on certain legal aspects of information society services, in particular electronic commerce, in the Internal Market (Directive on electronic commerce) OJ L 178, 17.7.2000. (2000)
49. CZ.NIC—Honeypot as a Service. https://haas.nic.cz/
50. Dionaea.: https://github.com/DinoTools/dionaea
51. Report with recommendations to the Commission on Civil Law Rules on Robotics (2015/2103(INL)). http://www.europarl.europa.eu/doceo/ document/A-8-2017-0005_EN.html
52. Communication from the Commission to the European parliament, the European Council, the Council, the European economic and social committee and the Committee of the regions- Artificial Intelligence for Europe (Com/2018/237 final). https://eur-lex.europa.eu/legal-content/EN/TXT/?uri=COM%3A2018%3A237%3AFIN
53. European Group on Tort Law.: http://www.egtl.org/

54. Spier, J., Busnelli, F.: Unification of tort law: Causation. Kluwer Law International (2000)
55. Koch, B.A.: The principles of European tort law. ERA Forum. **8**, 107–124 (2007). https://doi. org/10.1007/s12027-007-0003-x
56. Directive 2009/140/EC of the European Parliament and of the Council of 25 November 2009 amending Directives 2002/21/EC on a common regulatory framework for electronic communications networks and services, 2002/19/EC on access to, and interconnection of, electronic communications networks and associated facilities, and 2002/20/EC on the authorisation of electronic communications networks and services (Text with EEA relevance) OJ L 337, 18.12.2009, pp. 37–69

Correction to: Cybersecurity Threats, Vulnerability and Analysis in Safety Critical Industrial Control System (ICS)

Xinxin Lou and Asmaa Tellabi

Correction to:
Chapter "Cybersecurity Threats, Vulnerability and Analysis in Safety Critical Industrial Control System (ICS)" in: E. Pricop et al. (eds.), *Recent Developments on Industrial Control Systems Resilience*, Studies in Systems, Decision and Control 255, https://doi.org/10.1007/978-3-030-31328-9_4

The original version of the book was inadvertently published with an incorrect figure 2 in Chapter "Cybersecurity Threats, Vulnerability and Analysis in Safety Critical Industrial Control System (ICS)". The figure 2 is updated with the correct figure given below and statement has been updated at page number 78 as "As for the safety critical system, we tend to consider the integrity as a first". The chapter and book have been updated with the changes.

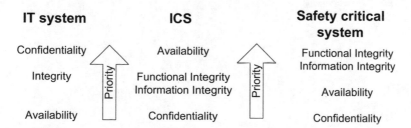

Fig. 2 Comparison of security objectives priority among various systems

---

The updated version of this chapter can be found at
https://doi.org/10.1007/978-3-030-31328-9_4

© Springer Nature Switzerland AG 2020
E. Pricop et al. (eds.), *Recent Developments on Industrial Control Systems Resilience*, Studies in Systems, Decision and Control 255,
https://doi.org/10.1007/978-3-030-31328-9_14

Printed in the United States
By Bookmasters